Factorial Designs

Probability and Mathematical Statistics (Continued)

WILLIAMS • Diffusions, Markov Processes, and Martingales, Volume I: Foundations

ZACKS • Theory of Statistical Inference

Applied Probability and Statistics

ANDERSON, AUQUIER, HAUCK, OAKES, VANDAELE, and WEISBERG • Statistical Methods for Comparative Studies

ARTHANARI and DODGE • Mathematical Programming in Statistics

BAILEY • The Elements of Stochastic Processes with Applications to the Natural Sciences

BAILEY • Mathematics, Statistics and Systems for Health

BARNETT • Interpreting Multivariate Data

BARNETT and LEWIS • Outliers in Statistical Data

BARTHOLOMEW • Stochastic Models for Social Processes, *Second Edition*

BARTHOLOMEW and FORBES • Statistical Techniques for Manpower Planning

BECK and ARNOLD • Parameter Estimation in Engineering and Science

BELSLEY, KUH, and WELSCH • Regression Diagnostics: Identifying Influential Data and Sources of Collinearity

BENNETT and FRANKLIN • Statistical Analysis in Chemistry and the Chemical Industry

BHAT • Elements of Applied Stochastic Processes

BLOOMFIELD • Fourier Analysis of Time Series: An Introduction

BOX • R. A. Fisher, The Life of a Scientist

BOX and DRAPER • Evolutionary Operation: A Statistical Method for Process Improvement

BOX, HUNTER, and HUNTER • Statistics for Experimenters: An Introduction to Design, Data Analysis, and Model Building

BROWN and HOLLANDER • Statistics: A Biomedical Introduction

BROWNLEE • Statistical Theory and Methodology in Science and Engineering, *Second Edition*

BURY • Statistical Models in Applied Science

CHAMBERS • Computational Methods for Data Analysis

CHATTERJEE and PRICE • Regression Analysis by Example

CHERNOFF and MOSES • Elementary Decision Theory

CHOW • Analysis and Control of Dynamic Economic Systems

CHOW • Econometric Analysis by Control Methods

CLELLAND, BROWN, and deCANI • Basic Statistics with Business Applications, *Second Edition*

COCHRAN • Sampling Techniques, *Third Edition*

COCHRAN and COX • Experimental Designs, *Second Edition*

CONOVER • Practical Nonparametric Statistics, *Second Edition*

CORNELL • Experiments with Mixtures: Designs, Models and The Analysis of Mixture Data

COX • Planning of Experiments

DANIEL • Biostatistics: A Foundation for Analysis in the Health Sciences, *Second Edition*

DANIEL • Applications of Statistics to Industrial Experimentation

DANIEL and WOOD • Fitting Equations to Data: Computer Analysis of Multifactor Data, *Second Edition*

DAVID • Order Statistics, *Second Edition*

DEMING • Sample Design in Business Research

DODGE and ROMIG • Sampling Inspection Tables, *Second Edition*

DRAPER and SMITH • Applied Regression Analysis, *Second Edition*

DUNN • Basic Statistics: A Primer for the Biomedical Sciences, *Second Edition*

DUNN and CLARK • Applied Statistics: Analysis of Variance and Regression

ELANDT-JOHNSON • Probability Models and Statistical Methods in Genetics

ELANDT-JOHNSON and JOHNSON • Survival Models and Data Analysis

continued on back

Factorial Designs

B. L. RAKTOE

Professor of Mathematics
University of Petroleum and Minerals
Dhahran, Saudi Arabia

A. HEDAYAT

Professor of Mathematics
University of Illinois
Chicago, Illinois

W. T. FEDERER

Liberty Hyde Bailey
Professor of Biological Statistics
Cornell University
Ithaca, New York

JOHN WILEY & SONS

New York Chichester Brisbane Toronto Singapore

Library of Congress Cataloging in Publication Data:

Raktoe, B L 1932–
Factorial designs.

(Wiley series in probability and mathematical statistics)
Includes bibliographies and index.
1. Factorial experiment designs.
I. Hedayat, A., 1937– joint author. II. Federer, Walter Theodore, 1915– joint author.
III. Title.

QA279.R35 519 80-28905
ISBN 0-471-09040-9

Printed in the United States of America

10 9 8 7 6 5 4 3 2 1

To

MADHURI RAKTOE

BATOOL HEDAYAT

and to

THE MEMORY OF LILLIAN FEDERER

Preface

The purpose of this book is to fill a void that currently exists in published literature regarding factorial design. The literature, though voluminous, does not contain a systematic treatise on the subject. This book provides a place, then, for the person interested in the pursuit of knowledge about the *concepts* of factorial design. It is not a review of published literature, but it provides a comprehensive and self-contained treatment of the basics of the theory. The material is presented in an organized manner and at a level suitable for persons with knowledge of the theory of linear models. This book is also a place to which statisticians and mathematicians may turn in order to become acquainted with the subject. Teachers of experiment and treatment designs will find in it the necessary concepts, definitions, and results for an introduction to factorial design.

In presenting the topics on the subject of factorial design, we have departed from the traditional approaches, notations, and definitions. We did this for several reasons. First, it is necessary to be precise and to rigorously define all concepts involved. This is essential for mathematicians and statisticians who for the first time would like to become acquainted with the subject. For example, the term "factor" in factorial experiments may not be in the mathematician's vocabulary. Another reason for departing from the traditional approach is that we wish the reader to be unencumbered by previous notions about factorial design for comprehension of the concepts. Further, traditional notation usually breaks down in a more general setting, whereas our notation allows for generality and preciseness. Moreover, current concepts of fractional factorial designs relate solely to vaguely defined regular fractions. Finally, current concepts of factorial designs pertain to n-way classifications with an equal number of observations per combination.

The traditional approach in experiment and treatment design work has been to construct classes of designs and then to describe the properties pertaining to the designs. We do not use this approach; instead, we first define the concepts involved in factorial design in combinatorial and statistical terms. Then we discuss restraints and objectives as related to criteria for "goodness." We then present some characterizations and constructions using the developed concepts. Only best linear unbiased estimation of parameters or linear parametric functions is considered throughout the book.

This book originated from lecture notes prepared by the authors for an advanced class in the design and analysis of experiments during the 1972 spring term at Cornell University and subsequently used and modified for several courses at various universities.

We wish to express our gratitude to all graduate students in the course, especially to J. Eccleston, P. Farquhar, and J. Joiner, for their suggestions and comments during the preparation of the book. We are indebted to Professor Jack C. Kiefer, who presented several of the lectures and did a critical reading of most of the chapters of the book. His comments led to a considerable improvement throughout. Needless to say, the authors assume complete responsibility for any errors.

B. L. Raktoe
A. Hedayat
W. T. Federer

Dhahran, Saudi Arabia
Chicago, Illinois
Ithaca, New York
June 1981

Contents

Factorial Designs

CHAPTER 1

Introduction

1.1 INTRODUCTORY REMARKS

Many factors or variables often affect the responses for a phenomenon under experimentation. For studying a particular phenomenon, an experimenter could attempt to vary only one factor at a time, but this procedure is not always desirable or practical. Combinations of levels of two or more factors are utilized in experimentation and have been for more than several decades. Several advantages related to statistical precision and inference, accrue from using all possible combinations of the levels of the t factors under consideration, that is, a *complete factorial.* In some situations, however, it may be necessary or desirable to use only a subset of a complete factorial, that is, a *fractional factorial.*

When the number of factors t and/or the number of levels for the factors increase, the total number of combinations becomes large. In experimentation it may not be possible to obtain sufficient homogeneous material for using all combinations in the experiment, but it may be possible to obtain groups (blocks) of relatively homogeneous material. The question of how to allocate fractions to the blocks then arises. For example, consider an experiment in which five types of vitamins (factors) each with two levels, presence and absence of a vitamin, are to be used in an experiment to study their effects on weight of rats in a nutrition study. There are 2^5 combinations of vitamin levels. Further suppose that the experimenter had groups of 8 relatively homogeneous rats and that one set of 32 rats could be studied at one time. The question then is which 8 of 32 combinations should be included in each group. Further, suppose that 10 sets of 32 rats were available, how should the combinations be allocated to the groups of 8 for each set of 32 rats? The theory of factorial design deals with such complex questions as well as many others. Another question that arises is if 6 rats are available, how should one select the fraction for experimentation?

Many problems in factorial design theory turn out to have a geometric, an algebraic, or a combinatorial flavor. As a consequence, finite mathematical structures such as groups, rings, fields, Euclidean, and projective geometries have been used successfully in clarifying, extending, and resolving many issues related to factorial design. Also, the subject of combinatorics has been used as a major tool in the enumeration aspects of fractional factorials. Some aspects of these are discussed and illustrated throughout the book.

The purpose of the present work is to present a unified, cohesive, and general account of factorial design concepts. The literature on factorial designs is mostly specific in nature with emphasis on a particular topic such as symmetrical k^t factorials, where each of the t factors is at k levels, a particular type of fractional factorial design, or computer generation of fractional factorial designs. We deal with the general $k_1 \times k_2 \times \ldots \times k_t$ asymmetrical factorial, with general fractions, with equal or unequally replicated combinations, with estimation of linear combinations of parameters, and with general properties and constructions. In this book we attempt to present a systematic and unified approach to the subject of factorial design. New notations and definitions were developed to cope with such a treatment of the subject. Chapter 2 is devoted to developing a set of precise definitions and notation as a foundation for the remaining chapters. A linear model with an additive error structure was selected since a more general model often becomes intractable and little theory is available. In this chapter and thereafter we do not use the unfortunate misnomer "*full rank model*" but focus attention on designs that, in conjunction with a given model, might produce a *full rank design matrix*.

In Chapter 3, the relationship of an n-way classification with missing subclasses to a factorial design is noted. It is usual in customary linear model theory to treat the topics separately even though a one-to-one relation exists. Also, genetic experiments involving a subset of all possible crosses of a set of parents are related to fractional factorial designs. A discussion of estimable linear parametric functions using a factorial design is given and related to general linear model theory. Unbiased estimation is discussed in this context.

In Chapter 4 we develop the full polynomial model and its corresponding orthogonal polynomial model using the Gram-Schmidt matrix orthogonalization procedure. Explicit formulas are given for the general case as well as an illustration for equally spaced levels up to a fifth degree polynomial. Best linear unbiased estimation of the full parametric vector is developed along with estimation of linear functions of the parametric vector under the assumption that some parameters are zero.

For any given fraction of a complete factorial, there usually are many different designs. The problems of selecting a particular factorial design that is best in some sense, is discussed in Chapter 5. Eight criteria for selecting a design from the set of possible designs are described; the first five criteria are solely variance-covariance related properties while the last three deal with mean square error, alias optimality, and alias balanced properties.

The problem of finding a factorial design with the minimum number of combinations for obtaining an unbiased estimator of an arbitrary linear parametric function is unresolved. A characterization of such designs is formulated and discussed in Chapter 6. Also, some results on the problem are presented. This type of estimation problem is not a part of standard linear model theory.

In Chapter 7, the total parametric vector $\boldsymbol{\beta}_\rho$ for a complete factorial is partitioned into three parts in our consideration of factorial designs. The first part, $\boldsymbol{\beta}_1$, contains parameters of primary interest to the experimenter. The second part, $\boldsymbol{\beta}_2$ contains parameters of secondary interest to the experimenter, which may or may not be small enough to be neglected. The third part, $\boldsymbol{\beta}_3$, consists of parameters that are assumed to be small and therefore can be neglected. Within this framework, minimal designs are formulated to obtain unbiased estimators of the parameters in $\boldsymbol{\beta}_1$. Two general classes of factorial designs, that is, the class of odd resolution designs and the class of even resolution designs, are considered. Four different situations involving different assumptions about the parameters in $\boldsymbol{\beta}_2$ and $\boldsymbol{\beta}_3$ are discussed. In this chapter the important notion of confounding is also discussed and illustrated.

Two particular properties of factorial designs, orthogonality and balancedness, are considered in Chapter 8. These properties can be useful in selecting a design in a manner that will make it optimal under a specified criterion. Variance balancedness and covariance balancedness, as defined in Chapter 5, are applied to the four cases of partitioning of the total parametric vector of Chapter 7.

In Chapter 9, we consider the problem of unbiased estimation for the parameters in $\boldsymbol{\beta}_1$ when the parameters in $\boldsymbol{\beta}_3$ are not zero. The theory of randomized designs is used in this context. A class of designs, generated by the group of level permutations of the factors, is produced to illustrate how randomization theory can be utilized. In addition, the concept of a regular fractional factorial design of the prime powered factorial is developed and the same group of level permutations is used to obtain a class of regular designs. The development in this chapter complements the theory of unbiased estimation of $\boldsymbol{\beta}_1$ based on a fixed design.

In Chapters 10 and 11, special cases of odd and even resolution factorial designs, that is designs of resolutions III, IV, and V, are discussed in some detail. Resolution III designs allow unbiased estimation of main effect parameters when all interaction parameters are zero. Resolution IV designs allow unbiased estimation of all main effect parameters in the presence of two factor interactions. Resolution V designs allow unbiased estimation of all main effect and all two factor interaction parameters, given that the remaining interactions are zero. Since these are the situations that have received considerable attention in the literature, it is appropriate to consider them in our setting, notation, and generality. The problems associated with the construction of resolution III, IV, and V factorial designs are discussed and illustrated.

Chapter 12 develops the relatively new area of search factorial designs. Along with the formulation, some unsolved problems are mentioned. Several examples are given to illustrate the basic ideas.

In Chapter 13, many construction methods for factorial designs are listed and several of the methods are used for illustration purposes. A special discussion is provided of the methods using orthogonal and balanced arrays, Latin squares and orthogonal Latin squares, Hadamard matrices, and the direct product of two designs.

Whenever there is a need to clarify concepts, definitions, and results, examples are given. At the end of each chapter, we provide a list of references with recommendations for additional reading on the contents of the chapter. Finally, to make the book relatively self-contained we present in the Appendix the basic notions of certain finite mathematical structures, such as groups, rings, fields, and geometries.

A historical development of the statistical work on factorial designs may be obtained from the references listed below. Before the 1930s factorial experimentation was called *complex experimentation*; after the 1930s the use of this term was discontinued.

1.2 SELECTED ADDITIONAL READING

1. Bose, R. C. (1947). Mathematical theory of the symmetrical factorial design. *Sankhyā*, **8**, 107–166.
2. Connor, W. S. (1960). Construction of fractional factorial designs of the mixed $2^m\ 3^n$ series. In *Contributions in Probability and Statistics*. Stanford University Press, Stanford.
3. Finney, D. J. (1945). The fractional replication of factorial experiments. *Ann. Eugen.*, **12**, 291–301 (correction **15**, 276).
4. Fisher, R. A. (1926). The arrangement of field experiments. *J. Agric.*, **33**, 503–513.
5. Fisher, R. A. (1935). *The Design of Experiments*, 1st ed. Oliver and Boyd, Edinburgh (1966, 8th ed.).
6. Fisher, R. A. (1942). The theory of confounding in factorial experiments in relation to the theory of groups. *Ann. Eugen.*, **11**, 341–353.
7. Fisher, R. A. (1945). A system of confounding for factors with more than two alternatives, giving completely orthogonal cubes and higher powers. *Ann. Eugen.*, **12**, 283–290.

8. Li, J. C. R. (1944). Design and statistical analysis of some confounded factorial experiments. *Iowa Agric. Expt. Sta. Res. Bull.*, No. **333**, 452–492.

9. Morrison, M. (1956). Fractional replication for mixed series. *Biometrics*, **12**, 1–19.

10. Yates, F. (1933). Principles of orthogonality and confounding in replicated experiments. *J. Agric. Sci.*, **23**, 108–145.

11. Yates, F. (1935). Complex experiments. *J. Royal Stat. Soc. Suppl.*, **2**, 181–247.

12. Yates, F. (1937). The design and analysis of factorial experiments. *Imp. Bur. Soil Sci. Tech. Commun.*, **35**, 1–95.

CHAPTER 2

Preliminaries and Notation

In the first section of this chapter we present the basic concepts and definitions associated with factorial design. The second section is mostly concerned with some statistical concepts and definitions that are required for the ensuing developments.

2.1 COMBINATORIAL ASPECTS AND DEFINITIONS

In this book a *set* is a collection of distinct elements, that is, a listing without replications. If replications are present, the term *collection* will be used.

Definition 2.1. Consider t nonempty, not necessarily distinct sets $G_1, G_2, \ldots, G_t$ with cardinalities $k_1, k_2, \ldots, k_t$, respectively, such that $k_i > 1$ for all i. With each set G_i, we shall associate a formal symbol F_i, which will be denoted as the *ith factor*.

The reader may question the reason for using two symbols F_i and G_i. The usefulness of this becomes apparent in the cases where the elements of the G_i are labeled with the same names. Example 2.2 below is a case in point.

Definition 2.2. The elements of G_i when associated with the formal symbol F_i will be called the *levels of the ith factor*.

The levels of a factor are the possible levels specified by the experimenter at which an experiment can be conducted. This does not mean all of them are used in a particular experiment.

Practitioners distinguish four types of labelings of the levels of a factor:

(i) The elements of G_i are called *nominal levels* (or versions) if there is neither an order relationship nor a measure of distance among the levels. Typical examples of factors with nominal levels are sex, race, and religion.

(ii) The elements of G_i are called *ordinal levels* if there is order relationship among the levels but there is no meaningful measure of distance among them. A typical example of a factor with ordinal levels is learning speed of a pupil with use of the three categories "slow," "medium," and "fast" as levels.

(iii) The elements of G_i are called *interval levels* if there is both an order relationship and a measure of distance among them. However, there is no meaningful origin of measurements (such as zero) for the purpose of comparison among the levels. A typical example is temperature. There are both an order relationship and a measure of distance between say 10°and 20°C. But we cannot say that at 20° the temperature is twice as hot as at 10°C.

(iv) The elements of G_i are called *ratio levels* if there is an order relationship, a measure of distance, and a meaningful origin for comparison. Typical examples are length, distance, and time.

A factor is called a *qualitative factor* if its levels are nominal levels. Otherwise, it is called a *quantitative factor*.

Remark. The levels of a qualitative factor can be conveniently coded with numbers for the purpose of identification.

Let G be the Cartesian product of the G_i, that is, $G = X_{i=1}^{t} G_i$, where the symbol X denotes the Cartesian product. The set G together with F_i are often referred to as the $k_1 \times k_2 \times \ldots \times k_t$ *factorial or* $k_1 \times k_2 \times \ldots \times k_t$ *t-way crossed classification*, and G itself is called the *factor space*. An element of G is called a *treatment*.

Hereafter, whenever we discuss G, it is to be understood that we mean the set G together with the associated factors $F_1, F_2, \ldots, F_t$. Also, the terms "combinations," "treatment combinations," "runs," "assemblies," and "subclasses" are frequently used in the literature as names for the elements of G. In this setting, we allow $t = 1$, but we should note that most statistical literature is confined to the case $t > 1$.

Before proceeding further we present two examples to illustrate the concepts defined thus far.

Example 2.1. Consider an experiment where the effects of temperature and humidity on the keeping quality of potatoes held in storage are being studied. The range of experimentation for temperature is from 1 to 6°C while that for humidity is from 60 to 95%. Represent temperature as the

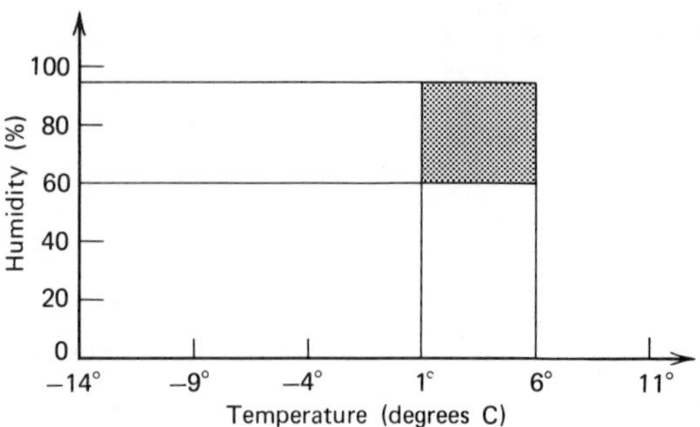

Figure 2.1. Combinations of temperature and humidity with shaded area representing the area of experimental interest.

horizontal axis and humidity as the vertical axis in the Cartesian plane. The shaded area in Figure 2.1 represents the combinations of temperature and humidity that are of experimental interest.

Identifying temperature with F_1 and humidity with F_2, we see that $G_1 = [1°\text{C}, 6°\text{C}]$ and $G_2 = [60\%, 95\%]$. For example, a level of F_1 is 3°C and a level of F_2 is 72.3%. The shaded area is G, and hence a treatment is a point in G, for instance (4°C,90%). In this example both factors are quantitative.

Example 2.2. A sociologist was interested in studying the influence of religious affiliation and income on attitude of individuals within a certain city toward extramarital sexual relations. The three denominations that could be studied were Protestant (P), Catholic (C), and Jewish (J); the income levels specified were low (L), medium (M), and high (H) as defined by the experimenter. The factors in this example are religion (F_1) and income (F_2) and the levels of the factors are given by $G_1 = \{\text{P}, \text{C}, \text{J}\}$ and $G_2 = \{\text{L}, \text{M}, \text{H}\}$. The factor space G consists of nine treatment combinations, that is

$$G = \{(P,L),(P,M),(P,H),(C,L),(C,M),\\ (C,H),(J,L),(J,M),(J,H)\}$$

In this example the factor religion is qualitative and the factor income is quantitative. The levels of religion are nominal while those of income are ordinal.

Often levels in such an experiment are coded by $\{0,1,2\}$ for each G_i. With such nomenclature, it is clear why "level one of F_2" is a useful designation, that is, why it was useful to introduce the symbol F_i. In coded form the factor space $G = \{(0,0), (0,1), (0,2), (1,0), (1,1), (1,2), (2,0), (2,1), (2,2)\}$ for the preceding example.

Definition 2.3. A *countable* set is a set that either has a finite number of elements or can be put into one-to-one correspondence with the set of positive integers. An *uncountable* set is a set that is not countable.

Since G is the Cartesian product of the G_i, it follows that G will be countable if and only if every G_i is countable. Also G will be uncountable if and only if at least one G_i is uncountable. In Example 2.1 the factor space G is uncountably infinite while that of Example 2.2 is finite and hence countable. Although there are many settings (e.g., response surface design problems) in which uncountably infinite factor spaces lead to interesting problems and applications, we will (unless otherwise stated) restrict the development to the case where G is *finite*. For this reason, let N be the cardinality of G and let G be indexed by a suitable index set.

Definition 2.4. A *factorial design* (or *factorial arrangement*) with parameters $k_1, k_2, \ldots, k_t, m, n, r_1, \ldots, r_N$, $m > 0$ and $n = \sum_{j=1}^{N} r_j > 0$, is a collection of n treatments of G such that the jth treatment in G has multiplicity $r_j \geq 0$ and m is the number of nonzero r_j. We denote such a factorial design by the symbol FD $(k_1, \ldots, k_t; m; n; r_1, \ldots, r_N)$.

This notation, although somewhat redundant, is useful. When there is no confusion the short notation FD will be used.

In a statistical setting the multiplicity r_j is referred to as *replication number* of the jth treatment. The statistician should observe that this definition is in agreement with the definition of a general t-way cross-classification with r_j observations on the jth treatment as used in statistical literature.

Definition 2.5. A factorial design is a *complete factorial design* or a *complete replicate* [denoted by CFD $(k_1, k_2, \ldots, k_t)$ or simply CFD if it is clear from the context] if $r_j > 0$ for all j, and a complete factorial design is said to be *minimal* if $r_j = 1$ for all j. A minimal complete factorial design will be denoted by MCFD $(k_1, \ldots, k_t)$ or simply MCFD if there is no ambiguity.

Note that an MCFD is a single copy of the factor space G and for this reason it is often referred to as a *single complete replicate of the* $k_1 \times k_2 \times \ldots \times k_t$ *factorial.* A complete factorial such that $r_j = r$, for all j, is said to consist of r *complete replicates of the* $k_1 \times k_2 \times \ldots \times k_t$ *factorial.*

Definition 2.6. An FD is *symmetrical* if $k_i = s$ for all i; otherwise it is *asymmetrical* or *mixed*. An FD is *prime powered* if $k_i = p_i^{u_i}$, such that for each i, p_i is a prime and u_i is a natural number greater than or equal to 1.

It follows that an FD can be *symmetrical prime powered* or *mixed prime powered*.

Since this book is mainly devoted to the family of fractional factorial designs, a formal definition of this is required.

Definition 2.7. A factorial design is said to be an *incomplete factorial design*, a *fractional factorial design* or, more simply, a *fractional replicate*, if some but not all $r_j > 0$. We denote a fractional replicate by FFD $(k_1,\ldots, k_t; m; n; r_1,\ldots, r_N)$ or by FFD if the context is clear.

We now illustrate the preceding definitions in the following two examples.

Example 2.3. Let $G_1 = \{0,1\}$ and $G_2 = \{0,1,2\}$ for two factors. The set of treatments $G = G_1 \times G_2 = \{(x_1, x_2), x_i \in G_i\} = \{(0,0), (0,1), (0,2), (1,0), (1,1), (1,2)\}$. Since $k_1 = 2$ and $k_2 = 3$, all the FD in this example are mixed prime powered. The factor space G is a minimal complete factorial design. An example of a complete factorial design that is not minimal is the collection $\{(0,0), (0,0), (0,1), (0,2), (0,2), (0,2), (1,0), (1,1), (1,2)\}$ $=$ FD$(2,3;6;9;2,1,3,1,1,1)$. An example of a fractional replicate is $\{(0,0),(0,0),(1,2),(1,2),(1,2),(1,2)\} =$ FFD $(2,3;2;6;2,0,0,0,0,4)$.

Example 2.4. In order to design an effective breakwater to protect a harbor from the forces of waves, an engineer measured the heights of the waves in the harbor area using a small scale model. The three specified lengths of breakwater were designated as d_1, d_2, and d_3, the two specified heights of breakwater were designated as h_1 and h_2, and the four feasible angles of the direction of force of the waves to the breakwater were designated as a_1, a_2, a_3, and a_4. Setting $G_1 = \{d_1, d_2, d_3\}, G_2 = \{h_1, h_2\}$, and $G_3 = \{a_1, a_2, a_3, a_4\}$, results in the following set of possible treatment combinations $G = G_1 \times G_2 \times G_3 = \{(d_1, h_1, a_1), (d_1, h_1, a_2), (d_1, h_1, a_3), (d_1, h_1, a_4), (d_1, h_2, a_1), (d_1, h_2, a_2), \ldots, (d_3, h_2, a_3), (d_3, h_2, a_4)\}$. These treatments may be reordered using the natural numbers 1,2,3,...,24 consecutively. Suppose now that the engineer could not conduct an experiment using the minimal factorial design G but due to cost limitations he or she was forced to use the fractional replicate FFD $(3, 2, 4; 8; 9; 2,0,0,0,1,0,1,1,1,0,0,0,0,1,0,0,0,0,0,0,0,0,1,1)$. In the original ordering this fractional replicate is $\{(d_1, h_1, a_1), (d_1, h_1, a_1), (d_1, h_2, a_1), (d_1, h_2, a_3),$

$(d_1,h_2,a_4),(d_2,h_1,a_1),(d_2,h_2,a_2),(d_3,h_2,a_3),(d_3,h_2,a_4)\}$, which may be shortened by using only the subscripts to obtain $\{(111), (111), (121), (123), (124),(211),(222),(323),(324)\}$.

From the preceding example the usefulness of an abbreviated notation becomes apparent. For this purpose the elements of G, that is, the g themselves, can be used as subscripts so that after deleting all the r_g that are equal to zero from the notation we obtain a more compact notation. This might, however, become cumbersome in the case of a large number of factors. Thus in the preceding fractional replicate we have FFD $(3,2,4;8;9;r_{(111)}=2,r_{(121)}=1,r_{(123)}=1,r_{(124)}=1,r_{(211)}=1,r_{(222)}=1,r_{(323)}=1, r_{(324)}=1)$. Finally if this notation is adopted there will appear exactly m such subscripts in the notation of a fractional factorial design, which then reflect the m distinct treatments appearing in the fractional factorial design.

With each treatment g in a factorial design we associate a *random variable* Y_g, which is called an *observation* or *response* or *measurement*. We omit the details of the customary measure-theoretic structure used in defining a random variable. For our purposes a random variable will always take on values in a finite dimensional Euclidean set. This is so because each Y_g will be finite dimensional and $n<\infty$ for any factorial design. Let Γ be an arbitrary factorial design and consider, unless otherwise stated, the univariate setting for each Y_g. That is, we assume each Y_g to be a one-dimensional random variable. Then with Γ we associate the $n\times 1$ observation vector $\mathbf{Y}_\Gamma$, whose gth element is Y_g. Let $F_{\mathbf{Y}_\Gamma}$ be the probability distribution of $\mathbf{Y}_\Gamma$, which is a (possibly unknown) member of a specified class $\mathcal{F}^*$ of distributions.

Although various other models can be postulated in certain situations the one we will consider throughout this book is the following:

Definition 2.8. By a *linear model* associated with a factorial design Γ we mean a relationship of the form

$$E[Y_g]=\sum_{i=0}^{k-1} f_i(g)\theta_i=\boldsymbol{f}'(g)\boldsymbol{\theta}$$

for each g in G, where $\boldsymbol{f}'(\cdot)=(f_0(\cdot),f_1(\cdot),\ldots,f_{k-1}(\cdot))$ is a vector of k real known functions defined on G and $\boldsymbol{\theta}'=(\theta_0,\theta_1,\ldots,\theta_{k-1})$ is a vector of k unknown parameters.

In matrix notation the linear model for any factorial design Γ may be written as

$$E[\mathbf{Y}_\Gamma]=\mathbf{W}_\Gamma\boldsymbol{\theta}, \tag{2.1}$$

where the element in the gth row and jth column of $\mathbf{W}_\Gamma$ is equal to $f_j(g)$ and $E(\cdot)$ is the expectation operator. The matrix $\mathbf{W}_\Gamma$ plays an important role in statistical analysis of data based on Γ. Therefore, it is appropriate to name this matrix. Throughout this book the $n \times k$ matrix $\mathbf{W}_\Gamma$ will be referred to as the *design matrix* associated with Γ relative to the given linear model or simply as the design matrix of Γ when the context is clear.

The reader may occasionally find in the literature the term design matrix being used for the design Γ itself. The majority of statisticians do not adhere to this practice and we will refrain from this also.

Linear models are often used for the purpose of statistical analysis. Two particular linear models frequently used are the *orthogonal polynomial* model and the *Helmert polynomial* model. These models are described in Chapter 4.

Example 2.5. An engineer working in material sciences experimented to determine electrical conductivity using two types of materials and both direct and alternating currents in a cold chamber. For this example material type is factor F_1 and current type is factor F_2. Let $G_1 = \{0,1\} = G_2$. Then $G = G_1 \times G_2 = \{(0,0),(0,1),(1,0),(1,1)\}$. Suppose that $E(Y_{(0,0)}, Y_{(0,1)}, Y_{(1,0)}, Y_{(1,1)}) = (\mu_{(0,0)}, \mu_{(0,1)}, \mu_{(1,0)}, \mu_{(1,1)})$. Assume that for $g = (x, y)$ the experimenter postulated the following relations between $E[Y_g]$ and the parameters θ_0, θ_1, θ_2, and θ_3:

$$E[Y_g] = \mu_{(x,y)} = \sum_{i=0}^{3} f_i(g)\theta_i$$

$$= \theta_0 + x\theta_1 + y\theta_2 + \left[\frac{(2x-1)(2y-1)+1}{2}\right]\theta_3.$$

In this linear model $f_0(g) = f_0[(x, y)] = 1$, $f_1(g) = f_1[(x, y)] = x$, $f_2[(x, y)] = y$, and $f_3(g) = f_3[(x, y)] = [(2x-1)(2y-1)+1]/2$. For the minimal complete factorial design in this experiment the model in matrix form is equal to

$$\begin{bmatrix} \mu_{(0,0)} \\ \mu_{(0,1)} \\ \mu_{(1,0)} \\ \mu_{(1,1)} \end{bmatrix} = \begin{bmatrix} 1 & 0 & 0 & 1 \\ 1 & 0 & 1 & 0 \\ 1 & 1 & 0 & 0 \\ 1 & 1 & 1 & 1 \end{bmatrix} \begin{bmatrix} \theta_0 \\ \theta_1 \\ \theta_2 \\ \theta_3 \end{bmatrix}.$$

Remark. If in the previous example it was assumed that $\theta_3 = 0$ and the experimenter had considered the fractional factorial design $\{(0,0),(0,1),(1,1)\}$, then the implied model for this experiment would have been

$$\begin{bmatrix} \mu_{(0,0)} \\ \mu_{(0,1)} \\ \mu_{(1,1)} \end{bmatrix} = \begin{bmatrix} 1 & 0 & 0 \\ 1 & 0 & 1 \\ 1 & 1 & 1 \end{bmatrix} \begin{bmatrix} \theta_0 \\ \theta_1 \\ \theta_2 \end{bmatrix}.$$

We could have considered functional relationships of the form $E[Y_g^p] = h(g, \theta_0, \theta_1, \ldots, \theta_{k-1})$, where for each $g \in \Gamma$, the function h could be quite complicated. Our linear model corresponds to the case $p = 1$ and the fact that h is linear in the parameters. The case $p = 1$ with the assumption that h is nonlinear in the parameters is known in the literature as a *nonlinear model*.

2.2 TREATMENT DESIGN AND STATISTICAL ESTIMATION PROBLEM

In practical applications an experimenter specifies his model from both theoretical and experimental considerations of the phenomenon under study. He conducts an experiment and makes observations in order to obtain estimates of the parameters in the model and to obtain evidence of the appropriateness of the proposed model. In experimentation, a treatment g is applied to an entity called an *experimental unit*. A response Y_g is observed and estimates of the parameters of the model are obtained using the Y_g; inferences may then be made from these estimates concerning the appropriateness of the specified model. The entire process of formulating and stating a model, taking observations, fitting the proposed model to the observations, and possibly conjecturing a new model until the discovery of a model that sufficiently describes the phenomenon under study is called *model building*.

We will not be concerned with all aspects of model building in this book since we are going to limit ourselves to the linear model $E[\mathbf{Y}_g] = \boldsymbol{\theta}' \boldsymbol{f}(g)$, where the set of parameters $\theta_0, \theta_1, \ldots, \theta_{k-1}$ will be referred to as the set of *factorial effects*. Essentially, the vector $\boldsymbol{\theta}$ reflects the behavior of $E[\mathbf{Y}_\Gamma]$ with respect to changes in the levels of the factors. Our designation of effects also includes the classical definition of effects in factorial experiments.

Suppose now that the experimenter's interest lies in obtaining information on $\boldsymbol{\theta}$ using a factorial design Γ. In typical applications the number of treatments in Γ depends on the number of unknown parameters in $\boldsymbol{\theta}$ and might also be dictated by economic and physical constraints. For given $\boldsymbol{\theta}$ and the possible choices of m distinct treatments, there may be many choices for the treatments in Γ such that information can be obtained on the parameters.

Explicitly, for a given linear model on G there is a family of c competing factorial designs $\Gamma_1, \Gamma_2, \ldots, \Gamma_c$. The matrix equations for these are

$$E[\mathbf{Y}_{\Gamma_1}] = \mathbf{W}_{\Gamma_1}\boldsymbol{\theta},\ E[\mathbf{Y}_{\Gamma_2}] = \mathbf{W}_{\Gamma_2}\boldsymbol{\theta}, \ldots, E[\mathbf{Y}_{\Gamma_c}] = \mathbf{W}_{\Gamma_c}\boldsymbol{\theta}.$$

In actual experimentation we will be dealing with the observational vector $\mathbf{Y}_\Gamma$ rather than $E[\mathbf{Y}_\Gamma]$ when the factorial design Γ is used. Denote by $\mathbf{Y}_\Gamma - E[\mathbf{Y}_\Gamma]$ the *deviation* or *error vector* $\boldsymbol{\epsilon}_\Gamma$. The previous equations can be written as

$$\begin{aligned} \mathbf{Y}_{\Gamma_1} &= \mathbf{W}_{\Gamma_1}\boldsymbol{\theta} + \boldsymbol{\epsilon}_{\Gamma_1} \\ \mathbf{Y}_{\Gamma_2} &= \mathbf{W}_{\Gamma_2}\boldsymbol{\theta} + \boldsymbol{\epsilon}_{\Gamma_2} \\ &\vdots \\ \mathbf{Y}_{\Gamma_c} &= \mathbf{W}_{\Gamma_c}\boldsymbol{\theta} + \boldsymbol{\epsilon}_{\Gamma_c}, \end{aligned}$$

where the $\boldsymbol{\epsilon}_{\Gamma_i}$ have zero expectation. Each equation is capable of providing some information on $\boldsymbol{\theta}$. This information depends on the treatments on which measurements have been made and the methods of estimating $\boldsymbol{\theta}$. Selecting the treatments in Γ_i is the *treatment design problem*, while selecting a method of estimating $\boldsymbol{\theta}$ is a *statistical estimation problem*. Ideally these problems should be considered together, but it is often too difficult to accomplish this and it is an aid to comprehension to consider them separately. Since for a given model the experimenter is usually confronted with a class of possible designs and a class of possible estimators, we clearly need criteria for selecting a particular treatment design and a particular method of estimation. In the following chapters we delve into these aspects further.

Example 2.6. Consider a $2\times2\times2$ factorial with the factors being camera brand, film type and filter type in an experiment with flashbulbs. Let the coded levels for the ith factor be indicated by $G_i = \{0,1\}, i = 1,2,3$. The complete replicate consists of $G = \{(0,0,0), (1,0,0), (0,1,0), (0,0,1), (1,1,0), (1,0,1), (0,1,1), (1,1,1)\}$. Consider the factorial design $\Gamma^* = \{(0,0,0),(1,1,0),(1,0,1),(0,1,1)\}$ and the linear model based on the four functions

$$f_0(g) = f_0[(x_1, x_2, x_3)] = 1$$

and

$$f_j(g) = f_j[(x_1, x_2, x_3)] = 2x_j - 1, \qquad j = 1,2,3.$$

Then the four-parameter linear model for Γ^* in the matrix notation $\mathbf{Y}_{\Gamma^*} =$

$\mathbf{X}_{\Gamma^*}\boldsymbol{\theta} + \boldsymbol{\epsilon}_{\Gamma^*}$ is explicitly

$$\begin{bmatrix} Y_{(0,0,0)} \\ Y_{(1,1,0)} \\ Y_{(1,0,1)} \\ Y_{(0,1,1)} \end{bmatrix} = \begin{bmatrix} 1 & -1 & -1 & -1 \\ 1 & 1 & 1 & -1 \\ 1 & 1 & -1 & 1 \\ 1 & -1 & 1 & 1 \end{bmatrix} \begin{bmatrix} \theta_0 \\ \theta_1 \\ \theta_2 \\ \theta_3 \end{bmatrix} + \begin{bmatrix} \epsilon_{(0,0,0)} \\ \epsilon_{(1,1,0)} \\ \epsilon_{(1,0,1)} \\ \epsilon_{(0,1,1)} \end{bmatrix}.$$

Consider the class $\mathcal{D}_4$ of designs such that each design consists of $m = n = 4$ distinct treatment combinations. Since G contains eight treatments, it follows that the cardinality of $\mathcal{D}_4$ is equal to $c = C(8,4) = 70$ and hence there are 70 competing designs. Each of the 70 designs is capable of providing some information relative to $\boldsymbol{\theta}$, and Γ^* is just one of these. The treatment design problem here consists of selecting a design in $\mathcal{D}_4$ with respect to some criterion and the statistical estimation problem consists of selecting a method of estimation of $\boldsymbol{\theta}$ with respect to a criterion.

The class $\mathcal{D}_4$ was considered in this example as an illustration. It is obvious that one may consider other classes, for example, by taking $m \neq 4$ and $m \neq n$, and then formulate the problems relative to these classes.

Remark. In the literature, selecting a method of estimation of $E[\mathbf{Y}_\Gamma]$ is known as the *response surface estimation problem* and the selection of treatments in Γ_i is known as the *response surface design problem*. Criteria for selecting a design are developed in Chapter 5.

2.3 SELECTED ADDITIONAL READING

Starred references are recommended for a first reading.

1. Anderson, V. L., and McLean, R. A. (1974). *Design of Experiments: A Realistic Approach*. Marcel Dekker, New York.
2. Bliss, C. I. (1960). *Statistics in Biology*, Vol. 2. McGraw-Hill, New York.
3. Bose, R. C. (1947). Mathematical theory of the symmetrical factorial design. *Sankhyā*, **8**, 107–166.
*4. Bose, R. C., and Srivastava, J. N. (1964). Mathematical theory of factorial designs I, analysis II, construction. *Bull. Int. Stat. Inst.*, **40**, 780–794.
5. Box, G. E. P., and Hunter, J. S. (1957). Multi-factor experimental designs for exploring response surfaces. *Ann. Math. Stat.*, **28**, 195–241.
6. Carrol, M. B., and Dykstra, O. (1958). The application of fractional factorials in a food research laboratory. In *Experimental Designs in Industry*, V. Chew, Ed. Wiley, New York, pp. 224–234.
*7. Chinloy, T., Innes, R. F., and Finney, D. J. (1953). An example of fractional replication in an experiment on sugar cane manuring. *J. Agric. Sci.*, **43**, 1–11.

*8. Cochran, W. G., and Cox, G. M. (1957). *Experimental Designs*, 2nd ed. Wiley, New York.

9. Connor, W. S. (1958). A fractional factorial experiment of the mixed $2^m 3^n$ series. *Trans. 14th Annu. Clin. Rochester Soc. Qual. Control*, 59–70.

*10. Daniel, C. (1957). Fractional replication in industrial experimentation. *Am. Soc. Qual. Control Annu. Conv. Trans.*, **11**, 229–233.

11. Davies, O. L. (1954). *The Design and Analysis of Industrial Experiments*. Oliver and Boyd, London.

12. Federer, W. T. (1955). *Experimental Design: Theory and Application*. Macmillan, New York.

13. Federer, W. T., and Balaam, L. N. (1972). *Bibliography on Experiment and Treatment Design. Pre-1968.* Oliver and Boyd, Edinburgh.

*14. Finney, D. J. (1945). The fractional replication of factorial experiments. *Ann. Eugen.*, **12**, 291–301.

15. Finney, D. J. (1960). *An Introduction to the Theory of Experimental Design*. Chicago University Press, Chicago.

16. Fisher, R. A. (1935). *The Design of Experiments*, 1st ed. Oliver and Boyd, Edinburgh. (1966, 8th ed.)

17. Horton, W. H. (1958). Experiences in fractional factorials. In *Experimental Designs in Industry*. V. Chew, Ed. Wiley, New York, pp. 207–223.

*18. John, P. W. M. (1971). *Statistical Design and Analysis of Experiments*. Macmillan, New York.

19. Kempthorne, O. (1947). A simple approach to confounding and fractional replication in factorial experiments. *Biometrika*, **34**, 255–272.

20. Kempthorne, O. (1952). *The Design and Analysis of Experiments*. Wiley, New York.

21. Kempthorne, O., and Folks, L. (1971). *Probability, Statistics and Data Analysis*. Iowa State University Press, Ames.

*22. Kempthorne, O., and Tischer, R. G. (1953). An example of the use of fractional replication. *Biometrics*, **9**, 295–303.

23. Kirk, R. E. (1968). *Experimental Design: Procedures for the Behavioural Sciences*. Brooks-Cole, Belmont, Calif.

24. Kishen, K. (1948). On fractional replication of the general symmetric factorial designs. *J. Indian Soc. Agric. Stat.*, **1**, 91–106.

25. Kurkjian, B., and Zelen, M. (1962). A calculus for factorial arrangements. *Ann. Math. Stat.*, **33**, 600–619.

26. Ogawa, J. (1975). *Statistical Theory of the Analysis of Experimental Designs*. Marcel Dekker, New York.

27. Raghavarao, D. (1971). *Construction and Combinatorial Problems in Design of Experiments*. Wiley, New York.

28. Rao, C. R. (1950). The theory of fractional replication in factorial experiments. *Sankhyā*, **10**, 81–86.

29. Winer, B. J. (1962). *Statistical Principles in Experimental Design*. McGraw-Hill, New York.

*30. Yates, F. (1937). The design and analysis of factorial experiments. *Imp. Bur. Soil Sci. Tech. Commun.*, **35**, 1–95.

CHAPTER 3

Some Facets of Factorial Design

The theory of factorial design is related to many topics in statistics and mathematics. The chapter on construction techniques utilizes some of these relationships. The purpose of the present chapter is to discuss some facets so that at least a deeper perspective about the scope of factorial design is initiated. For further exploration the reader is referred to the literature list at the end of the chapter.

3.1 BALANCED AND UNBALANCED CROSSED CLASSIFICATIONS AND FACTORIAL DESIGNS

In the previous chapter we equated the terms "$k_1 \times k_2 \times \ldots \times k_t$ factorial" with "$k_1 \times k_2 \times \ldots \times k_t$ t-way crossed classification." If it is clear from the context, then $k_1 \times k_2 \times \ldots \times k_t$ is deleted and we simply say *t-way crossed classification*. This term comes from the fact that in many investigations the resulting data can be displayed according to the t factors in a t-way table. It is customary in such situations to refer to the level combinations as *subclasses* or *cells* rather than treatments.

Example 3.1. Suppose that 12 uniformly soiled bath towels were washed with two brands of detergents using three different brands of washing machines such that every detergent was tested twice in each of the washing machines. If the measurements were washability scores, then the data can be displayed in a two-way classification table as

		Brand of washing machine		
		1	2	3
Brand of detergent	1	$Y^{(1)}_{(1,1)}$	$Y^{(1)}_{(1,2)}$	$Y^{(1)}_{(1,3)}$
		$Y^{(2)}_{(1,1)}$	$Y^{(2)}_{(1,2)}$	$Y^{(2)}_{(1,3)}$
	2	$Y^{(1)}_{(2,1)}$	$Y^{(1)}_{(2,2)}$	$Y^{(1)}_{(2,3)}$
		$Y^{(2)}_{(2,1)}$	$Y^{(2)}_{(2,2)}$	$Y^{(2)}_{(2,3)}$

Here $Y^{(h)}_{(x,y)}$ denotes the hth observation on treatment $g=(x,y)$. In the notation and definitions of the previous chapter, we have $G_1=\{1,2\}, G_2=\{1,2,3\}$, and the factor space $G=\{(1,1),(1,2),(1,3),(2,1),(2,2),(2,3)\}$. The six subclasses are simply the elements of G. The two-way classification displayed above consists of observations from a CFD with all $r_j=2$, $j=1,2,\ldots,6$.

In this example we have a case of what is known in the literature as an *equal numbers* or *balanced* two-way classification. In general, a t-way classification is said to be balanced if $r_j=r\geq 1$ for all $j=1,2,\ldots,N$. If all the r_j are not equal, then one speaks of the *unequal numbers* or *unbalanced* or *missing data* or *messy data* case. In the terminology of Chapter 2, the balanced t-way crossed classification corresponds to a CFD with $r_j=r$, and an unbalanced t-way classification may arise from a CFD or an FFD.

Example 3.2. Suppose that for some reason four scores were lost or were not recorded in the experiment of Example 3.1, resulting in the following table:

		Brand of washing machine		
		1	2	3
Brand of detergent	1	$Y^{(1)}_{(1,1)}$	$Y^{(1)}_{(1,2)}$	$Y^{(1)}_{(1,3)}$
		$Y^{(2)}_{(1,1)}$		
	2	$Y^{(1)}_{(2,1)}$		$Y^{(1)}_{(2,3)}$
		$Y^{(2)}_{(2,1)}$		$Y^{(2)}_{(2,3)}$

Here, $r_1=r_{(1,1)}=2, r_2=r_{(1,2)}=1, r_3=r_{(1,3)}=1, r_4=r_{(2,1)}=2, r_5=r_{(2,2)}=0$, and $r_6=r_{(2,3)}=2$, ana we have an example of an unbalanced two-way classification, which in factorial terminology corresponds to a fractional factorial design Γ from the 2×3 factorial.

One of the simpler linear models usually assumed for the $k_1\times k_2$ two-way classification, where the levels of the factors are labeled with integers, is given by the equation

$$E\left[Y_{(x,y)}\right]=f_0(x,y)\mu+\sum_i f_{1i}(x,y)\tau_i+\sum_j f_{2j}(x,y)\rho_j, \tag{3.1}$$

where $i,\ x=1,2,\ldots,k_1,\ j,\ y=1,2,\ldots,k_2$, and $f_0(x,y)=1$, for all $x,\ y$,

$$f_{1i}(x,y)=\begin{cases}1, & \text{if } x=i\\ 0, & \text{otherwise,}\end{cases}$$

and

$$f_{2j}(x,y)=\begin{cases}1, & \text{if } y=j\\ 0, & \text{otherwise.}\end{cases}$$

This model can be easily generalized to the t-way crossed classification. The parameter μ is called the *general mean*, τ_i is known as the *true effect of level i* of the first classification (brand of detergent), and ρ_j is known as the *true effect of level j* of the second classification (brand of washing machine). In matrix notation the model for the observations of Example 3.2 is written as

$$E[\mathbf{Y}_\Gamma]=E\begin{bmatrix}Y_{(1,1)}^{(1)}\\ Y_{(1,1)}^{(2)}\\ Y_{(1,2)}^{(1)}\\ Y_{(1,3)}^{(1)}\\ Y_{(2,1)}^{(1)}\\ Y_{(2,1)}^{(2)}\\ Y_{(2,3)}^{(1)}\\ Y_{(2,3)}^{(2)}\end{bmatrix}=\begin{bmatrix}1&1&0&1&0&0\\1&1&0&1&0&0\\1&1&0&0&1&0\\1&1&0&0&0&1\\1&0&1&1&0&0\\1&0&1&1&0&0\\1&0&1&0&0&1\\1&0&1&0&0&1\end{bmatrix}\begin{bmatrix}\mu\\ \tau_1\\ \tau_2\\ \rho_1\\ \rho_2\\ \rho_3\end{bmatrix}=\mathbf{W}_\Gamma\boldsymbol{\theta}.$$

The model described in (3.1) is a special case of the general linear model introduced in the previous chapter under Definition 2.8; hence the analysis of balanced or unbalanced t-way crossed classifications can be approached using factorial concepts.

Remark. The design and analysis of certain other well-known experiment designs may also be considered from the viewpoint of factorials and fractional replication. Readers familiar with the subject will realize that, for example, the *randomized complete block design* is a two-way classification design, and, depending on whether there are missing plots or not, it is an FFD or CFD from a two-factor factorial. Likewise the *general incomplete block design* may be viewed as a factorial design from a two-factor factorial. The *Latin square design* and the *Youden design* are three-way classification designs and as such the FFD from a three-factor factorial. Finally the reader is invited to consider the *nested classification design* from a factorial viewpoint.

3.2 DIALLEL CROSSING DESIGNS AND FACTORIAL DESIGNS

In genetic and breeding experiments various types of genetic crosses can be made between pairs of k lines. The cross of two lines is denoted as a *diallel cross* and the set of crosses to be made is known as a *diallel crossing plan*. If all possible crosses including crossing a line with itself (selfs) are made among k lines, then the resulting k^2 *crosses* are in one-to-one correspondence with the treatments of a two-factor factorial with k levels for each factor. Denote this diallel crossing design as Plan I. Note that a diallel crossing design could involve k_1 lines crossed in all combinations with k_2 other lines resulting in the $k_1 \times k_2$ factorial. Of course this can be generalized to all possible crosses when t sets of lines are considered and thus give rise to the $k_1 \times k_2 \times \ldots \times k_t$ factorial.

In Plan I if line x as the male parent is crossed with line y as the female parent then the cross may be denoted as (x, y). Now, if the cross is (y, x), that is, if the male and female parents are interchanged, then this is called the *reciprocal cross* of the original cross (x, y). This is the reason that there are $k(k-1)/2 + k(k-1)/2 + k = k^2$ crosses giving rise to the k^2 factorial.

In certain situations, the experimenters' interest may center on:

Plan II	All possible crosses (x, y), $x < y$ to produce $k(k-1)/2$ crosses, or
Plan III	$k(k-1)/2$ crosses plus the k selfs, that is, all crosses (x, y), $x \leq y$, to produce $k(k+1)/2$ crosses, or
Plan IV	$k(k-1)/2$ crosses plus $k(k-1)/2$ reciprocal crosses, that is, all crosses (x, y), $x \neq y$, to produce $k^2 - k = k(k-1)$ crosses.

Plan I corresponds to an MCFD of the k^2 factorial and Plans II–IV correspond to the FFD from the same factorial. If k is large, the fraction of crosses investigated may be much smaller than the $k(k-1)/2$ crosses of Plan II, thus giving rise to different FFD.

Example 3.3. Suppose that the yield is measured on $9(9-1)/2=36$ crosses (Plan II) among nine varieties of corn, which are planted on four plots each, that is, $r_j=4, j=1,2,\ldots,36$. Denoting the total yield of the cross (x, y) by $Y_{(x,y)}$, we obtain Table 3.1.

The data displayed in Table 3.1 may be viewed as coming from an FFD with $m=36$, $r_j=4$ for all j, and $n=144$. For a yield measurement $Y_{(x,y)}$ a typical linear model assumed is

$$E\left[Y_{(x,y)}\right]=f_0(x,y)\mu+\sum_i f_{1i}(x,y)\alpha_i+\sum_j f_{2j}(x,y)\beta_j+\sum_{i,j} f_{3ij}(x,y)\gamma_{ij}, \tag{3.2}$$

where $i, x=1,2,\ldots,k-1$, $j, y=2,3,\ldots,k$, $i<j$, $x<y$, and

$$f_0(x,y)=1, \qquad \text{for all } x, y,$$

$$f_{1i}(x,y)=\begin{cases}1, & \text{if } x=i\\ 0, & \text{otherwise,}\end{cases}$$

$$f_{2j}(x,y)=\begin{cases}1, & \text{if } y=j\\ 0, & \text{otherwise,}\end{cases}$$

$$f_{3ij}(x,y)=\begin{cases}1, & \text{if } x=i \text{ and } y=j\\ 0, & \text{otherwise.}\end{cases}$$

The parameter μ is called the *grand mean*, α_i is called the *true effect of line i*, β_j is called the *true effect of line j*, and γ_{ij} is known as the *true effect due to the cross of lines i and j.*

Table 3.1 Total Yield of Corn for 36 Crosses (Plan II) Among Nine Corn Varieties

Parent number	Parent number (y)								
(x)	1	2	3	4	5	6	7	8	9
1	—	$Y_{(1,2)}$	$Y_{(1,3)}$	$Y_{(1,4)}$	$Y_{(1,5)}$	$Y_{(1,6)}$	$Y_{(1,7)}$	$Y_{(1,8)}$	$Y_{(1,9)}$
2		—	$Y_{(2,3)}$	$Y_{(2,4)}$	$Y_{(2,5)}$	$Y_{(2,6)}$	$Y_{(2,7)}$	$Y_{(2,8)}$	$Y_{(2,9)}$
3			—	$Y_{(3,4)}$	$Y_{(3,5)}$	$Y_{(3,6)}$	$Y_{(3,7)}$	$Y_{(3,8)}$	$Y_{(3,9)}$
4				—	$Y_{(4,5)}$	$Y_{(4,6)}$	$Y_{(4,7)}$	$Y_{(4,8)}$	$Y_{(4,9)}$
5					—	$Y_{(5,6)}$	$Y_{(5,7)}$	$Y_{(5,8)}$	$Y_{(5,9)}$
6						—	$Y_{(6,7)}$	$Y_{(6,8)}$	$Y_{(6,9)}$
7							—	$Y_{(7,8)}$	$Y_{(7,9)}$
8								—	$Y_{(8,9)}$
9									—

The reader is invited to write out in matrix notation the model for the 36 observations. Note that the linear model given previously is a special case of the linear model given under Definition 2.8 and hence the analysis of diallel crossing experiments can be performed along factorial lines. When parent numbers x and parent numbers y represent the same set of parents, the linear model needs to be adjusted accordingly. The FFD described as Plan II has uses in many areas of study other than breeding and genetics. Some of these are tournament designs, cockfighting games, designs for paired comparisons, designs for successive teaching methods, designs for repeated measurements, designs for communication studies in psychology, and designs for competition in mixtures of plant and animal strains.

3.3 ESTIMABLE LINEAR PARAMETRIC FUNCTIONS IN FACTORIAL DESIGN

Let Γ be a given factorial design from the general $k_1 \times k_2 \times \ldots \times k_t$ factorial. In Chapter 2 we saw that the linear model for Γ in matrix notation is written as

$$E[\mathbf{Y}_\Gamma] = \mathbf{W}_\Gamma \boldsymbol{\theta}, \tag{3.3}$$

where the element in the gth row and jth column of $\mathbf{W}_\Gamma$ is equal to $f_j(g)$. If the $n \times k$ design matrix $\mathbf{W}_\Gamma$ in expression (3.3) is of full rank, that is, if the rank is equal to k, then we shall refer to the pair (Γ, model) that generated $\mathbf{W}_\Gamma$ as a *full rank setting*. We emphasize the dependency of the rank of $\mathbf{W}_\Gamma$ on both the model and the design Γ. For brevity a design in a full rank setting will be called a *full rank design*.

Example 3.4. Consider again the $2 \times 2 \times 2$ or 2^3 factorial introduced in Example 2.6 and the class $\mathfrak{D}_4$ of designs such that each design consists of $m = n = 4$ treatment combinations. It is a nice exercise to verify that among the 70 possible designs there are exactly 12 less than full rank designs and 58 full rank designs. Examples of such designs based on model (2.2) are, respectively, $\Gamma_1 = \{(0,0,0), (0,1,0), (0,0,1), (0,1,1)\}$ and $\Gamma_2 = \{(0,0,0), (1,1,0), (1,0,1), (0,1,1)\}$. By the way, Example 3.4 is related to the problem of *degenerate and nondegenerate simplexes of the t-cube*. In this example $t = 3$.

An interesting special case of (3.3) with $k = N$, which is treated extensively in the subsequent chapters, starts out with the minimal complete factorial design ρ consisting of $N = \prod k_i$ treatment combinations and the model

$$E[\mathbf{Y}_\rho] = \mathbf{W}_\rho \boldsymbol{\theta}, \tag{3.4}$$

where $\boldsymbol{\theta}$ is an $N \times 1$ vector of factorial effects and rank of $\mathbf{W}_\rho$ is equal to N. The polynomial model and orthogonal polynomial models as introduced in Chapter 4 belong to this setting. If Γ is an FFD, then it is implied that the $n \times N$ design matrix $\mathbf{W}_\Gamma$ has rank m, since Γ has m independent rows. Hence an FFD in this setting always leads to a less than full rank design.

Example 3.5. Consider the 2×2 or 2^2 factorial introduced in Example 2.5 with the following model for the minimal complete factorial design ρ:

$$E\begin{bmatrix} Y_{(0,0)} \\ Y_{(0,1)} \\ Y_{(1,0)} \\ Y_{(1,1)} \end{bmatrix} = \begin{bmatrix} 1 & 0 & 0 & 1 \\ 1 & 0 & 1 & 0 \\ 1 & 1 & 0 & 0 \\ 1 & 1 & 1 & 1 \end{bmatrix} \begin{bmatrix} \theta_0 \\ \theta_1 \\ \theta_2 \\ \theta_3 \end{bmatrix}. \tag{3.5}$$

Based on this model the FFD $\Gamma = \{(0,0),(0,0),(0,1),(1,0)\}$ has as design matrix

$$\mathbf{W}_\Gamma = \begin{bmatrix} 1 & 0 & 0 & 1 \\ 1 & 0 & 0 & 1 \\ 1 & 0 & 1 & 0 \\ 1 & 1 & 0 & 0 \end{bmatrix},$$

which has rank three, since Γ contains exactly three distinct treatment combinations. Hence Γ is a less than full rank design.

Now, let Γ be an FD of the $k_1 \times k_2 \times \dots \times k_t$ factorial and let $\mathbf{Y}_\Gamma^*$ be an $n \times 1$ observation vector with the underlying assumptions

$$\begin{aligned} \mathbf{Y}_\Gamma^* &= \mathbf{W}_\Gamma^* \boldsymbol{\theta} + \boldsymbol{\epsilon}_\Gamma^*, \\ \operatorname{Cov}(\boldsymbol{\epsilon}_\Gamma^*) &= E[\boldsymbol{\epsilon}_\Gamma^* \boldsymbol{\epsilon}_\Gamma^{*\prime}] = \sigma^2 \mathbf{V}_\Gamma, \end{aligned} \tag{3.6}$$

where $\mathbf{W}_\Gamma^*$ has rank less than or equal to k and $\mathbf{V}_\Gamma$ is a known and positive definite matrix of rank n. From a well-known theorem in matrix algebra, we know that there exists a nonsingular matrix $\mathbf{P}_\Gamma$ such that $\mathbf{P}_\Gamma' \mathbf{V}_\Gamma \mathbf{P}_\Gamma = \mathbf{I}_n$. On premultiplying the first equation in (3.6) by $\mathbf{P}_\Gamma'$, we obtain

$$\begin{aligned} \mathbf{Y}_\Gamma &= \mathbf{W}_\Gamma \boldsymbol{\theta} + \boldsymbol{\epsilon}_\Gamma, \\ \operatorname{Cov}(\boldsymbol{\epsilon}_\Gamma) &= \mathbf{P}_\Gamma' \mathbf{V}_\Gamma \mathbf{P}_\Gamma = \sigma^2 \mathbf{I}_n, \end{aligned} \tag{3.7}$$

where $\mathbf{Y}_\Gamma = \mathbf{P}'_\Gamma \mathbf{Y}^*_\Gamma$, $\mathbf{W}_\Gamma = \mathbf{P}'_\Gamma \mathbf{W}^*_\Gamma$, and $\epsilon_\Gamma = \mathbf{P}'_\Gamma \epsilon^*_\Gamma$. Hence without loss of generality we restrict ourselves below to the assumptions in (3.7).

Definition 3.1. A parametric function is said to be a *linear parametric function* of $\boldsymbol{\theta}' = (\theta_1, \theta_2, \ldots, \theta_k)$ if it is of the form

$$\psi = \mathbf{C}'\boldsymbol{\theta} = \sum_{i=1}^{k} c_i \theta_i, \tag{3.8}$$

where the $\mathbf{C}' = (c_1, c_2, \ldots, c_k)$ is a vector of known coefficients.

Definition 3.2. Two linear parametric functions $\psi_1 = \mathbf{C}'_1 \boldsymbol{\theta}$ and $\psi_2 = \mathbf{C}'_2 \boldsymbol{\theta}$ are said to be (*algebraically*) *independent* if $\mathbf{C}_1$ cannot be written as a scalar multiple of $\mathbf{C}_2$.

Definition 3.3. Two linear parametric functions $\psi_1 = \mathbf{C}'_1 \boldsymbol{\theta}$ and $\psi_2 = \mathbf{C}'_2 \boldsymbol{\theta}$ are said to be (*algebraically*) *orthogonal* if $\mathbf{C}'_1 \mathbf{C}_2 = \sum_{i=1}^{k} c_{1i} c_{2i} = 0$.

Definition 3.4. A linear parametric function $\psi = \mathbf{C}'\boldsymbol{\theta}$ is a *contrast* if $\sum_{i=1}^{k} c_i = 0$ and it is a *normalized contrast* if in addition $\sum_{i=1}^{k} c_i^2 = 1$.

Remark. Some linear parametric functions are often referred to in the statistical literature as *comparisons* and orthogonal contrasts are called *orthogonal comparisons*. The notions of algebraic independence and orthogonality are generalized in the usual way from two comparisons to a set of comparisons.

Example 3.6. Consider the 4×1 parametric vector $\boldsymbol{\theta}$ of Example 3.5; then it is easily verified that

$$\psi_1 = (\tfrac{1}{2} \quad -\tfrac{1}{2} \quad -\tfrac{1}{2} \quad \tfrac{1}{2})\boldsymbol{\theta},$$

$$\psi_2 = (1 \quad 1 \quad -1 \quad -1)\boldsymbol{\theta},$$

and

$$\psi_3 = (1 \quad -1 \quad 1 \quad -1)\boldsymbol{\theta}$$

form a set of three algebraically orthogonal contrasts, of which the first one is normalized. In matrix notation this set of three comparisons is written as

$$\boldsymbol{\psi}=\begin{bmatrix}\psi_1\\ \psi_2\\ \psi_3\end{bmatrix}=\begin{bmatrix}\frac{1}{2} & -\frac{1}{2} & -\frac{1}{2} & \frac{1}{2}\\ 1 & 1 & -1 & -1\\ 1 & -1 & 1 & -1\end{bmatrix}\begin{bmatrix}\theta_0\\ \theta_1\\ \theta_2\\ \theta_3\end{bmatrix}=\mathbf{C}'\boldsymbol{\theta},$$

where $\mathbf{C}'$ is known as the *contrast matrix*.

Example 3.7. For the two-way unbalanced crossed classification of Example 3.2, let

$$\begin{aligned}\psi_1&=(1\ 1\ 0\ 0\ 0\ 0)\boldsymbol{\theta}=\mu+\tau_1,\\ \psi_2&=(1\ 0\ 1\ 0\ 0\ 0)\boldsymbol{\theta}=\mu+\tau_2,\\ \psi_3&=(1\ 0\ 0\ 1\ 0\ 0)\boldsymbol{\theta}=\mu+\rho_1,\\ \psi_4&=(1\ 0\ 0\ 0\ 1\ 0)\boldsymbol{\theta}=\mu+\rho_2,\\ \psi_5&=(1\ 0\ 0\ 0\ 0\ 1)\boldsymbol{\theta}=\mu+\rho_3.\end{aligned}$$

In matrix form these five linear functions are written as

$$\boldsymbol{\psi}=\begin{bmatrix}1&1&0&0&0&0\\ 1&0&1&0&0&0\\ 1&0&0&1&0&0\\ 1&0&0&0&1&0\\ 1&0&0&0&0&1\end{bmatrix}\begin{bmatrix}\mu\\ \tau_1\\ \tau_2\\ \rho_1\\ \rho_2\\ \rho_3\end{bmatrix}=\mathbf{C}'\boldsymbol{\theta}.$$

Since the rank of $\mathbf{C}'$ is five it follows that $\boldsymbol{\psi}$ is a vector of algebraically independent linear combinations. None of the combinations are contrasts, and it can be verified that they form an algebraically nonorthogonal set.

Definition 3.5. A function $t(\mathbf{Y}_\Gamma)$ of the observations $\mathbf{Y}_\Gamma$ is said to be an *unbiased estimator* of a parametric function if $E[t(\mathbf{Y}_\Gamma)]$ is equal to the parametric function.

In Example 3.2 the function $t(\mathbf{Y}_\Gamma)=Y_{11}^{(1)}$ is an unbiased estimator of $h(\boldsymbol{\theta})=\mu+\tau_1+\rho_1$, since $E[Y_{11}^{(1)}]=(1\,1\,0\,1\,0\,0)\boldsymbol{\theta}=\mu+\tau_1+\rho_1$.

Example 3.8. Consider a single factor factorial with $G=\{x_1,x_2,\ldots,x_N\}$, where the x_i are concentrations of a drug in a toxicity experiment. Let the linear model be

$$\mathbf{Y}_G = \begin{bmatrix} Y_{x_1} \\ Y_{x_2} \\ \vdots \\ Y_{x_N} \end{bmatrix} = \begin{bmatrix} 1 & x_1 \\ 1 & x_2 \\ \vdots & \vdots \\ 1 & x_N \end{bmatrix} \begin{bmatrix} \theta_0 \\ \theta_1 \end{bmatrix} + \begin{bmatrix} \epsilon_{x_1} \\ \epsilon_{x_2} \\ \vdots \\ \epsilon_{x_N} \end{bmatrix} = \mathbf{W}_G \boldsymbol{\theta} + \boldsymbol{\epsilon}_G$$

and $E[\boldsymbol{\epsilon}_G \boldsymbol{\epsilon}_G'] = \sigma^2 \mathbf{I}_N$. For $N > 2$, the reader should verify that an unbiased estimator of the vector of parametric functions

$$\mathbf{h} = \begin{bmatrix} h_1(\boldsymbol{\theta}) \\ h_2(\boldsymbol{\theta}) \\ h_3(\sigma^2) \end{bmatrix} = \begin{bmatrix} \theta_0 \\ \theta_1 \\ \sigma^2 \end{bmatrix}$$

is given by

$$\mathbf{t} = \begin{bmatrix} t_1(\mathbf{Y}_G) \\ t_2(\mathbf{Y}_G) \\ t_3(\mathbf{Y}_G) \end{bmatrix} = \begin{bmatrix} \hat{\theta}_0 \\ \hat{\theta}_1 \\ \hat{\sigma}^2 \end{bmatrix} = \begin{bmatrix} \bar{Y} - \hat{\theta}_1 \bar{x} \\ \left[\sum_{i=1}^{N} (x_i - \bar{x})(Y_{x_i} - \bar{Y}) \right] \Big/ \left[\sum_{i=1}^{N} (x_i - \bar{x})^2 \right] \\ \sum_{i=1}^{N} \left(Y_{x_i} - \hat{\theta}_0 - \hat{\theta}_1 x_i \right)^2 / (N-2) \end{bmatrix},$$

where $\bar{Y} = (1/N) \sum_{i=1}^{N} Y_{x_i}$ and $\bar{x} = (1/N) \sum_{i=1}^{N} x_i$. These estimators can be written compactly as

$$\hat{\boldsymbol{\theta}} = \begin{bmatrix} \hat{\theta}_0 \\ \theta_1 \end{bmatrix} = (\mathbf{W}_G' \mathbf{W}_G)^{-1} \mathbf{W}_G' \mathbf{Y}_G, \tag{3.9}$$

and

$$\hat{\sigma}^2 = \left[(\mathbf{Y}_G - \mathbf{W}_G \hat{\boldsymbol{\theta}})' (\mathbf{Y}_G - \mathbf{W}_G \hat{\boldsymbol{\theta}}) \right] / (N-2). \tag{3.10}$$

Definition 3.6. A function of parameters is said to be *estimable under the design setting (model,* Γ) if there exists an unbiased estimator for it based on the observation vector $\mathbf{Y}_\Gamma$.

Remark. Henceforth if a function of parameters is estimable under (model, Γ), we shall briefly say it is *estimable under* Γ if the underlying model is clear from the context.

Example 3.9. In Example 3.2 the parametric function $\psi=\rho_1-\rho_2$ is estimable, since

$$\begin{aligned} E\left[Y_{(1,1)}^{(2)}-Y_{(1,2)}^{(1)}\right] &= (1\ \ 1\ \ 0\ \ 1\ \ 0\ \ 0)\boldsymbol{\theta}-(1\ \ 1\ \ 0\ \ 0\ \ 1\ \ 0)\boldsymbol{\theta} \\ &= (0\ \ 0\ \ 0\ \ 1\ \ -1\ \ 0)\boldsymbol{\theta} \\ &= \rho_1-\rho_2. \end{aligned}$$

That is, an unbiased estimator of the contrast $\rho_1-\rho_2$ is $Y_{(1,1)}^{(2)}-Y_{(1,2)}^{(1)}$.

Definition 3.7. A function of parameters is said to be *linearly estimable* under Γ if there exists an unbiased estimator for the function that can be written as a linear combination of the observations in $\mathbf{Y}_\Gamma$.

Example 3.10. The function $h(\boldsymbol{\theta})=\rho_1-\rho_2$ in Example 3.9 is linearly estimable since an unbiased estimator is $Y_{(1,1)}^{(2)}-Y_{(1,2)}^{(1)}=(0\ \ 1\ \ -1\ \ 0\ \ 0\ \ 0\ \ 0)\mathbf{Y}_\Gamma$, which is a linear combination of the observations.

Throughout this book, unless specifically stated otherwise, we mean linearly estimable when we say estimable. The following theorems are standard results in the theory of linear estimation.

Theorem 3.1. The linear parametric function $\psi=\mathbf{C}'\boldsymbol{\theta}$ is estimable under Γ if and only if $\mathbf{C}'$ lies in the row space of $\mathbf{W}_\Gamma$, that is, if and only if there exists a vector $\mathbf{A}'$ such that $\mathbf{C}'=\mathbf{A}'\mathbf{W}_\Gamma$.

Definition 3.8. A linear unbiased estimator $t^*(\mathbf{Y}_\Gamma)=\mathbf{A}^{*\prime}\mathbf{Y}_\Gamma$ of the estimable function $\psi=\mathbf{C}'\boldsymbol{\theta}$ is said to be a *best linear unbiased estimator* (BLUE) of ψ if $\mathrm{Var}[t^*(\mathbf{Y}_\Gamma)]=\sigma^2\mathbf{A}^{*\prime}\mathbf{A}^*$ is minimum in the class of all linear unbiased estimators of ψ.

Definition 3.9. A vector of functions $\hat{\boldsymbol{\theta}}'=(\hat{\theta}_1=\hat{\theta}_1(\mathbf{Y}_\Gamma), \hat{\theta}_2=\hat{\theta}_2(\mathbf{Y}_\Gamma), \ldots, \hat{\theta}_k=\hat{\theta}_k(\mathbf{Y}_\Gamma))$ of the observations under Γ is said to be a vector of *least squares estimators* (LS estimators) of the parameters in $\boldsymbol{\theta}'=(\theta_1,\theta_2,\ldots,\theta_k)$ if the values $\theta_j=\hat{\theta}_j$, $j=1,2,\ldots,k$, minimize $(\mathbf{Y}_\Gamma-\mathbf{W}_\Gamma\boldsymbol{\theta})'(\mathbf{Y}_\Gamma-\mathbf{W}_\Gamma\boldsymbol{\theta})$.

The following results relate to estimation of an estimable function under a design Γ.

Theorem 3.2. Every estimable function $\psi=\mathbf{C}'\boldsymbol{\theta}$ has a unique linear unbiased estimator $\hat{\psi}$, which has minimum variance in the class of all linear unbiased estimators. The estimator may be obtained as $\hat{\psi}=\mathbf{C}'\hat{\boldsymbol{\theta}}$, where $\hat{\boldsymbol{\theta}}$ is a least squares estimator of $\boldsymbol{\theta}$.

The estimator $\hat{\psi}$ above is called the least squares estimator of ψ and by Definition 3.8 it is the best linear unbiased estimator (BLUE) of ψ. In practice we may calculate $\hat{\boldsymbol{\theta}}$ by utilizing the *normal equations*, which are obtained by minimizing $\boldsymbol{\epsilon}'_\Gamma\boldsymbol{\epsilon}_\Gamma = (\mathbf{Y}_\Gamma - \mathbf{W}_\Gamma\boldsymbol{\theta})'(\mathbf{Y}_\Gamma - \mathbf{W}_\Gamma\boldsymbol{\theta})$ with respect to $\boldsymbol{\theta}$. These equations are

(3.11) $$\mathbf{W}'_\Gamma\mathbf{W}_\Gamma\boldsymbol{\theta} = \mathbf{W}'_\Gamma\mathbf{Y}_\Gamma,$$

and a least squares estimator of $\boldsymbol{\theta}$ is given by

(3.12) $$\hat{\boldsymbol{\theta}} = (\mathbf{W}'_\Gamma\mathbf{W}_\Gamma)^- \mathbf{W}'_\Gamma\mathbf{Y}_\Gamma,$$

where $(\mathbf{W}'_\Gamma\mathbf{W}_\Gamma)^-$ is a *generalized inverse* of $\mathbf{W}'_\Gamma\mathbf{W}_\Gamma$. If $\mathbf{W}_\Gamma$ is of full rank then $\mathbf{W}'_\Gamma\mathbf{W}_\Gamma$ has an inverse and the least squares estimator $\hat{\boldsymbol{\theta}} = (\mathbf{W}'_\Gamma\mathbf{W}_\Gamma)^{-1}\mathbf{W}'_\Gamma\mathbf{Y}_\Gamma$ is unique.

Theorem 3.3. If $\psi_1 = \mathbf{C}'_1\boldsymbol{\theta}$ and $\psi_2 = \mathbf{C}'_2\boldsymbol{\theta}$ are estimable functions with the BLUE $\hat{\psi}_1 = \mathbf{C}'_1\hat{\boldsymbol{\theta}}$ and $\hat{\psi}_2 = \mathbf{C}'_2\hat{\boldsymbol{\theta}}$, then

(3.13) $$\begin{aligned} \operatorname{Var}(\mathbf{C}'_1\hat{\boldsymbol{\theta}}) &= \sigma^2\mathbf{C}'_1(\mathbf{W}'_\Gamma\mathbf{W}_\Gamma)^- \mathbf{C}_1; \\ \operatorname{Cov}(\mathbf{C}'_1\hat{\boldsymbol{\theta}}, \mathbf{C}'_2\hat{\boldsymbol{\theta}}) &= \sigma^2\mathbf{C}'_1(\mathbf{W}'_\Gamma\mathbf{W}_\Gamma)^- \mathbf{C}_2. \end{aligned}$$

Before providing examples we generalize the preceding results to a vector of parametric functions.

Theorem 3.4. Let $\boldsymbol{\psi} = \mathbf{C}'\boldsymbol{\theta}$ be a vector of estimable parametric functions. Then the BLUE of $\boldsymbol{\psi}$ is

(3.14) $$\hat{\boldsymbol{\psi}} = \mathbf{C}'(\mathbf{W}'_\Gamma\mathbf{W}_\Gamma)^- \mathbf{W}'_\Gamma\mathbf{Y}_\Gamma$$

and

(3.15) $$\operatorname{Cov}(\hat{\boldsymbol{\psi}}) = \sigma^2\mathbf{C}'(\mathbf{W}'_\Gamma\mathbf{W}_\Gamma)^- \mathbf{C}.$$

In particular, the BLUE of the response surface $E[\mathbf{Y}_\Gamma] = \mathbf{W}_\Gamma\boldsymbol{\theta}$ is given by $\widehat{E[\mathbf{Y}_\Gamma]} = \mathbf{W}_\Gamma(\mathbf{W}'_\Gamma\mathbf{W}_\Gamma)^-\mathbf{W}'_\Gamma\mathbf{Y}_\Gamma$.

Observe that the matrix $(\mathbf{W}'_\Gamma\mathbf{W}_\Gamma)$ plays a major role in statistical analysis of data. Due to its crucial importance in (3.14) and (3.15) we shall call it the *information matrix* of Γ relative to $\boldsymbol{\theta}$ under a given model. More specifically, let $\mathbf{Y}_\Gamma$ be the vector of n observations obtained under Γ. If the linear model associated with $\mathbf{Y}_\Gamma$ is

$$\mathbf{Y}_\Gamma = \mathbf{W}_\Gamma \boldsymbol{\theta} + \boldsymbol{\epsilon}_\Gamma, \qquad \mathrm{Cov}(\boldsymbol{\epsilon}_\Gamma) = \sigma^2 \mathbf{I}_n$$

then the matrix $(\mathbf{W}'_\Gamma \mathbf{W}_\Gamma)$ is called the *information matrix* of Γ with respect to $\boldsymbol{\theta}$ for the given model.

Definition 3.10. A vector of estimable functions $\boldsymbol{\psi} = \mathbf{C}'\boldsymbol{\theta}$ is said to be *statistically orthogonal* if under Γ the matrix $\mathbf{C}'(\mathbf{W}'_\Gamma \mathbf{W}_\Gamma)^- \mathbf{C}$ is diagonal.

Note that if $\boldsymbol{\psi} = \mathbf{C}'\boldsymbol{\theta}$ is statistically orthogonal, then this does not necessarily imply that $\mathbf{C}'\boldsymbol{\theta}$ is algebraically orthogonal and vice versa. The same comment applies to a vector of contrasts. However, in many important "balanced" settings a vector of contrasts are statistically orthogonal if and only if they are algebraically orthogonal.

In the analysis of data from experiments, and in particular from factorial experiments, we are usually interested in statistically orthogonal contrasts. This is in general a fairly difficult task. In some settings the equivalence of algebraic and statistical orthogonality leads to a simpler analysis, because algebraic orthogonality is easy to construct (e.g., the Gram-Schmidt process, see Chapter 4) and easy to verify. The concept of algebraic orthogonality is illustrated in Example 3.11.

Hereafter, unless otherwise stated, when we use the terms "independent" and "orthogonal," we mean them in the algebraic sense.

The following theorem tells us how many linearly independent functions can be estimated from a given design Γ.

Theorem 3.5. If the rank of the design matrix $\mathbf{W}_\Gamma$ is equal to s, then there are exactly s linearly independent estimable functions.

Example 3.11. Consider the $2 \times 2 \times 2$ factorial with $G_1 = G_2 = G_3 = \{0,1\}$ and the following linear model for the FFD $\Gamma = \{(0,0,0),(0,0,0),(0,1,0),(0,1,1),(0,1,1)\}$:

$$\begin{bmatrix} Y^{(1)}_{(0,0,0)} \\ Y^{(2)}_{(0,0,0)} \\ Y^{(1)}_{(0,1,0)} \\ Y^{(1)}_{(0,1,1)} \\ Y^{(2)}_{(0,1,1)} \end{bmatrix} = \begin{bmatrix} 1 & -1 & -1 \\ 1 & -1 & -1 \\ 1 & -1 & 1 \\ 1 & -1 & -1 \\ 1 & -1 & -1 \end{bmatrix} \begin{bmatrix} \theta_0 \\ \theta_1 \\ \theta_2 \end{bmatrix} + \begin{bmatrix} \epsilon^{(1)}_{(0,0,0)} \\ \epsilon^{(2)}_{(0,0,0)} \\ \epsilon^{(1)}_{(0,1,0)} \\ \epsilon^{(1)}_{(0,1,1)} \\ \epsilon^{(2)}_{(0,1,1)} \end{bmatrix},$$

where $\mathrm{Cov}(\boldsymbol{\epsilon}_\Gamma)=\sigma^2\mathbf{I}$. Here the design matrix $\mathbf{W}_\Gamma$ is determined by $f_0[(x_1,x_2,x_3)]=1$, $f_1[(x_1,x_2,x_3)]=2x_1-1$, and $f_2[(x_1,x_2,x_3)]=(2x_1-1)(2x_2-1)(2x_3-1)$. The rank of $\mathbf{W}_\Gamma$ is equal to two, so that by Theorem 3.5 exactly two linearly independent functions of $\boldsymbol{\theta}$ can be estimated. Let these two functions be given by

$$\boldsymbol{\psi}=\mathbf{C}'\boldsymbol{\theta}=\begin{bmatrix}1 & -1 & -1\\ 0 & 0 & 1\end{bmatrix}\begin{bmatrix}\theta_0\\ \theta_1\\ \theta_2\end{bmatrix}.$$

The vector $\boldsymbol{\psi}$ is estimable, since $\mathbf{C}'=\mathbf{K}'\mathbf{W}_\Gamma$, where

$$\mathbf{K}'=\begin{bmatrix}1 & -1 & 0 & -1 & 2\\ \frac{1}{2} & \frac{1}{2} & \frac{1}{2} & -\frac{1}{2} & -1\end{bmatrix}.$$

To obtain the BLUE of $\boldsymbol{\psi}$ we use the generalized inverse

$$(\mathbf{W}_\Gamma'\mathbf{W}_\Gamma)^-=\frac{1}{16}\begin{bmatrix}5 & 0 & 3\\ 0 & 0 & 0\\ 3 & 0 & 5\end{bmatrix}.$$

That $(\mathbf{W}_\Gamma'\mathbf{W}_\Gamma)^-$ is a generalized inverse may be verified by calculating $(\mathbf{W}_\Gamma'\mathbf{W}_\Gamma)(\mathbf{W}_\Gamma'\mathbf{W}_\Gamma)^-(\mathbf{W}_\Gamma'\mathbf{W}_\Gamma)$ and seeing whether this is equal to $\mathbf{W}_\Gamma'\mathbf{W}_\Gamma$. Doing this we first calculate

$$(\mathbf{W}_\Gamma'\mathbf{W}_\Gamma)=\begin{bmatrix}5 & -5 & -3\\ -5 & 5 & 3\\ -3 & 3 & 5\end{bmatrix}.$$

Next,

$$\begin{aligned}&(\mathbf{W}_\Gamma'\mathbf{W}_\Gamma)(\mathbf{W}_\Gamma'\mathbf{W}_\Gamma)^-(\mathbf{W}_\Gamma'\mathbf{W}_\Gamma)\\ &=\frac{1}{16}\begin{bmatrix}5 & -5 & -3\\ -5 & 5 & 3\\ -3 & 3 & 5\end{bmatrix}\begin{bmatrix}5 & 0 & 3\\ 0 & 0 & 0\\ 3 & 0 & 5\end{bmatrix}\begin{bmatrix}5 & -5 & -3\\ -5 & 5 & 3\\ -3 & 3 & 5\end{bmatrix}\\ &=\begin{bmatrix}1 & 0 & 0\\ -1 & 0 & 0\\ 0 & 0 & 1\end{bmatrix}\begin{bmatrix}5 & -5 & -3\\ -5 & 5 & 3\\ -3 & 3 & 5\end{bmatrix}\\ &=\begin{bmatrix}5 & -5 & -3\\ -5 & 5 & 3\\ -3 & 3 & 5\end{bmatrix}=(\mathbf{W}_\Gamma'\mathbf{W}_\Gamma).\end{aligned}$$

Hence $(\mathbf{W}_\Gamma'\mathbf{W}_\Gamma)^-$ is a legitimate generalized inverse, so that the BLUE of $\boldsymbol{\psi}$

and its covariance by Theorem 3.4 are equal to

$$\hat{\psi}=\begin{bmatrix}1 & -1 & -1\\ 0 & 0 & 1\end{bmatrix}\frac{1}{16}\begin{bmatrix}5 & 0 & 3\\ 0 & 0 & 0\\ 3 & 0 & 5\end{bmatrix}\begin{bmatrix}1 & 1 & 1 & 1 & 1\\ -1 & -1 & -1 & -1 & -1\\ -1 & -1 & 1 & -1 & -1\end{bmatrix}\mathbf{Y}_\Gamma$$

$$=\frac{1}{16}\begin{bmatrix}2 & 0 & -2\\ 3 & 0 & 5\end{bmatrix}\begin{bmatrix}1 & 1 & 1 & 1 & 1\\ -1 & -1 & -1 & -1 & -1\\ -1 & -1 & 1 & -1 & -1\end{bmatrix}\mathbf{Y}_\Gamma$$

$$=\frac{1}{16}\begin{bmatrix}4 & 4 & 0 & 4 & 4\\ -2 & -2 & 8 & -2 & -2\end{bmatrix}\mathbf{Y}_\Gamma$$

$$=\begin{bmatrix}\left(Y^{(1)}_{(0,0,0)}+Y^{(2)}_{(0,0,0)}+Y^{(1)}_{(0,1,1)}+Y^{(2)}_{(0,1,1)}\right)/4\\ \left(-Y^{(1)}_{(0,0,0)}-Y^{(2)}_{(0,0,0)}+4Y^{(1)}_{(0,1,0)}-Y^{(1)}_{(0,1,1)}-Y^{(2)}_{(0,1,1)}\right)/8\end{bmatrix}$$

and

$$\operatorname{Cov}(\hat{\psi})=\sigma^2\begin{bmatrix}1 & -1 & -1\\ 0 & 0 & 1\end{bmatrix}\frac{1}{16}\begin{bmatrix}5 & 0 & 3\\ 0 & 0 & 0\\ 3 & 0 & 5\end{bmatrix}\begin{bmatrix}1 & 0\\ -1 & 0\\ -1 & 1\end{bmatrix}$$

$$=\frac{\sigma^2}{16}\begin{bmatrix}2 & 0 & -2\\ 3 & 0 & 5\end{bmatrix}\begin{bmatrix}1 & 0\\ -1 & 0\\ -1 & 1\end{bmatrix}$$

$$=\frac{\sigma^2}{16}\begin{bmatrix}4 & -2\\ -2 & 5\end{bmatrix}.$$

Instead of working with the vector of independent contrasts

$$\psi=\mathbf{C}'\boldsymbol{\theta}=\begin{bmatrix}1 & -1 & -1\\ 0 & 0 & 1\end{bmatrix}\begin{bmatrix}\theta_0\\ \theta_1\\ \theta_2\end{bmatrix}$$

we may use the Gram-Schmidt matrix $\mathbf{H}^*$ and row-wise orthogonalize the matrix $\mathbf{C}'$ and obtain

$$\psi^*=\mathbf{H}^*\mathbf{C}'\boldsymbol{\theta}=\mathbf{C}^{*\prime}\boldsymbol{\theta}=\begin{bmatrix}1 & 0\\ \frac{1}{3} & 1\end{bmatrix}\begin{bmatrix}1 & -1 & -1\\ 0 & 0 & 1\end{bmatrix}\begin{bmatrix}\theta_0\\ \theta_1\\ \theta_2\end{bmatrix}$$

$$=\begin{bmatrix}1 & -1 & -1\\ \frac{1}{3} & -\frac{1}{3} & \frac{2}{3}\end{bmatrix}\begin{bmatrix}\theta_0\\ \theta_1\\ \theta_2\end{bmatrix}.$$

This has provided us with two algebraically orthogonal comparisons, which are estimable, since

$$\mathbf{C}^{*\prime} = \mathbf{H}^*\mathbf{C}' = \mathbf{H}^*\mathbf{K}'\mathbf{W}_\Gamma = \mathbf{K}^{*\prime}\mathbf{W}_\Gamma.$$

Note however that ψ^* is not statistically orthogonal, since

$$\mathbf{C}^{*\prime}(\mathbf{W}_\Gamma'\mathbf{W}_\Gamma)^- \mathbf{C}^* = \frac{1}{16}\begin{bmatrix} 1 & -1 & -1 \\ \frac{1}{3} & -\frac{1}{3} & \frac{2}{3} \end{bmatrix}\begin{bmatrix} 5 & 0 & 3 \\ 0 & 0 & 0 \\ 3 & 0 & 5 \end{bmatrix}\begin{bmatrix} 1 & \frac{1}{3} \\ -1 & -\frac{1}{3} \\ -1 & \frac{2}{3} \end{bmatrix}$$

$$= \frac{1}{144}\begin{bmatrix} 36 & -6 \\ -6 & 37 \end{bmatrix}.$$

It is an interesting exercise to construct two estimable statistically orthogonal comparisons for the preceding example.

The next theorem provides an unbiased estimator of the variance σ^2 of the model in Equation (3.7).

Theorem 3.6. An unbiased estimator of σ^2 is given by the expression

$$\hat{\sigma}^2 = \frac{[(\mathbf{Y}_\Gamma - \mathbf{W}_\Gamma\hat{\boldsymbol{\theta}})'(\mathbf{Y}_\Gamma - \mathbf{W}_\Gamma\hat{\boldsymbol{\theta}})]}{(n-s)}, \tag{3.16}$$

where $\hat{\boldsymbol{\theta}}$ is a least squares estimator of $\boldsymbol{\theta}$ and s is the rank of the design matrix $\mathbf{W}_\Gamma$.

3.4 SELECTION OF ESTIMABLE SUBSETS OF PARAMETERS IN FACTORIAL DESIGN

In Equation (3.4) we considered a linear model associated with the minimal complete factorial ρ, that is, $E[\mathbf{Y}_\rho] = \mathbf{W}_\rho\boldsymbol{\theta}$, where the rank of $\mathbf{W}_\rho$ is equal to N. This type of model arises frequently in factorial experimentation and may be deduced as follows.

Let the treatment combinations in ρ be indexed from 1 to N, and set $E[\mathbf{Y}_\rho] = \boldsymbol{\mu} = (E[Y_1] = \mu_1, E[Y_2] = \mu_2, \ldots, E[Y_N] = \mu_N)'$. Associate with $\boldsymbol{\mu}$ a vector of N independent linear combinations of $\boldsymbol{\mu}$ given by $\boldsymbol{\theta} = \mathbf{W}_\rho\boldsymbol{\mu}$, where the $N \times N$ matrix $\mathbf{W}_\rho$ has rank N. If $\boldsymbol{\theta}$ is a vector of orthogonal linear combinations, then $\mathbf{W}_\rho$ is orthogonal in the sense that $\mathbf{W}_\rho'\mathbf{W}_\rho$ is a diagonal matrix. If the elements of $\boldsymbol{\theta}$ are orthonormal linear combinations, then $\mathbf{W}_\rho$ is orthonormal, $\mathbf{W}_\rho'\mathbf{W}_\rho = \mathbf{I}$. In many settings the first element of $\boldsymbol{\theta}$ is of the form $(c, c, \ldots, c)\boldsymbol{\mu}$ and the rest of them form a set of $N-1$ orthogonal contrasts. These contrasts should be meaningful comparisons among the treatment means μ_i and should be dictated by the goals of the experimenter. When the factors are quantitative and the goal is to fit an orthogonal polynomial model, then the orthogonal matrix $\mathbf{W}_\rho$ has meaning, and we have devoted Chapter 4 to this important case.

Throughout this section we associate the following assumptions with the minimal complete factorial design ρ of the $k_1 \times k_2 \times \ldots \times k_t$ factorial

$$\mathbf{Y}_\rho = \mathbf{W}_\rho\boldsymbol{\theta} + \boldsymbol{\epsilon}_\rho, \quad \operatorname{Cov}(\boldsymbol{\epsilon}_\rho) = \sigma^2\mathbf{I}_N, \tag{3.17}$$

where $N = \prod k_i$, and $\mathbf{W}_\rho$ is an $N \times N$ matrix of rank N. The parametric vector $\boldsymbol{\theta}$ could have the meaning described previously. For any fractional factorial design Γ, we have

$$\mathbf{Y}_\Gamma = \mathbf{W}_\Gamma\boldsymbol{\theta} + \boldsymbol{\epsilon}_\Gamma, \quad \operatorname{Cov}(\boldsymbol{\epsilon}_\Gamma) = \sigma^2\mathbf{I}_n, \tag{3.18}$$

where the $n \times N$ matrix $\mathbf{W}_\Gamma$ is simply read off from $\mathbf{W}_\rho$ taking repetitions of treatment combinations in Γ into account. Since Γ contains m distinct treatment combinations, it follows that the rank of $\mathbf{W}_\Gamma$ is m and hence exactly m linearly independent parametric functions of $\boldsymbol{\theta}$ are estimable and we already know how to find the BLUEs.

Since $m < N$ in a fractional factorial design, the experimenter frequently partitions the parametric vector $\boldsymbol{\theta}$ as

$$\boldsymbol{\theta}' = \left(\boldsymbol{\theta}_1' \vdots \boldsymbol{\theta}_2'\right) \tag{3.19}$$

where $\boldsymbol{\theta}_1$ is an $N_1 \times 1$ vector to be estimated under the *knowledge* that the $N_2 \times 1 = (N - N_1) \times 1$ vector $\boldsymbol{\theta}_2$ is *negligible*. (A more general partitioning of $\boldsymbol{\theta}$ is discussed in Chapter 7.) This leads to the model

$$\mathbf{Y}_\Gamma = \mathbf{W}_{\Gamma 1}\boldsymbol{\theta}_1 + \boldsymbol{\epsilon}_\Gamma, \quad \operatorname{Cov}(\boldsymbol{\epsilon}_\Gamma) = \sigma^2\mathbf{I}_n, \tag{3.20}$$

where the $n \times N_1$ matrix $\mathbf{W}_{\Gamma 1}$ is obtained from $\mathbf{W}_\Gamma$ or $\mathbf{W}_\rho$ and has rank s. A necessary condition for the estimability of $\boldsymbol{\theta}_1$ is that $m \geq N_1$. Assume now that this is the case. Further, if $s = N_1$, then $\boldsymbol{\theta}_1$ is estimable and its BLUE and covariance are immediately found from expressions (3.14) and (3.15).

That is,

$$\hat{\boldsymbol{\theta}}_1 = (\mathbf{W}'_{\Gamma 1}\mathbf{W}_{\Gamma 1})^{-1}\mathbf{W}'_{\Gamma 1}\mathbf{Y}_\Gamma, \tag{3.21}$$
$$\mathrm{Cov}(\hat{\boldsymbol{\theta}}_1) = \sigma^2(\mathbf{W}'_{\Gamma 1}\mathbf{W}_{\Gamma 1})^{-1}.$$

On the other hand, if $s < N_1$, then in light of the development in the previous section we know that exactly s independent linear functions of $\boldsymbol{\theta}_1$ are estimable.

If the experimenter is unable to specify which specific parameters are negligible and simply wants to know what N_1 subvectors $\boldsymbol{\theta}_1$ are estimable under the *assumption* that each time $\boldsymbol{\theta}_2$ is negligible, then we are immediately confronted with a formidable problem. There are N parameters in $\boldsymbol{\theta}$ and N_1 of these can be selected in $\mathrm{C}(N, N_1)$ ways. Thus there are $\mathrm{C}(N, N_1)$ possible selections of $\boldsymbol{\theta}_1$ leading to $\mathrm{C}(N, N_1)$ possible design matrices $\mathbf{W}_{\Gamma 1}$. Some of these design matrices will be less than full rank (i.e., $\mathbf{W}'_{\Gamma 1}\mathbf{W}_{\Gamma 1}$ will be singular), so that $\boldsymbol{\theta}_1$ will not be estimable. For the general $k_1 \times k_2 \times \ldots \times k_t$ factorial and a given design Γ, the problem of determining the cardinality of the class of singular selections of $\boldsymbol{\theta}_1$ is not resolved at present. Neither is resolved a procedure to determine the singularity of Γ relative to $\boldsymbol{\theta}_1$ by inspecting the treatment combinations themselves and not $\mathbf{W}_{\Gamma 1}$. This means that for a given design Γ there is no analytical procedure to generate a catalogue of nonsingular selections of $\boldsymbol{\theta}_1$. Of course, for given k_i and Γ the computer can always be brought to our aid and this has been done in practice for some cases.

The problem of selecting estimable subvectors $\boldsymbol{\theta}_1$ may be viewed as a strict matrix algebra problem. For a given design Γ the design matrix relative to the whole vector $\boldsymbol{\theta}$ is the $n \times N$ matrix W_Γ of Equation (3.18). Selection of an $N_1 \times 1$ vector $\boldsymbol{\theta}_1$ implies the selection of N_1 columns of $\mathbf{W}_\Gamma$, leading to $\mathrm{C}(N, N_1)$ matrices which are the design matrices $\mathbf{W}_{\Gamma 1}$. The estimability of $\boldsymbol{\theta}_1$ is then determined by the rank of $\mathbf{W}_{\Gamma 1}$. It is clear that in the calculation of the rank of $\mathbf{W}_{\Gamma 1}$ we may ignore repetitions of rows, that is, restrict ourselves to the m distinct treatment combinations in Γ and hence to the reduced $m \times N_1$ matrix $\mathbf{W}_{\Gamma 1}$.

The selection problem of $\boldsymbol{\theta}_1$ may be viewed as the *dual* of the design selection problem for given $m \times 1$ vector $\boldsymbol{\theta}_1$ under the assumption that the $(N-m) \times 1$ vector $\boldsymbol{\theta}_2$ is negligible. In this setting expression (3.17) reduces to

$$\mathbf{Y}_\rho = \mathbf{W}_{\rho 1}\boldsymbol{\theta}_1 + \boldsymbol{\epsilon}_\rho, \tag{3.22}$$
$$\mathrm{Cov}(\boldsymbol{\epsilon}_\rho) = \sigma^2\mathbf{I}_N,$$

where $\mathbf{W}_{\rho 1}$ is an $N \times m$ matrix of rank N_1. Suppose now that the problem consists of selecting an FFD consisting of N_1 distinct treatment combinations. This can be done in $C(N, N_1)$ ways; assume that $N_1 \geq m$. Selecting a design Γ of N_1 treatment combinations implies the selection of N_1 rows from among N rows of $\mathbf{W}_{\rho 1}$. The estimability of $\boldsymbol{\theta}_1$ is then determined by the rank of the $N_1 \times m$ matrix $\mathbf{W}_{\Gamma 1}$. It thus follows that the design selection problem is tied to the parameter selection problem. This essentially comes about from the fact that the roles of treatments and parameters can be interchanged.

Example 3.12. Consider a 2^3 factorial with each $G_i = \{0, 1\}$. Assume the following orthogonal model for the minimal complete factorial design ρ:

$$\begin{bmatrix} Y_{(0,0,0)} \\ Y_{(1,0,0)} \\ Y_{(0,1,0)} \\ Y_{(0,0,1)} \\ Y_{(1,1,0)} \\ Y_{(1,0,1)} \\ Y_{(0,1,1)} \\ Y_{(1,1,1)} \end{bmatrix} = \begin{bmatrix} 1 & -1 & -1 & -1 & 1 & 1 & 1 & -1 \\ 1 & 1 & -1 & -1 & -1 & -1 & 1 & 1 \\ 1 & -1 & 1 & -1 & -1 & 1 & -1 & 1 \\ 1 & -1 & -1 & 1 & 1 & -1 & -1 & 1 \\ 1 & 1 & 1 & -1 & 1 & -1 & -1 & -1 \\ 1 & 1 & -1 & 1 & -1 & 1 & -1 & -1 \\ 1 & -1 & 1 & 1 & -1 & -1 & 1 & -1 \\ 1 & 1 & 1 & 1 & 1 & 1 & 1 & 1 \end{bmatrix}$$

$$\times \begin{bmatrix} \alpha_0 \\ \alpha_1 \\ \alpha_2 \\ \alpha_3 \\ \alpha_4 \\ \alpha_5 \\ \alpha_6 \\ \alpha_7 \end{bmatrix} + \begin{bmatrix} \epsilon_{(0,0,0)} \\ \epsilon_{(1,0,0)} \\ \epsilon_{(0,1,0)} \\ \epsilon_{(0,0,1)} \\ \epsilon_{(1,1,0)} \\ \epsilon_{(1,0,1)} \\ \epsilon_{(0,1,1)} \\ \epsilon_{(1,1,1)} \end{bmatrix}, \quad \text{that is,}$$

$$\mathbf{Y}_\rho = \mathbf{W}_\rho \boldsymbol{\theta} + \boldsymbol{\epsilon}_\rho,$$

where $\mathrm{Cov}(\boldsymbol{\epsilon}_\rho) = \sigma^2 \mathbf{I}_8$. In this example $\boldsymbol{\theta} = \mathbf{W}_\rho^{-1} \boldsymbol{\mu} = \frac{1}{8} \mathbf{W}'_\rho \boldsymbol{\mu}$. Consider the design $\Gamma = \{(0,0,0), (1,1,0), (1,0,1), (0,1,1)\}$, then according to Equation

(3.18) we have

$$E\begin{bmatrix} Y_{(0,0,0)} \\ Y_{(1,1,0)} \\ Y_{(1,0,1)} \\ Y_{(0,1,1)} \end{bmatrix} = \begin{bmatrix} 1 & -1 & -1 & -1 & 1 & 1 & 1 & -1 \\ 1 & 1 & 1 & -1 & 1 & -1 & -1 & -1 \\ 1 & 1 & -1 & 1 & -1 & 1 & -1 & -1 \\ 1 & -1 & 1 & 1 & -1 & -1 & 1 & -1 \end{bmatrix} \boldsymbol{\theta},$$

$$E[\mathbf{Y}_\Gamma] = \mathbf{W}_\Gamma \boldsymbol{\theta}.$$

Since the rank of $\mathbf{W}_\Gamma$ is equal to 4 it follows that a set of exactly four linearly independent functions of $\boldsymbol{\theta}$ is estimable. For example, the four linear functions given by $\boldsymbol{\psi} = \mathbf{W}_\Gamma \boldsymbol{\theta}$ itself are estimable and the BLUE is equal to $\hat{\boldsymbol{\psi}} = \mathbf{W}_\Gamma (\mathbf{W}_\Gamma' \mathbf{W}_\Gamma)^- \mathbf{W}_\Gamma' \mathbf{Y}_\Gamma$.

If the total parametric vector is partitioned as

$$\boldsymbol{\theta}' = \left(\boldsymbol{\theta}_1' \vdots \boldsymbol{\theta}_2'\right) = \left(\alpha_0, \alpha_1, \alpha_2, \alpha_3 \vdots \alpha_4, \alpha_5, \alpha_6, \alpha_7\right)$$

under the assumption that $\boldsymbol{\theta}_2$ is negligible, then according to Equation (3.20) we have the model

$$E\begin{bmatrix} Y_{(0,0,0)} \\ Y_{(1,1,0)} \\ Y_{(1,0,1)} \\ Y_{(0,1,1)} \end{bmatrix} = \begin{bmatrix} 1 & -1 & -1 & -1 \\ 1 & 1 & 1 & -1 \\ 1 & 1 & -1 & 1 \\ 1 & -1 & 1 & 1 \end{bmatrix} \begin{bmatrix} \alpha_0 \\ \alpha_1 \\ \alpha_2 \\ \alpha_3 \end{bmatrix},$$

$$E[\mathbf{Y}_\Gamma] = \mathbf{W}_\Gamma \boldsymbol{\theta}_1.$$

The design matrix $\mathbf{W}_{\Gamma 1}$ is of full rank and in fact orthogonal; that is, $\mathbf{W}_{\Gamma 1}' \mathbf{W}_{\Gamma 1} = 4\mathbf{I}_4$. The BLUE of $\boldsymbol{\theta}_1$ is obtained as

$$\hat{\boldsymbol{\theta}}_1 = (\mathbf{W}_{\Gamma 1}' \mathbf{W}_{\Gamma 1})^{-1} \mathbf{W}_{\Gamma 1}' \mathbf{Y}_\Gamma = \tfrac{1}{4} \mathbf{W}_{\Gamma 1}' \mathbf{Y}_\Gamma, \tag{3.23}$$

giving us explicitly the estimators

$$\hat{\alpha}_0 = \tfrac{1}{4}\left(Y_{(0,0,0)} + Y_{(1,1,0)} + Y_{(1,0,1)} + Y_{(0,1,1)}\right),$$

$$\hat{\alpha}_1 = \tfrac{1}{4}\left(-Y_{(0,0,0)} + Y_{(1,1,0)} + Y_{(1,0,1)} - Y_{(0,1,1)}\right),$$

$$\hat{\alpha}_2 = \tfrac{1}{4}\left(-Y_{(0,0,0)} + Y_{(1,1,0)} - Y_{(1,0,1)} + Y_{(0,1,1)}\right),$$

and

$$\hat{\alpha}_3 = \tfrac{1}{4}\left(-Y_{(0,0,0)} - Y_{(1,1,0)} + Y_{(1,0,1)} + Y_{(0,1,1)}\right).$$

If the experimenter cannot indicate which parameters are negligible and is, say, simply interested in knowing which four parameters among the eight are estimable, then the problem is to determine which of the C(8,4) = 70 possible selections of $\boldsymbol{\theta}_1$ lead to nonsingular matrices $\mathbf{W}_{\Gamma 1}$. In this particular example, the reader is invited to determine (with or without the aid of a computer) which of the 70 selections result in estimable $\boldsymbol{\theta}_1$ and which lead to nonestimable $\boldsymbol{\theta}_1$ and then see how this ties up with the dual problem illustrated at the end of the second chapter and in Example 3.4.

3.5 SELECTED ADDITIONAL READING

Starred references are recommended for a first reading.

1. Banerjee, K. S. (1975). *Weighing Designs*. Marcel Dekker, New York.

*2. Bose, R. C., and Carter, R. L. (1962). Response model coefficients and the individual degrees of freedom of a factorial design. *Biometrics*, **18**, 160–171.

3. Box, G. E. P., and Youle, P. V. (1955). The exploration and exploitation of response surfaces: An example of the link between the fitted surface and the basic mechanism of the system. *Biometrics*, **11**, 287–322.

4. Bradley, H. E. (1968). Multiple classification analysis for arbitrary experimental arrangements. *Technometrics*, **10**, 13–27.

5. Daniel, C. (1959). Use of half-normal plots in interpreting factorial two-level experiments. *Technometrics*, **1**, 311–341.

6. Das, M. N., and Jain, R. C. (1970). On component analysis of factorial and fractional factorial experiments. *Biometrics*, **26**, 823–833.

7. Draper, N. R., and Smith, H. (1966). *Applied Regression Analysis*. Wiley, New York.

*8. Federer, W. T. (1967). Diallel cross designs and their relation to fractional replication. *Der Züchter*, **37**, 174–178.

9. Federer, W. T., and Zelen, M. (1966). Analysis of multifactor classifications with unequal numbers of observations. *Biometrics*, **22**, 525–552.

10. Finney, D. J. (1952). *Statistical Method in Biological Assay*. Griffin, London.

*11. Fyfe, J. L., and Gilbert, N. (1963). Partial diallel crosses. *Biometrics*, **19**, 278–286.

12. Graybill, F. A. (1961). *An Introduction to Linear Statistical Models*, Vol. 1. McGraw-Hill, New York.

*13. Graybill, F. A. (1976). *Theory and Application of the Linear Model*. Duxbury Press, Belmont, Calif.

*14. Haberman, S. J. (1975). Direct products and linear models for complete factorial tables. *Ann. Stat.*, **3**, 314–333.

15. Hinkelmann, K. (1963). A commonly occurring incomplete multiple classification model. *Biometrics*, **19**, 105–117.

16. Hocking, R. R., and Speed, F. M. (1975). A full rank analysis of some linear model problems. *J. Am. Stat. Assoc.*, **70**, 706–712.

17. Kishen, K. (1942). On expressing any single degree of freedom for treatment in an s^m factorial arrangement in terms of its sets for main effects and interactions. *Sankhyā*, **6**, 133–140.

18. Krumbein, W. C., and Graybill, F. A. (1965). *An Introduction to Statistical Models in Geology*. McGraw-Hill, New York.

19. Mosteller, F., and Tukey, J. W. (1977). *Data Analysis and Regression*. Addison-Wesley, Reading, Mass.

20. Paik, U. B., and Federer, W. T. (1974). Analysis of nonorthogonal n-way classifications. *Ann. Stat.*, **5**, 1000–1021.

21. Rao, C. R. (1946). On the linear combinations of observations and the general theory of least squares. *Sankhyā*, **7**, 237–256.

*22. Rao, C. R. (1973). *Linear Statistical Inference and its Applications*. 2nd ed. Wiley, New York.

23. Rao, C. R., and Mitra, S. K. (1971). *Generalized Inverse of Matrices and its Applications*. Wiley, New York.

24. Reiersøl, O. (1963). Identifiability, estimability, phenorestricting specifications and zero Lagrange multipliers in the analysis of variance. *Skand. Aktuarietidskr.* **46**, 131–142.

*25. Scheffé, H. (1959). *The Analysis of Variance*. Wiley, New York.

26. Searle, S. R. (1971). *Linear Models*. Wiley, New York.

27. Tocher, K. D. (1952). The design and analysis of block experiments. *J. R. Stat. Soc.*, **B14**, 45–100.

28. Urquhart, N. S., Weeks, D. L., and Henderson, C. R. (1973). Estimation associated with linear models: a revisitation. *Commun. Stat.* **1**, 303–330.

29. Williams, E. J. (1959). *Regression Analysis*. Wiley, New York.

30. Yates, F. (1935). Complex experiments. *J. R. Stat. Soc.*, **B2**, 181–247.

31. Zelen, M., and Federer, W. T. (1965). Application of the calculus for factorial arrangements. III. Analysis of factorials with unequal numbers of observations. *Sankhyā*, **A27**, 383–400.

32. Zyskind, G., and Martin, F. B. (1969). On best linear estimation and a general Gauss-Markoff theorem in linear models with arbitrary non-negative covariance structure. *SIAM J. Appl. Math.*, **17**, 1190–1200.

CHAPTER 4

Orthogonal Polynomial Model and Estimation of its Parameters

In the previous chapter we stated the model $E[\mathbf{Y}_\rho]=\mathbf{W}_\rho\boldsymbol{\theta}$, where ρ is the minimal complete factorial design, $\boldsymbol{\theta}$ is the full parametric vector, and the $N\times N$ design matrix $\mathbf{W}_\rho$ is of full rank. For any factorial design Γ, this model then serves as a base, since the design matrix of Γ relative to $\boldsymbol{\theta}$ (or a subvector $\boldsymbol{\theta}_1$) can be simply read off from $\mathbf{W}_\rho$. Two popular models of this type are the orthogonal polynomial model and the Helmert orthogonal polynomial model. In this chapter we develop the orthogonal polynomial model and provide the least squares estimators of the parametric vector and of a vector of parametric functions with and without the assumption that some of the components of the parametric vector are negligible. The use of the Helmert orthogonal matrices or other contrast matrices is also pointed out.

4.1 THE ORTHOGONAL POLYNOMIAL MODEL

We shall assume throughout this chapter that the levels of all factors are quantified. In order to understand the orthogonal polynomial model for the t-factor factorial we first introduce this for the case $t=1$, that is, for the single factor experiment. Denote the levels of this single factor by $\{z_1, z_2, \ldots, z_v\}$ and let Y_{z_j} be the observation at the z_jth level.

Definition 4.1. The *full polynomial model* for the single factor factorial is given by the equation

$$E[Y_{z_j}] = \theta_0 p_0(z_j) + \theta_1 p_1(z_j) + \cdots + \theta_{v-1} p_{v-1}(z_j), \tag{4.1}$$

where $p_w(z_j) = z_j^w$, $j = 1,2,\ldots,v$, $w = 0,1,2,\ldots,v-1$.

As in regression theory, θ_0 is called the *intercept*, θ_1 is the *linear regression coefficient*, θ_2 is the *quadratic regression coefficient*, and so on. In matrix form, model (4.1) is written as

$$E[\mathbf{Y}] = \mathbf{P}\boldsymbol{\theta}, \tag{4.2}$$

where $\mathbf{P}$ is a $v \times v$ matrix with the (j,w)th entry being equal to $p_w(z_j) = z_j^w$, $w = 0,1,2,\ldots,v-1$, $j = 1,2,\ldots,v$.

Example 4.1. Suppose that a variable is measured at five levels $\{z_1, z_2,\ldots, z_5\}$ of the factor temperature. Then the polynomial model (4.2) is explicitly equal to,

$$E\begin{bmatrix} Y_1 \\ Y_2 \\ Y_3 \\ Y_4 \\ Y_5 \end{bmatrix} = \begin{bmatrix} 1 & z_1 & z_1^2 & z_1^3 & z_1^4 \\ 1 & z_2 & z_2^2 & z_2^3 & z_2^4 \\ 1 & z_3 & z_3^2 & z_3^3 & z_3^4 \\ 1 & z_4 & z_4^2 & z_4^3 & z_4^4 \\ 1 & z_5 & z_5^2 & z_5^3 & z_5^4 \end{bmatrix} \begin{bmatrix} \theta_0 \\ \theta_1 \\ \theta_2 \\ \theta_3 \\ \theta_4 \end{bmatrix}.$$

Remark. It is not necessary to have the full polynomial model and in practical applications one or more of the θ_w could be zero or negligible. For the full model the matrix $\mathbf{P}$ is a Van der Monde matrix of rank v with determinant equal to $\Pi_{i<j}(z_j - z_i)$. When r of the θ_w are deleted, then the resulting design matrix has rank $v - r$.

Let $\mathbf{H}$ be the triangular Gram-Schmidt transformation matrix that orthonormalizes the columns of $\mathbf{P}$ from left to right. It follows that Equation (4.2) can be rewritten as

$$E[\mathbf{Y}] = \mathbf{P}\mathbf{H}\mathbf{H}^{-1}\boldsymbol{\theta} = \mathbf{M}\boldsymbol{\phi}, \tag{4.3}$$

where $\mathbf{M} = \mathbf{P}\mathbf{H}$ and $\boldsymbol{\phi} = \mathbf{H}^{-1}\boldsymbol{\theta}$. This model is called the *single factor orthogonal polynomial model*. Indicate the columns of $\mathbf{P}$ and $\mathbf{M}$ by $\mathbf{P}_0, \mathbf{P}_1, \ldots, \mathbf{P}_{v-1}$ and $\mathbf{M}_0, \mathbf{M}_1, \ldots, \mathbf{M}_{v-1}$, respectively. Then the Gram-Schmidt process is defined by

$$
\begin{aligned}
&\mathbf{W}_0 = \mathbf{P}_0,\\
&\mathbf{W}_1 = \mathbf{P}_1 - \frac{\mathbf{W}_0 \cdot \mathbf{P}_1}{\mathbf{W}_0 \cdot \mathbf{W}_0}\mathbf{W}_0,\\
(4.4)\qquad &\mathbf{W}_2 = \mathbf{P}_2 - \frac{\mathbf{W}_1 \cdot \mathbf{P}_2}{\mathbf{W}_1 \cdot \mathbf{W}_1}\mathbf{W}_1 - \frac{\mathbf{W}_0 \cdot \mathbf{P}_2}{\mathbf{W}_0 \cdot \mathbf{W}_0}\mathbf{W}_0,\\
&\cdots\cdots\cdots\cdots\\
&\mathbf{W}_{v-1} = \mathbf{P}_{v-1} - \frac{\mathbf{W}_{v-2} \cdot \mathbf{P}_{v-1}}{\mathbf{W}_{v-2} \cdot \mathbf{W}_{v-2}}\mathbf{W}_{v-2} - \cdots - \frac{\mathbf{W}_0 \cdot \mathbf{P}_{v-1}}{\mathbf{W}_0 \cdot \mathbf{W}_0}\mathbf{W}_0,
\end{aligned}
$$

and

$$\mathbf{M}_i = \frac{\mathbf{W}_i}{\|\mathbf{W}_i\|}, \qquad i = 0, 1, \ldots, v-1.$$

Here $\mathbf{U}\cdot\mathbf{V}$ means the usual *dot product* $\Sigma u_i v_i$ of the vectors $\mathbf{U}$ and $\mathbf{V}$, and $\|\mathbf{U}\| = (\Sigma u_i^2)^{1/2}$ is the *length* of the vector $\mathbf{U}$.

Explicitly, the (j, h)th entry of $\mathbf{M}$ is given by

$$(4.5)\qquad m_{jh} = m_h(z_j) = \frac{1}{c_h}\left[z_j^h - \sum_{a=0}^{h-1} m_a(z_j) \sum_{b=1}^{v} z_b^h m_a(z_b)\right],$$

$j = 1, 2, \ldots, v$, $h = 1, 2, \ldots, v-1$, where c_h is the normalization constant

$$(4.6)\qquad c_h = \left\{\sum_{j=1}^{v}\left[z_j^h - \sum_{a=0}^{h-1} m_a(z_j) \sum_{b=1}^{v} z_b^h m_a(z_b)\right]^2\right\}^{1/2}$$

and the $(j,0)$th entry m_{j0} in $\mathbf{M}$ is $v^{-1/2}$ for all j. It then follows that $\mathbf{M}'\mathbf{M} = \mathbf{I}_v$. If we denote the elements of $\boldsymbol{\phi}$ by $\phi^0, \phi^1, \ldots, \phi^{v-1}$, then in regression theory ϕ^0 is called the *intercept*, ϕ^1 is the *linear regression coefficient eliminating the intercept and ignoring all higher degree terms*, ϕ^2 is the *quadratic regression coefficient eliminating the intercept and the linear regression coefficient and ignoring all higher degree terms*, and so on. This order of eliminating and ignoring regression coefficients is due to the left-to-right orthonormalization of the matrix $\mathbf{P}$.

In the case where the levels of the factor are equispaced and are coded as $1, 2, \ldots, v$, tables have been prepared that provide the matrix in orthogonal form and from which the orthonormal form is obtained by dividing each column-vector by its length. Some examples of these tables are given in Example 4.4 where the equispaced case is developed. Selected tables for

$v \leq 9$ are given in Table 4.1. Additional tables are given in the relevent references at the end of the chapter.

The triangular Gram-Schmidt transformation matrix in (4.3) will be needed in estimating $\boldsymbol{\theta}$ once an estimator of $\boldsymbol{\phi}$ is obtained. The reader is invited to describe the elements of $\mathbf{H}$ explicitly.

In many settings the normalization of the $\mathbf{W}_i$ is not carried out. In such cases the model is written as $E[\mathbf{Y}] = \mathbf{M}\boldsymbol{\phi}$, where $\boldsymbol{\phi} = \mathbf{H}^{*-1}\boldsymbol{\theta}$ and the Gram-Schmidt transformation matrix $\mathbf{H}^*$ is then equal to $\mathbf{H}$ without the lengths as divisors. This model is often referred to as the *columnwise orthogonal polynomial model*, since $\mathbf{W}'\mathbf{W}$ is not equal to the identity matrix but equal to a diagonal matrix.

Example 4.2. Suppose that the levels of a factor are $z_1 = 0$, $z_2 = 1$, $z_3 = 3$, and $z_4 = 4$. Then the second degree polynomial model for the observations is

$$E\begin{bmatrix} Y_0 \\ Y_1 \\ Y_3 \\ Y_4 \end{bmatrix} = \begin{bmatrix} 1 & 0 & 0 \\ 1 & 1 & 1 \\ 1 & 3 & 9 \\ 1 & 4 & 16 \end{bmatrix} \begin{bmatrix} \theta_0 \\ \theta_1 \\ \theta_2 \end{bmatrix}.$$

Using (4.4) we first obtain the $\mathbf{W}_i$ and then the $\mathbf{M}_i$, that is,

$$\mathbf{W}'_0 = \mathbf{P}'_0 = (1,1,1,1),$$

$$\mathbf{W}'_1 = \mathbf{P}'_1 - \frac{\mathbf{W}_0 \cdot \mathbf{P}_1}{\mathbf{W}_0 \cdot \mathbf{W}_0}\mathbf{W}'_0 = (0,1,3,4) - 2(1,1,1,1) = (-2,-1,1,2),$$

$$\mathbf{W}'_2 = \mathbf{P}'_2 - \frac{\mathbf{W}_1 \cdot \mathbf{P}_2}{\mathbf{W}_1 \cdot \mathbf{W}_1}\mathbf{W}'_1 - \frac{\mathbf{W}_0 \cdot \mathbf{P}_2}{\mathbf{W}_0 \cdot \mathbf{W}_0}\mathbf{W}'_0$$

$$= (0,1,9,16) - 4(-2,-1,1,2) - \tfrac{13}{2}(1,1,1,1)$$

$$= \left(\frac{3}{2}, -\frac{3}{2}, -\frac{3}{2}, \frac{3}{2}\right).$$

Hence

$$\mathbf{M}'_0 = \frac{\mathbf{W}'_0}{\|\mathbf{W}_0\|} = \frac{1}{2}(1,1,1,1) = \left(\frac{1}{2}, \frac{1}{2}, \frac{1}{2}, \frac{1}{2}\right),$$

$$\mathbf{M}'_1 = \frac{\mathbf{W}'_1}{\|\mathbf{W}_1\|} = \frac{1}{\sqrt{10}}(-2,\ -1,\ 1,\ 2) = \left(\frac{-2}{\sqrt{10}}, \frac{-1}{\sqrt{10}}, \frac{1}{\sqrt{10}}, \frac{2}{\sqrt{10}}\right),$$

$$\mathbf{M}'_2 = \frac{\mathbf{W}'_2}{\|\mathbf{W}_2\|} = \frac{1}{3}\left(\frac{3}{2}, \frac{-3}{2}, \frac{-3}{2}, \frac{3}{2}\right) = \left(\frac{1}{2}, -\frac{1}{2}, -\frac{1}{2}, \frac{1}{2}\right).$$

Therefore, the quadratic orthogonal polynomial model is equal to

$$E\begin{bmatrix} Y_0 \\ Y_1 \\ Y_3 \\ Y_4 \end{bmatrix} = \begin{bmatrix} \frac{1}{2} & \frac{-2}{\sqrt{10}} & \frac{1}{2} \\ \frac{1}{2} & \frac{-1}{\sqrt{10}} & \frac{-1}{2} \\ \frac{1}{2} & \frac{1}{\sqrt{10}} & \frac{-1}{2} \\ \frac{1}{2} & \frac{2}{\sqrt{10}} & \frac{1}{2} \end{bmatrix} \begin{bmatrix} \phi_0 \\ \phi_1 \\ \phi_2 \end{bmatrix} = \begin{bmatrix} m_{10} & m_{11} & m_{12} \\ m_{20} & m_{21} & m_{22} \\ m_{30} & m_{31} & m_{32} \\ m_{40} & m_{41} & m_{42} \end{bmatrix} \boldsymbol{\phi}.$$

Let us verify the elements in the second column of **M** by calculating them directly via formulas (4.5) and (4.6):

$$\begin{aligned} m_{11} &= \frac{1}{c_1}\{z_1 - m_0(z_1)[z_1 m_0(z_1) + z_2 m_0(z_2) + z_3 m_0(z_3) + z_4 m_0(z_4)]\} \\ &= \frac{1}{c_1}\left\{0 - \frac{1}{2}\left[(0)\left(\frac{1}{2}\right) + (1)\left(\frac{1}{2}\right) + (3)\left(\frac{1}{2}\right) + (4)\left(\frac{1}{2}\right)\right]\right\} \\ &= \frac{1}{c_1}\{0 - 2\} = -\frac{2}{c_1}, \\ m_{21} &= \frac{1}{c_1}\{z_2 - m_0(z_2)[z_1 m_0(z_1) + z_2 m_0(z_2) + z_3 m_0(z_3) + z_4 m_0(z_4)]\} \\ &= \frac{1}{c_1}\left\{1 - \frac{1}{2}\left[(0)\left(\frac{1}{2}\right) + (1)\left(\frac{1}{2}\right) + (3)\left(\frac{1}{2}\right) + (4)\left(\frac{1}{2}\right)\right]\right\} \\ &= \frac{1}{c_1}\{1 - 2\} = -\frac{1}{c_1}, \\ m_{31} &= \frac{1}{c_1}\{z_3 - m_0(z_3)[z_1 m_0(z_1) + z_2 m_0(z_2) + z_3 m_0(z_3) + z_4 m_0(z_4)]\} \\ &= \frac{1}{c_1}\left\{3 - \frac{1}{2}\left[(0)\left(\frac{1}{2}\right) + (1)\left(\frac{1}{2}\right) + (3)\left(\frac{1}{2}\right) + (4)\left(\frac{1}{2}\right)\right]\right\} \\ &= \frac{1}{c_1}\{3 - 2\} = \frac{1}{c_1}, \\ m_{41} &= \frac{1}{c_1}\{4 - m_0(z_4)[z_1 m_0(z_1) + z_2 m_0(z_2) + z_3 m_0(z_3) + z_4 m_0(z_4)]\} \\ &= \frac{1}{c_1}\left\{4 - \frac{1}{2}\left[(0)\left(\frac{1}{2}\right) + (1)\left(\frac{1}{2}\right) + (3)\left(\frac{1}{2}\right) + (4)\left(\frac{1}{2}\right)\right]\right\} \\ &= \frac{1}{c_1}\{4 - 2\} = \frac{2}{c_1}. \end{aligned}$$

Now, $c_1 = [(-2)^2 + (-1)^2 + (1)^2 + (2)^2]^{1/2} = \sqrt{10}$, so that $\mathbf{M}'_1 = (1/c_1)(-2,-1,1,2) = (-2/\sqrt{10}, -1/\sqrt{10}, 1/\sqrt{10}, 2/\sqrt{10})$, which verifies the earlier result.

Definition 4.2. A real n-tuple $(x_1, x_2, \dots, x_n)$ is said to be *less than* a real n-tuple $(y_1, y_2, \dots, y_n)$ if and only if for the first u such that $x_u \neq y_u$ we have $x_u < y_u, 1 \leq u \leq n$.

Definition 4.3. A set of real n-tuples is said to be *lexicographically ordered* if it is ordered according to Definition 4.2.

Definition 4.4. The *left Kronecker product* of two matrices $\mathbf{A}_{m\times n} = (a_{ij})$ and $\mathbf{B}_{r\times s} = (b_{ij})$ is equal to the $mr \times ns$ matrix

$$\mathbf{A} \otimes \mathbf{B} = \begin{bmatrix} a_{11}\mathbf{B} & a_{12}\mathbf{B} & \dots & a_{1n}\mathbf{B} \\ a_{21}\mathbf{B} & a_{22}\mathbf{B} & \dots & a_{2n}\mathbf{B} \\ \cdots & \cdots & \cdots & \cdots \\ a_{m1}\mathbf{B} & a_{m2}\mathbf{B} & \dots & a_{mn}\mathbf{B} \end{bmatrix}. \tag{4.7}$$

Definition 4.5. The *symbolic left Kronecker product* of two vectors $\mathbf{X}_{m\times 1}$ and $\mathbf{Y}_{n\times 1}$ is equal to the $mn \times 1$ vector

$$\begin{aligned} \mathbf{X} \circledast \mathbf{Y} = [&(x_1, y_1), (x_1, y_2), \dots, (x_1, y_n), \dots, \\ &(x_m, y_1), (x_m, y_2), \dots, (x_m, y_n)]'. \end{aligned} \tag{4.8}$$

This is often simply written as

$$\mathbf{X} \circledast \mathbf{Y} = (x_1y_1, x_1y_2, \dots, x_1y_n, \dots, x_my_1, x_my_2, \dots, x_my_n)'. \tag{4.9}$$

We are now ready to generalize the orthogonal polynomial model to the case $t \geq 2$. Let the levels of each factor be ordered in increasing order and, as before, let ρ be the minimal complete factorial design and $\mathbf{Y}_\rho$ be the corresponding observation vector. The t-factor orthogonal polynomial model, which will be adopted throughout the ensuing developments is given by

$$E[\mathbf{Y}_\rho] = \mathbf{X}_\rho \boldsymbol{\beta}_\rho \tag{4.10}$$

where the subscripts of $\mathbf{Y}_\rho$ are lexicographically ordered,

$$\mathbf{X}_\rho = \mathbf{M}_1 \otimes \mathbf{M}_2 \otimes \cdots \otimes \mathbf{M}_t, \tag{4.11}$$

and

$$\boldsymbol{\beta}_\rho = \boldsymbol{\phi}_1 \circledast \boldsymbol{\phi}_2 \circledast \cdots \circledast \boldsymbol{\phi}_t. \tag{4.12}$$

The matrix $\mathbf{M}_i$ and the vector $\boldsymbol{\phi}_i$ are the design matrix and parametric vector for the ith factor after left to right orthonormalization (or orthogonalization) as described earlier in (4.4). The Kronecker product $\mathbf{M}_1 \otimes \mathbf{M}_2 \otimes \cdots \otimes \mathbf{M}_t$ and the symbolic Kronecker product $\boldsymbol{\phi}_1 \circledast \boldsymbol{\phi}_2 \circledast \cdots \circledast \boldsymbol{\phi}_t$ are obtained as the usual generalizations of Definitions 4.4 and 4.5. It follows that the superscripts of the elements of $\boldsymbol{\beta}_\rho$ are also lexicographically ordered. The elements of $\boldsymbol{\beta}_\rho$, also called *factorial effects*, have been traditionally named in the following manner: $\phi_1^0\phi_2^0\ldots\phi_t^0$ is called the *general mean* or *intercept*, $\phi_1^0\phi_2^0\ldots\phi_q^p\ldots\phi_t^0$ is called the *pth main effect of the qth factor*, $\phi_1^{v_1}\phi_2^{v_2}\phi_3^0\ldots\phi_t^0$ is called the *v_1th degree of factor F_1 by v_2th degree of factor F_2 interaction effect*, and so on. Also, an effect $\phi_1^{i_1}\phi_2^{i_2}\ldots\phi_t^{i_t}$ is said to be of *degree or order k* if k of the exponents $i_1, i_2, \ldots, i_t$ are nonzero.

One may in the same way generalize the nonorthogonal polynomial model of Definition 4.1 to the case of t factors. In this case the design matrix $\mathbf{P} = \mathbf{P}_1 \otimes \mathbf{P}_2 \otimes \cdots \otimes \mathbf{P}_t$ and the parametric vector is $\boldsymbol{\theta}_\rho = \boldsymbol{\theta}_1 \circledast \boldsymbol{\theta}_2 \circledast \cdots \circledast \boldsymbol{\theta}_t$.

As stated before there is an error vector associated with the observation vector $\mathbf{Y}_\rho$ and model (4.10) may be rewritten as

$$\mathbf{Y}_\rho = \mathbf{X}_\rho \boldsymbol{\beta}_\rho + \boldsymbol{\epsilon}_\rho, \tag{4.13}$$

where the error vector $\boldsymbol{\epsilon}_\rho$ is assumed to have

$$\begin{aligned} E[\boldsymbol{\epsilon}_\rho] &= \mathbf{0}, \\ \operatorname{Cov}(\boldsymbol{\epsilon}_\rho) &= \sigma^2 \mathbf{I}_N. \end{aligned} \tag{4.14}$$

Example 4.3. Consider a two-factor factorial with $G_1 = \{0, 2, 5\}$ and $G_2 = \{0, 1, 3, 6\}$. The two Van der Monde matrices are then

$$\mathbf{P}_1 = \begin{bmatrix} 1 & z_{11} & z_{11}^2 \\ 1 & z_{12} & z_{12}^2 \\ 1 & z_{13} & z_{13}^2 \end{bmatrix} = \begin{bmatrix} 1 & 0 & 0 \\ 1 & 2 & 4 \\ 1 & 5 & 25 \end{bmatrix}$$

and

$$\mathbf{P}_2 = \begin{bmatrix} 1 & z_{21} & z_{21}^2 & z_{21}^3 \\ 1 & z_{22} & z_{22}^2 & z_{22}^3 \\ 1 & z_{23} & z_{23}^2 & z_{23}^3 \\ 1 & z_{24} & z_{24}^2 & z_{24}^3 \end{bmatrix} = \begin{bmatrix} 1 & 0 & 0 & 0 \\ 1 & 1 & 1 & 1 \\ 1 & 3 & 9 & 27 \\ 1 & 6 & 36 & 216 \end{bmatrix}.$$

On orthonormalizing the columns of $\mathbf{P}_1$ and $\mathbf{P}_2$ as described above, we obtain

$$\mathbf{M}_1 = \begin{bmatrix} \frac{+1}{\sqrt{3}} & \frac{-7}{\sqrt{114}} & \frac{+3}{\sqrt{38}} \\ \frac{+1}{\sqrt{3}} & \frac{-1}{\sqrt{114}} & \frac{-5}{\sqrt{38}} \\ \frac{+1}{\sqrt{3}} & \frac{+8}{\sqrt{114}} & \frac{+2}{\sqrt{38}} \end{bmatrix},$$

$$\mathbf{M}_2 = \begin{bmatrix} \frac{+1}{\sqrt{4}} & \frac{-5}{\sqrt{84}} & \frac{+9}{\sqrt{308}} & \frac{-5}{\sqrt{132}} \\ \frac{+1}{\sqrt{4}} & \frac{-3}{\sqrt{84}} & \frac{-3}{\sqrt{308}} & \frac{+9}{\sqrt{132}} \\ \frac{+1}{\sqrt{4}} & \frac{+1}{\sqrt{84}} & \frac{-13}{\sqrt{308}} & \frac{-5}{\sqrt{132}} \\ \frac{+1}{\sqrt{4}} & \frac{+7}{\sqrt{84}} & \frac{+7}{\sqrt{308}} & \frac{+1}{\sqrt{132}} \end{bmatrix}.$$

The two parametric vectors for this example are

$$\boldsymbol{\phi}'_1 = (\phi_1^0, \phi_1^1, \phi_1^2), \qquad \boldsymbol{\phi}'_2 = (\phi_2^0, \phi_2^1, \phi_2^2, \phi_2^3).$$

Hence the design matrix $\mathbf{X}_\rho$ and the parametric vector $\boldsymbol{\beta}_\rho$ are obtained as

$$\mathbf{X}_\rho = \mathbf{M}_1 \otimes \mathbf{M}_2 = \begin{bmatrix} \frac{+1}{\sqrt{3}}\mathbf{M}_2 & \frac{-7}{\sqrt{114}}\mathbf{M}_2 & \frac{+3}{\sqrt{38}}\mathbf{M}_2 \\ \frac{+1}{\sqrt{3}}\mathbf{M}_2 & \frac{-1}{\sqrt{114}}\mathbf{M}_2 & \frac{-5}{\sqrt{38}}\mathbf{M}_2 \\ \frac{+1}{\sqrt{3}}\mathbf{M}_2 & \frac{+8}{\sqrt{114}}\mathbf{M}_2 & \frac{+2}{\sqrt{38}}\mathbf{M}_2 \end{bmatrix}$$

and

$$\boldsymbol{\beta}'_\rho = (\boldsymbol{\phi}_1 \circledast \boldsymbol{\phi}_2)' = (\phi_1^0\phi_2^0, \phi_1^0\phi_2^1, \phi_1^0\phi_2^2, \phi_1^0\phi_2^3, \phi_1^1\phi_2^0, \phi_1^1\phi_2^1,$$
$$\phi_1^1\phi_2^2, \phi_1^1\phi_2^3, \phi_1^2\phi_2^0, \phi_1^2\phi_2^1, \phi_1^2\phi_2^2, \phi_1^2\phi_2^3).$$

In the model $E[\mathbf{Y}_\rho] = \mathbf{X}_\rho \boldsymbol{\beta}_\rho$ the observation vector $\mathbf{Y}_\rho$ is lexicographically ordered and is compatible with $\boldsymbol{\beta}'_\rho$ above; that is,

$$\mathbf{Y}'_\rho = (Y_{(0,0)}, Y_{(0,1)}, Y_{(0,2)}, Y_{(0,3)}, Y_{(1,0)}, Y_{(1,1)}, Y_{(1,2)}, Y_{(1,3)},$$
$$Y_{(2,0)}, Y_{(2,1)}, Y_{(2,2)}, Y_{(2,3)}).$$

Example 4.4. Suppose that the v levels $z_1^*, z_2^*, \ldots, z_v^*$ of a factor are *equispaced*; that is, $z_{i+1}^* - z_i^* = d$, (say) for $i = 1, 2, \ldots, v-1$. Using the transformation $z = (1/d)(z^* - \bar{z}^*) + (v+1)/2$, where $\bar{z}^* = (1/v)\Sigma z_i^*$, we obtain the coded levels $z_1 = 1$, $z_2 = 2, \ldots, z_v = v$. The kth degree polynomial model for an observation Y_z at the level z is given by

$$E[Y_z] = \theta_0 + \theta_1 z + \theta_2 z^2 + \cdots + \theta_{v-1} z^k, \qquad k \le v-1,$$

which in terms of orthogonal polynomials may be rewritten as

$$E[Y_z] = \phi_0 + \phi_1 \xi_1(z) + \phi_2 \xi_2(z) + \cdots + \phi_{v-1} \xi_k(z),$$

where $\xi_i(z)$ is a polynomial of degree i in z, and $\Sigma_{z=1}^{v} \xi_i(z)\xi_j(z) = 0$ and $\xi_0(z) = 1$. Note that this approach fits the earlier discussed nonnormalized orthogonal model. The v values of the ξ for the first five polynomials are determined for $z = 1, 2, \ldots, v$ by

$$\xi_1(z) = \lambda_1(z - \bar{z}),$$

$$\xi_2(z) = \lambda_2\left[(z - \bar{z})^2 - \frac{v^2 - 1}{12}\right],$$

$$\xi_3(z) = \lambda_3\left[(z - \bar{z})^3 - (z \quad \bar{z})\left(\frac{3v^2 - 7}{20}\right)\right],$$

$$\xi_4(z) = \lambda_4\left[(z - \bar{z})^4 - (z - \bar{z})^2\left(\frac{3v^2 - 13}{14}\right) + \frac{3(v^2 - 1)(v^2 - 9)}{560}\right],$$

$$\xi_5(z) = \lambda_5\left[(z - \bar{z})^5 - (z - \bar{z})^3\left(\frac{5v^2 - 35}{18}\right)\right.$$
$$\left. + (z - \bar{z})\left(\frac{15v^4 - 230v^2 + 407}{1008}\right)\right],$$

where $\bar{z}=(1/v)\Sigma z_i$ and the λ are so chosen that the values of the ξ evaluated at the v values of z are integers reduced to lowest terms.

As an illustration suppose that it is desired to fit a third degree orthogonal polynomial to $v=5$ equispaced levels of a factor. Then

$$\bar{z}=\frac{1}{5}\Sigma z_i=\frac{v+1}{2}=3,$$
$$\xi_1(z)=\lambda_1(z-3),$$
$$\xi_2(z)=\lambda_2\left[(z-3)^2-2\right],$$
$$\xi_3(z)=\lambda_3\left[(z-3)^3-(z-3)\tfrac{17}{5}\right].$$

Hence the values of the ξ evaluated at $z=1,2,3,4,5$ are

z	$\xi_1(z)$	$\xi_2(z)$	$\xi_3(z)$
1	$-2\lambda_1=-2$	$+2\lambda_2=+2$	$-\frac{6}{5}\lambda_3=-1$
2	$-\lambda_1=-1$	$-\lambda_2=-1$	$+\frac{12}{5}\lambda_3=+2$
3	$0=0$	$-2\lambda_2=-2$	$0=0$
4	$+\lambda_1=+1$	$-\lambda_2=-1$	$-\frac{12}{5}\lambda_3=-2$
5	$+2\lambda_1=+2$	$2\lambda_2=+2$	$+\frac{6}{5}\lambda_3=+1$

where $\lambda_1=+1$, $\lambda_2=+1$ and $\lambda_3=\frac{5}{6}$. In matrix notation the (nonnormalized) orthogonal polynomial model for this case is

$$E\begin{bmatrix}Y_1\\Y_2\\Y_3\\Y_4\\Y_5\end{bmatrix}=\begin{bmatrix}1&-2&2&-1\\1&-1&-1&2\\1&0&-2&0\\1&1&-1&-2\\1&2&2&1\end{bmatrix}\begin{bmatrix}\phi_0\\\phi_1\\\phi_2\\\phi_3\end{bmatrix},$$

and the reader may verify that the information matrix is diagonal with entries (5, 10, 14, 10).

As remarked elsewhere extended design matrices in integer form are available for fitting orthogonal polynomials to equispaced factors. In Table 4.1 we present selected matrices for values of $v\leq 9$.

Let us now illustrate expression (4.10) for two factors with equispaced levels $G_1=\{1,2\}$ and $G_2=\{1,2,3\}$. From any table of orthogonal polynomials or by direct calculation from above we find the nonnormalized

Table 4.1 Selected Design Matrices for Orthogonal Polynomials

$v=2$

ξ_0	ξ_1
1	−1
1	1

$v=3$

ξ_0	ξ_1	ξ_2
1	−1	1
1	0	−2
1	1	1

$v=4$

ξ_0	ξ_1	ξ_2	ξ_3
1	−3	1	−1
1	−1	−1	3
1	1	−1	−3
1	3	1	1

$v=5$

ξ_0	ξ_1	ξ_2	ξ_3	ξ_4
1	−2	2	−1	1
1	−1	−1	2	−4
1	0	−2	0	6
1	1	−1	−2	−4
1	2	2	1	1

$v=6$

ξ_0	ξ_1	ξ_2	ξ_3	ξ_4	ξ_5
1	−5	5	−5	1	−1
1	−3	−1	7	−3	5
1	−1	−4	4	2	−10
1	1	−4	−4	2	10
1	3	−1	−7	−3	−5
1	5	5	5	1	1

$v=7$

ξ_0	ξ_1	ξ_2	ξ_3	ξ_4	ξ_5	ξ_6
1	−3	5	−1	3	−1	1
1	−2	0	1	−7	4	−6
1	−1	−3	1	1	−5	15
1	0	−4	0	6	0	−20
1	1	−3	−1	1	5	15
1	2	0	−1	−7	−4	−6
1	3	5	1	3	1	1

$v=8$

ξ_0	ξ_1	ξ_2	ξ_3	ξ_4	ξ_5
1	−7	7	−7	7	−7
1	−5	1	5	−13	23
1	−3	−3	7	−3	−17
1	−1	−5	3	9	−15
1	1	−5	−3	9	15
1	3	−3	−7	−3	17
1	5	1	−5	−13	−23
1	7	7	7	7	7

$v=9$

ξ_0	ξ_1	ξ_2	ξ_3	ξ_4	ξ_5
1	−4	28	−14	14	−4
1	−3	7	7	−21	11
1	−2	−8	13	−11	−4
1	−1	−17	9	9	−9
1	0	−20	0	18	0
1	1	−17	−9	9	9
1	2	−8	−13	−11	4
1	3	7	−7	−21	−11
1	4	28	14	14	4

Note. The tables for $v=8$ and $v=9$ go to orthogonal polynomials of degree ≤ 5.

orthogonal matrices

$$\mathbf{M}_1=\begin{bmatrix}1 & -1\\ 1 & 1\end{bmatrix},\qquad \mathbf{M}_2=\begin{bmatrix}1 & -1 & 1\\ 1 & 0 & -2\\ 1 & 1 & 1\end{bmatrix},$$

so that the design matrix $\mathbf{X}_\rho$ in nonnormalized form is equal to

$$\mathbf{X}_\rho=\mathbf{M}_1\otimes\mathbf{M}_2=\begin{bmatrix}1 & -1 & 1 & -1 & 1 & -1\\ 1 & 0 & -2 & -1 & 0 & 2\\ 1 & 1 & 1 & -1 & -1 & -1\\ 1 & -1 & 1 & 1 & -1 & 1\\ 1 & 0 & -2 & 1 & 0 & -2\\ 1 & 1 & 1 & 1 & 1 & 1\end{bmatrix}.$$

4.2 BEST LINEAR UNBIASED ESTIMATION OF β_ρ FROM THE MINIMAL COMPLETE FACTORIAL DESIGN UNDER THE ORTHOGONAL POLYNOMIAL MODEL

Applying the least squares procedure (introduced in Definition 3.9) to the model $\mathbf{Y}_\rho = \mathbf{X}_\rho \boldsymbol{\beta}_\rho + \boldsymbol{\epsilon}_\rho$, where $\mathbf{X}_\rho$ is in orthonormal form, we obtain the following best linear unbiased estimator (BLUE) for $\boldsymbol{\beta}_\rho$ from Equation (3.12):

$$\hat{\boldsymbol{\beta}}_\rho = \left[\mathbf{X}'_\rho \mathbf{X}_\rho\right]^{-1} \mathbf{X}'_\rho \mathbf{Y}_\rho = \mathbf{X}'_\rho \mathbf{Y}_\rho, \tag{4.15}$$

since $\mathbf{X}'_\rho \mathbf{X}_\rho = \mathbf{I}_N$. The covariance matrix of this estimator is equal to

$$\operatorname{Cov}\left(\hat{\boldsymbol{\beta}}_\rho\right) = \left[\mathbf{X}'_\rho \mathbf{X}_\rho\right]^{-1} \sigma^2 = \sigma^2 \mathbf{I}_N. \tag{4.16}$$

In the nonnormalized orthogonal case $\mathbf{X}'_\rho \mathbf{X}_\rho = \mathbf{D}_\rho =$ diagonal $(d_1, d_2, \ldots, d_N)$, so that $\hat{\boldsymbol{\beta}}_\rho = \mathbf{D}_\rho^{-1} \mathbf{X}'_\rho \mathbf{Y}_\rho$ and covariance matrix of $\hat{\boldsymbol{\beta}}_\rho$ is then equal to $\sigma^2 \mathbf{D}_\rho^{-1}$. From the expression in Equation (3.16) it follows that the minimal complete factorial arrangement does not provide an unbiased estimator of σ^2 and hence of $\operatorname{Cov}(\hat{\boldsymbol{\beta}}_\rho)$ under the orthonormalized (or nonnormalized orthogonal) polynomial model. If one or more treatments are replicated, that is, if we take a complete factorial Γ with $n > N$, then the resulting design is capable of providing unbiased estimators for both $\boldsymbol{\beta}_\rho$ and σ^2. The BLUE of $\boldsymbol{\beta}_\rho$ in this case is $\hat{\boldsymbol{\beta}}_\rho = (\mathbf{W}'_\Gamma \mathbf{W}_\Gamma)^{-1} \mathbf{W}'_\Gamma \mathbf{Y}_\Gamma$ and an unbiased estimator of σ^2 is $\hat{\sigma}^2 = [(\mathbf{Y}_\Gamma - \mathbf{W}_\Gamma \hat{\boldsymbol{\beta}}_\rho)'(\mathbf{Y}_\Gamma - \mathbf{W}_\Gamma \hat{\boldsymbol{\beta}}_\rho)]/(n - N)$.

Note that if $\operatorname{Cov}(\boldsymbol{\epsilon}_\rho) = \sigma^2 \mathbf{V}$, where $\mathbf{V}$ is a known positive definite matrix then the BLUE of $\boldsymbol{\beta}_\rho$ is

$$\hat{\boldsymbol{\beta}}_\rho = \left[\mathbf{X}'_\rho \mathbf{V}^{-1} \mathbf{X}_\rho\right]^{-1} \mathbf{X}'_\rho \mathbf{V}^{-1} \mathbf{Y}_\rho, \tag{4.17}$$

with

$$\operatorname{Cov}\left(\hat{\boldsymbol{\beta}}_\rho\right) = \left[\mathbf{X}'_\rho \mathbf{V}^{-1} \mathbf{X}_\rho\right]^{-1} \sigma^2. \tag{4.18}$$

In the case of the nonorthogonal polynomial model, the BLUE of $\boldsymbol{\theta}_\rho$ is obtained as $\hat{\boldsymbol{\theta}}_\rho = [\mathbf{P}'_\rho \mathbf{P}_\rho]^{-1} \mathbf{P}'_\rho \mathbf{Y}_\rho$ with $\operatorname{Cov}(\hat{\boldsymbol{\theta}}_\rho) = [\mathbf{P}'_\rho \mathbf{P}_\rho]^{-1} \sigma^2$.

Example 4.5. Considering the 2×3 equispaced factorial introduced in Example 4.3, the orthonormal polynomial model is

$$E\left[\mathbf{Y}_\rho\right] = \left\{ \begin{bmatrix} \frac{1}{\sqrt{2}} & \frac{-1}{\sqrt{2}} \\ \frac{1}{\sqrt{2}} & \frac{1}{\sqrt{2}} \end{bmatrix} \otimes \begin{bmatrix} \frac{1}{\sqrt{3}} & \frac{-1}{\sqrt{2}} & \frac{1}{\sqrt{6}} \\ \frac{1}{\sqrt{3}} & 0 & \frac{-2}{\sqrt{6}} \\ \frac{1}{\sqrt{3}} & \frac{1}{\sqrt{2}} & \frac{1}{\sqrt{6}} \end{bmatrix} \right\} \left\{ \boldsymbol{\phi}_1 \circledast \boldsymbol{\phi}_2 \right\},$$

that is,

$$E\begin{bmatrix} Y_{(0,0)} \\ Y_{(0,1)} \\ Y_{(0,2)} \\ Y_{(1,0)} \\ Y_{(1,1)} \\ Y_{(1,2)} \end{bmatrix} = \begin{bmatrix} \frac{1}{\sqrt{6}} & \frac{-1}{2} & \frac{1}{\sqrt{12}} & \frac{-1}{\sqrt{6}} & \frac{1}{2} & \frac{-1}{\sqrt{12}} \\ \frac{1}{\sqrt{6}} & 0 & \frac{-2}{\sqrt{12}} & \frac{-1}{\sqrt{6}} & 0 & \frac{2}{\sqrt{12}} \\ \frac{1}{\sqrt{6}} & \frac{1}{2} & \frac{1}{\sqrt{12}} & \frac{-1}{\sqrt{6}} & \frac{-1}{2} & \frac{-1}{\sqrt{12}} \\ \frac{1}{\sqrt{6}} & \frac{-1}{2} & \frac{1}{\sqrt{12}} & \frac{1}{\sqrt{6}} & \frac{-1}{2} & \frac{1}{\sqrt{12}} \\ \frac{1}{\sqrt{6}} & 0 & \frac{-2}{\sqrt{12}} & \frac{1}{\sqrt{6}} & 0 & \frac{-2}{\sqrt{12}} \\ \frac{1}{\sqrt{6}} & \frac{1}{2} & \frac{1}{\sqrt{12}} & \frac{1}{\sqrt{6}} & \frac{1}{2} & \frac{1}{\sqrt{12}} \end{bmatrix} \begin{bmatrix} \phi_1^0\phi_2^0 \\ \phi_1^0\phi_2^1 \\ \phi_1^0\phi_2^2 \\ \phi_1^1\phi_2^0 \\ \phi_1^1\phi_2^1 \\ \phi_1^1\phi_2^2 \end{bmatrix}.$$

Using Equation (4.15) we see that the BLUE of $\boldsymbol{\beta}_p$ is

$$\hat{\boldsymbol{\beta}}_p = \begin{bmatrix} \frac{1}{\sqrt{6}} & \frac{1}{\sqrt{6}} & \frac{1}{\sqrt{6}} & \frac{1}{\sqrt{6}} & \frac{1}{\sqrt{6}} & \frac{1}{\sqrt{6}} \\ \frac{-1}{2} & 0 & \frac{1}{2} & \frac{-1}{2} & 0 & \frac{1}{2} \\ \frac{1}{\sqrt{12}} & \frac{-2}{\sqrt{12}} & \frac{1}{\sqrt{12}} & \frac{1}{\sqrt{12}} & \frac{-2}{\sqrt{12}} & \frac{1}{\sqrt{12}} \\ \frac{-1}{\sqrt{6}} & \frac{-1}{\sqrt{6}} & \frac{-1}{\sqrt{6}} & \frac{1}{\sqrt{6}} & \frac{1}{\sqrt{6}} & \frac{1}{\sqrt{6}} \\ \frac{1}{2} & 0 & \frac{-1}{2} & \frac{-1}{2} & 0 & \frac{1}{2} \\ \frac{-1}{\sqrt{12}} & \frac{2}{\sqrt{12}} & \frac{-1}{\sqrt{12}} & \frac{1}{\sqrt{12}} & \frac{-2}{\sqrt{12}} & \frac{1}{\sqrt{12}} \end{bmatrix} \begin{bmatrix} Y_{(0,0)} \\ Y_{(0,1)} \\ Y_{(0,2)} \\ Y_{(1,0)} \\ Y_{(1,1)} \\ Y_{(1,2)} \end{bmatrix}.$$

4.3 BEST LINEAR UNBIASED ESTIMATION OF LINEAR PARAMETRIC FUNCTIONS USING AN ARBITRARY FACTORIAL DESIGN UNDER AN ORTHOGONAL POLYNOMIAL MODEL

Let Γ be an arbitrary factorial design $\mathrm{FD}(k_1, k_2, \ldots, k_t; m; n; r_1, r_2, \ldots, r_N)$. We associate a polynomial model with Γ in the following way:

(4.19) $$E[\mathbf{Y}_\Gamma] = \mathbf{X}_\Gamma \boldsymbol{\beta}_p,$$

where the observation vector $\mathbf{Y}_\Gamma$ is written as

$$(4.20)\qquad \mathbf{Y}'_\Gamma=\left(Y_{g_1}^{(1)},Y_{g_1}^{(2)},\ldots,Y_{g_1}^{(r_1)},Y_{g_2}^{(1)},Y_{g_2}^{(2)},\ldots,Y_{g_2}^{(r_2)},\ldots,\right.$$
$$\left.Y_{g_N}^{(1)},Y_{g_N}^{(2)},\ldots,Y_{g_N}^{(r_N)}\right).$$

As before, $Y_{g_i}^{(h)}$ refers to the observation associated with the hth repetition of the treatment g_i. The $n\times N$ design matrix $\mathbf{X}_\Gamma$ is obtained from the matrix $\mathbf{X}_\rho$ of the related minimal complete factorial arrangement under the orthogonal polynomial model taking repetitions of treatments into account. Explicitly, the design matrix $\mathbf{X}_\Gamma$ can be rewritten as

$$(4.21)\qquad \mathbf{X}_\Gamma=\begin{bmatrix}\mathbf{1}_{r_1} & \mathbf{X}_{g_1}\\ \mathbf{1}_{r_2} & \mathbf{X}_{g_2}\\ & \vdots \\ \mathbf{1}_{r_N} & \mathbf{X}_{g_N}\end{bmatrix},$$

where $\mathbf{X}_{g_i}$ is the ith row of $\mathbf{X}_\rho$ of the previous section, and $\mathbf{1}_{r_i}$ is a column vector of ones of order r_i. It is understood that whenever $r_i=0$, then the corresponding $\mathbf{1}_{r_i}\mathbf{X}_{g_i}$ in the matrix of (4.21) is not present. Finally $\boldsymbol{\beta}_\rho$ is the full parametric vector as in (4.12).

It should be noted that one can associate with the minimal complete factorial arrangement ρ and $\mathbf{X}_\rho^*\boldsymbol{\beta}_\rho^*$ where $\mathbf{X}_\rho^*$ is an arbitrary columnwise orthogonal or orthonormal matrix, and similar results will be obtained. Such a model has already been introduced at the beginning of Section 3.4, where $\mathbf{X}_\rho^*$ was a contrast matrix. For example, instead of using matrices $\mathbf{M}_i$ arising from orthogonal polynomials, we may use $\mathbf{R}_i$, known as *Helmert orthogonal matrices* given by

$$\mathbf{R}_i=\begin{bmatrix}1 & 1 & 1 & \cdots & 1\\ 1 & -1 & 1 & \cdots & 1\\ 1 & 0 & -2 & \cdots & 1\\ \vdots & \vdots & \vdots & \vdots & \vdots\\ 1 & 0 & 0 & \cdots & -(k_i-1)\end{bmatrix}.$$

One may wish to consider the $\mathbf{M}_i$ or $\mathbf{R}_i$ as a contrast matrix (see Chapter 3) for the ith factor. The $\mathbf{M}_i$, $\mathbf{R}_i$ or any other contrast matrix of interest to the experimenter may be used individually or in combination to form the design matrix $\mathbf{X}_\rho$ in (4.11).

Example 4.6. Refer to the equispaced 2×3 factorial of Example 4.5, and let $\Gamma=\{(0,0),(0,0),(1,0),(1,1),(1,1),(1,1),(1,2)\}$. Note that $r_1=r_{(0,0)}=2$,

$r_2 = r_{(0,1)} = 0$, $r_3 = r_{(0,2)} = 0$, $r_4 = r_{(1,0)} = 1$, $r_5 = r_{(1,1)} = 3$, and $r_6 = r_{(1,2)} = 1$. Hence the model for Γ is explicitly equal to

$$E[\mathbf{Y}_\Gamma] = \begin{bmatrix} \mathbf{1}_2 & \mathbf{X}_1 \\ \mathbf{1}_1 & \mathbf{X}_4 \\ \mathbf{1}_3 & \mathbf{X}_5 \\ \mathbf{1}_1 & \mathbf{X}_6 \end{bmatrix} \boldsymbol{\beta}_\rho,$$

that is,

$$E\begin{bmatrix} Y^{(1)}_{(0,0)} \\ Y^{(2)}_{(0,0)} \\ Y^{(1)}_{(1,0)} \\ Y^{(1)}_{(1,1)} \\ Y^{(2)}_{(1,1)} \\ Y^{(3)}_{(1,1)} \\ Y^{(1)}_{(1,2)} \end{bmatrix} = \begin{bmatrix} \frac{1}{\sqrt{6}} & \frac{-1}{2} & \frac{1}{\sqrt{12}} & \frac{-1}{\sqrt{6}} & \frac{1}{2} & \frac{-1}{\sqrt{12}} \\ \frac{1}{\sqrt{6}} & \frac{-1}{2} & \frac{1}{\sqrt{12}} & \frac{-1}{\sqrt{6}} & \frac{1}{2} & \frac{-1}{\sqrt{12}} \\ \frac{1}{\sqrt{6}} & \frac{-1}{2} & \frac{1}{\sqrt{12}} & \frac{1}{\sqrt{6}} & \frac{-1}{2} & \frac{1}{\sqrt{12}} \\ \frac{1}{\sqrt{6}} & 0 & \frac{-2}{\sqrt{12}} & \frac{1}{\sqrt{6}} & 0 & \frac{-2}{\sqrt{12}} \\ \frac{1}{\sqrt{6}} & 0 & \frac{-2}{\sqrt{12}} & \frac{1}{\sqrt{6}} & 0 & \frac{-2}{\sqrt{12}} \\ \frac{1}{\sqrt{6}} & 0 & \frac{-2}{\sqrt{12}} & \frac{1}{\sqrt{6}} & 0 & \frac{-2}{\sqrt{12}} \\ \frac{1}{\sqrt{6}} & \frac{1}{2} & \frac{1}{\sqrt{12}} & \frac{1}{\sqrt{6}} & \frac{1}{2} & \frac{1}{\sqrt{12}} \end{bmatrix} \begin{bmatrix} \phi_1^0\phi_2^0 \\ \phi_1^0\phi_2^1 \\ \phi_1^0\phi_2^2 \\ \phi_1^1\phi_2^0 \\ \phi_1^1\phi_2^1 \\ \phi_1^1\phi_2^2 \end{bmatrix}.$$

Suppose now that the experimenter is interested in estimating a vector $\boldsymbol{\psi}$ of v independent linear parametric functions specified by $\boldsymbol{\psi} = \mathbf{L}'\boldsymbol{\beta}_\rho$, where $\mathbf{L}'$ is a matrix of order $v \times N$ of rank $v \leq N$. We shall distinguish between the following two cases, which are treated successively.

Case 1. No specific *a priori* assumptions are made on the components of $\boldsymbol{\beta}_\rho$. Let Γ be a design such that $\mathbf{L}'\boldsymbol{\beta}_\rho$ is estimable; that is, there exists a matrix $\mathbf{K}'_\Gamma$ such that

$$\mathbf{L}' = \mathbf{K}'_\Gamma \mathbf{X}_\Gamma. \tag{4.22}$$

Then the BLUE of $\mathbf{L}'\boldsymbol{\beta}_\rho$ is immediately obtained from (3.14) of Chapter 3 and is equal to

$$\hat{\boldsymbol{\psi}} = \widehat{\mathbf{L}'\boldsymbol{\beta}_\rho} = \mathbf{L}'(\mathbf{X}'_\Gamma \mathbf{X}_\Gamma)^- \mathbf{X}'_\Gamma \mathbf{Y}'_\Gamma, \tag{4.23}$$

where $(\mathbf{A})^-$ indicates a generalized inverse of $\mathbf{A}$. The expected value and the covariance matrix of this estimator are, respectively,

$$E[\hat{\psi}] = E[\widehat{\mathbf{L}'\boldsymbol{\beta}_\rho}] = \mathbf{K}'_\Gamma\mathbf{X}_\Gamma(\mathbf{X}'_\Gamma\mathbf{X}_\Gamma)^-\mathbf{X}'_\Gamma\mathbf{X}_\Gamma\boldsymbol{\beta}_\rho \tag{4.24}$$
$$= \mathbf{K}'_\Gamma\mathbf{X}_\Gamma\boldsymbol{\beta}_\rho = \mathbf{L}'\boldsymbol{\beta}_\rho = \psi,$$

$$\operatorname{Cov}(\hat{\psi}) = \mathbf{L}'(\mathbf{X}'_\Gamma\mathbf{X}_\Gamma)^-\mathbf{L}\sigma^2 \tag{4.25}$$
$$= \mathbf{K}'_\Gamma\mathbf{X}_\Gamma(\mathbf{X}'_\Gamma\mathbf{X}_\Gamma)^-\mathbf{X}'_\Gamma\mathbf{K}_\Gamma\sigma^2.$$

It is well known that $\mathbf{X}_\Gamma(\mathbf{X}'_\Gamma\mathbf{X}_\Gamma)^-\mathbf{X}'_\Gamma$ is invariant under any choice of a generalized inverse for $\mathbf{X}'_\Gamma\mathbf{X}_\Gamma$.

Remark. Case 1 includes *response surface estimation* by setting $\mathbf{L}' = \mathbf{X}_\Gamma$, so that $\hat{\psi} = \widehat{\mathbf{X}_\Gamma\boldsymbol{\beta}_\rho} = \mathbf{X}_\Gamma(\mathbf{X}'_\Gamma\mathbf{X}_\Gamma)^-\mathbf{X}'_\Gamma\mathbf{Y}_\Gamma$. Since in this case $E[\hat{\psi}] = E[\widehat{\mathbf{X}_\Gamma\boldsymbol{\beta}_\rho}] = \mathbf{X}_\Gamma\boldsymbol{\beta}_\rho = E[\mathbf{Y}_\Gamma]$, the estimator $\widehat{E[\mathbf{Y}_\Gamma]}$ is often written as $\widehat{\mathbf{X}_\Gamma\boldsymbol{\beta}_\rho}$.

Example 4.7. Consider the 2^3 factorial with $G_1 = G_2 = G_3 = \{0,1\}$ and the nonnormalized orthogonal model

$$E[\mathbf{Y}_\rho] = [\mathbf{X}_1 \otimes \mathbf{X}_2 \otimes \mathbf{X}_3][\phi_1 \circledast \phi_2 \circledast \phi_3] = \mathbf{X}_\rho\boldsymbol{\beta}_\rho,$$

where

$$\mathbf{Y}'_\rho = (Y_{(0,0,0)}, Y_{(0,0,1)}, Y_{(0,1,0)}, Y_{(0,1,1)}, Y_{(1,0,0)}, Y_{(1,0,1)}, Y_{(1,1,0)}, Y_{(1,1,1)}),$$

$$\mathbf{X}_\rho = \begin{bmatrix} 1 & -1 \\ 1 & 1 \end{bmatrix} \otimes \begin{bmatrix} 1 & -1 \\ 1 & 1 \end{bmatrix} \otimes \begin{bmatrix} 1 & -1 \\ 1 & 1 \end{bmatrix}$$
$$= \begin{bmatrix} 1 & -1 & -1 & 1 & -1 & 1 & 1 & -1 \\ 1 & 1 & -1 & -1 & -1 & -1 & 1 & 1 \\ 1 & -1 & 1 & -1 & -1 & 1 & -1 & 1 \\ 1 & 1 & 1 & 1 & -1 & -1 & -1 & -1 \\ 1 & -1 & -1 & 1 & 1 & -1 & -1 & 1 \\ 1 & 1 & -1 & -1 & 1 & 1 & -1 & -1 \\ 1 & -1 & 1 & -1 & 1 & -1 & 1 & -1 \\ 1 & 1 & 1 & 1 & 1 & 1 & 1 & 1 \end{bmatrix},$$

and

$$\boldsymbol{\beta}'_\rho = (\phi_1 \circledast \phi_2 \circledast \phi_3)'$$
$$= (\phi_1^0\phi_2^0\phi_3^0, \phi_1^0\phi_2^0\phi_3^1, \phi_1^0\phi_2^1\phi_3^0, \phi_1^0\phi_2^1\phi_3^1, \phi_1^1\phi_2^0\phi_3^0,$$
$$\phi_1^1\phi_2^0\phi_3^1, \phi_1^1\phi_2^1\phi_3^0, \phi_1^1\phi_2^1\phi_3^1).$$

Let

$$\mathbf{L}' = \begin{bmatrix} 1 & 0 & 0 & 0 & 0 & 0 & 0 & 1 \\ 0 & 1 & 0 & 0 & 0 & 0 & 1 & 0 \\ 0 & 0 & 1 & 0 & 0 & 1 & 0 & 0 \\ 0 & 0 & 0 & 1 & 1 & 0 & 0 & 0 \end{bmatrix}.$$

Then

$$\mathbf{K}'_\Gamma = \mathbf{L}'\mathbf{X}_\Gamma^{-1} = \tfrac{1}{8}\mathbf{L}'\mathbf{X}'_\Gamma = \tfrac{1}{4}\begin{bmatrix} 0 & 1 & 1 & 0 & 1 & 0 & 0 & 1 \\ 0 & 1 & -1 & 0 & -1 & 0 & 0 & 1 \\ 0 & -1 & 1 & 0 & -1 & 0 & 0 & 1 \\ 0 & -1 & -1 & 0 & 1 & 0 & 0 & 1 \end{bmatrix},$$

so that $\mathbf{L}'\boldsymbol{\beta}_\rho$ is an estimable vector of linear parametric functions.

From (4.23) the BLUE of $\mathbf{L}'\boldsymbol{\beta}_\rho$ is obtained as

$$\begin{aligned} \widehat{\mathbf{L}'\boldsymbol{\beta}_\rho} &= \mathbf{L}'(\mathbf{X}'_\Gamma\mathbf{X}_\Gamma)^{-1}\mathbf{X}'_\Gamma\mathbf{Y}_\Gamma \\ &= \mathbf{K}'_\Gamma\mathbf{X}_\Gamma(\tfrac{1}{8}\mathbf{I}_8)\mathbf{X}'_\Gamma\mathbf{Y}_\Gamma \\ &= \mathbf{K}'_\Gamma\mathbf{Y}_\Gamma \end{aligned}$$

$$= \begin{bmatrix} 0 & 1 & 1 & 0 & 1 & 0 & 0 & 1 \\ 0 & 1 & -1 & 0 & -1 & 0 & 0 & 1 \\ 0 & -1 & 1 & 0 & -1 & 0 & 0 & 1 \\ 0 & -1 & -1 & 0 & 1 & 0 & 0 & 1 \end{bmatrix} \begin{bmatrix} Y_{(0,0,0)} \\ Y_{(0,0,1)} \\ Y_{(0,1,0)} \\ Y_{(0,1,1)} \\ Y_{(1,0,0)} \\ Y_{(1,0,1)} \\ Y_{(1,1,0)} \\ Y_{(1,1,1)} \end{bmatrix}$$

$$= \begin{bmatrix} Y_{(0,0,1)} + Y_{(0,1,0)} + Y_{(1,0,0)} + Y_{(1,1,1)} \\ Y_{(0,0,1)} - Y_{(0,1,0)} - Y_{(1,0,0)} + Y_{(1,1,1)} \\ -Y_{(0,0,1)} + Y_{(0,1,0)} - Y_{(1,0,0)} + Y_{(1,1,1)} \\ -Y_{(0,0,1)} - Y_{(0,1,0)} + Y_{(1,0,0)} + Y_{(1,1,1)} \end{bmatrix}.$$

Thus $\widehat{\mathbf{L}'\boldsymbol{\beta}_\rho}$ is obtained from only four of the eight observations and the reader is invited to modify $\mathbf{L}'$ so that the other four observations would be used.

The covariance of $\widehat{\mathbf{L}'\boldsymbol{\beta}_\rho}$ is computed from Equation (4.25) as

$$\operatorname{Cov}\left(\widehat{\mathbf{L}'\boldsymbol{\beta}_\rho}\right) = \mathbf{K}'_\Gamma \mathbf{K}_\Gamma \sigma^2$$

$$= \frac{\sigma^2}{4}\begin{bmatrix} 1 & 0 & 0 & 0 \\ 0 & 1 & 0 & 0 \\ 0 & 0 & 1 & 0 \\ 0 & 0 & 0 & 1 \end{bmatrix}.$$

That is, each element of $\widehat{\mathbf{L}'\boldsymbol{\beta}_\rho}$ has variance equal to $\sigma^2/4$ and the covariance between any two elements of $\widehat{\mathbf{L}'\boldsymbol{\beta}_\rho}$ is equal to 0. In this example no estimate of σ^2 is available, since $n-s=8-8=0$ and thus Equation (3.16) cannot be used. If any of the preceding eight treatments are repeated and the same orthogonal polynomial model is used, then an unbiased estimator of σ^2 is obtainable from (3.16).

Case 2. Suppose now that the experimenter has *a priori* knowledge of the exact values of some of the components of $\boldsymbol{\beta}_\rho$. We may assume without loss of generality that these values are zero and $\boldsymbol{\beta}'_\rho = [\boldsymbol{\beta}'_1 \vdots \boldsymbol{\beta}'_2] = [\boldsymbol{\beta}'_1 \vdots \mathbf{0}']$. In this case Model (4.19) reduces to

$$E[\mathbf{Y}_\Gamma] = \left[\mathbf{X}_{\Gamma 1} \vdots \mathbf{X}_{\Gamma 2}\right]\begin{bmatrix} \boldsymbol{\beta}_1 \\ \cdots \\ \mathbf{0} \end{bmatrix} = \mathbf{X}_{\Gamma 1}\boldsymbol{\beta}_1. \tag{4.26}$$

Let Γ be a design such that $\mathbf{L}'_1\boldsymbol{\beta}_1$ is estimable; that is, there exists a $\mathbf{K}'_\Gamma$ such that

$$\mathbf{L}'_1 = \mathbf{K}'_\Gamma \mathbf{X}_{\Gamma 1}. \tag{4.27}$$

The BLUE of $\mathbf{L}'_1\boldsymbol{\beta}_1$, together with its covariance matrix, are obtained from (3.14) and (3.15) and are now given by

$$\widehat{\mathbf{L}'_1\boldsymbol{\beta}_1} = \mathbf{L}'_1(\mathbf{X}'_{\Gamma 1}\mathbf{X}_{\Gamma 1})^{-}\mathbf{X}'_{\Gamma 1}\mathbf{Y}_\Gamma, \tag{4.28}$$

$$\operatorname{Cov}\left(\widehat{\mathbf{L}'_1\boldsymbol{\beta}_1}\right) = \mathbf{L}'_1(\mathbf{X}'_{\Gamma 1}\mathbf{X}_{\Gamma 1})^{-}\mathbf{L}_1\sigma^2. \tag{4.29}$$

Remark. We have assumed that $\boldsymbol{\beta}_2 = \mathbf{0}$ so that the BLUE $\widehat{\mathbf{L}'_1\boldsymbol{\beta}_1}$ is an unbiased estimator of $\boldsymbol{\beta}_1$. If in reality $\boldsymbol{\beta}_2 \neq \mathbf{0}$, then $\widehat{\mathbf{L}'_1\boldsymbol{\beta}_1}$ is no longer unbiased, since then

$$E\left[\widehat{\mathbf{L}'_1\boldsymbol{\beta}_1}\right] = \mathbf{L}'_1\boldsymbol{\beta}_1 + \mathbf{A}_\Gamma\boldsymbol{\beta}_2, \tag{4.30}$$

where

$$\mathbf{A}_\Gamma = \mathbf{L}_1'(\mathbf{X}_{\Gamma 1}'\mathbf{X}_{\Gamma 1})^- \mathbf{X}_{\Gamma 1}'\mathbf{X}_{\Gamma 2}. \tag{4.31}$$

The covariance of this biased estimator is clearly equal to (4.29). In this case the covariance matrix is not informative enough as a measure of inaccuracy, since it does not take into account the *bias* $\mathbf{A}_\Gamma\boldsymbol{\beta}_2$. The matrix $\mathbf{A}_\Gamma$ is known in the literature as the *alias matrix* of the design Γ relative to $\mathbf{L}_1'\boldsymbol{\beta}_1$ and $\boldsymbol{\beta}_2$. Another quantity sensitive to precision and bias called the *mean square error* (MSE) is given by

$$\begin{aligned} \text{MSE}(\widehat{\mathbf{L}_1'\boldsymbol{\beta}_1}) &= E[(\widehat{\mathbf{L}_1'\boldsymbol{\beta}_1} - \mathbf{L}_1'\boldsymbol{\beta}_1)(\widehat{\mathbf{L}_1'\boldsymbol{\beta}_1} - \mathbf{L}_1'\boldsymbol{\beta}_1)'] \\ &= \text{Cov}(\widehat{\mathbf{L}_1'\boldsymbol{\beta}_1}) + \mathbf{A}_\Gamma\boldsymbol{\beta}_2\boldsymbol{\beta}_2'\mathbf{A}_\Gamma'. \end{aligned} \tag{4.32}$$

Example 4.8. Suppose that in Example 4.7 the parametric vector $\boldsymbol{\beta}_\rho$ is partitioned as $\boldsymbol{\beta}_1' = (\phi_1^0\phi_2^0\phi_3^0, \phi_1^0\phi_2^0\phi_3^1, \phi_1^0\phi_2^1\phi_3^0, \phi_1^1\phi_2^0\phi_3^0)$ and $\boldsymbol{\beta}_2' = (\phi_1^0\phi_2^1\phi_3^1, \phi_1^1\phi_2^0\phi_3^1, \phi_1^1\phi_2^1\phi_3^0, \phi_1^1\phi_2^1\phi_3^1)$. Now if $\boldsymbol{\beta}_2 = \mathbf{0}$ and Γ is the minimal complete factorial ρ, then $\mathbf{X}_\Gamma\boldsymbol{\beta}_\rho = \mathbf{X}_{\Gamma 1}\boldsymbol{\beta}_1$, where $\mathbf{X}_{\Gamma 1}$ is simply the matrix consisting of columns 1,2,3, and 5 of $\mathbf{X}_\rho$, that is,

$$E\begin{bmatrix} Y_{(0,0,0)} \\ Y_{(0,0,1)} \\ Y_{(0,1,0)} \\ Y_{(0,1,1)} \\ Y_{(1,0,0)} \\ Y_{(1,0,1)} \\ Y_{(1,1,0)} \\ Y_{(1,1,1)} \end{bmatrix} = \begin{bmatrix} 1 & -1 & -1 & -1 \\ 1 & 1 & -1 & -1 \\ 1 & -1 & 1 & -1 \\ 1 & 1 & 1 & -1 \\ 1 & -1 & -1 & 1 \\ 1 & 1 & -1 & 1 \\ 1 & -1 & 1 & 1 \\ 1 & 1 & 1 & 1 \end{bmatrix} \begin{bmatrix} \phi_1^0\phi_2^0\phi_3^0 \\ \phi_1^0\phi_2^0\phi_3^1 \\ \phi_1^0\phi_2^1\phi_3^0 \\ \phi_1^1\phi_2^0\phi_3^0 \end{bmatrix}.$$

Since $\mathbf{X}_{\Gamma 1}$ is columnwise orthogonal and thus of full column rank, it follows that the BLUE of $\mathbf{L}_1'\boldsymbol{\beta}_1 = \mathbf{I}_4\boldsymbol{\beta}_1 = \boldsymbol{\beta}_1$ is given by $\widehat{\mathbf{L}_1'\boldsymbol{\beta}_1} = \frac{1}{8}\mathbf{X}_{\Gamma 1}'\mathbf{Y}_\Gamma = \hat{\boldsymbol{\beta}}_1$, that is,

$$\hat{\boldsymbol{\beta}}_1 = \frac{1}{8}\begin{bmatrix} 1 & 1 & 1 & 1 & 1 & 1 & 1 & 1 \\ -1 & 1 & -1 & 1 & -1 & 1 & -1 & 1 \\ -1 & -1 & 1 & 1 & -1 & -1 & 1 & 1 \\ -1 & -1 & -1 & -1 & 1 & 1 & 1 & 1 \end{bmatrix}\mathbf{Y}_\Gamma.$$

The covariance of $\hat{\boldsymbol{\beta}}_1$ is simply equal to $\frac{1}{8}\sigma^2\mathbf{I}_4$. In this example it is not necessary to calculate $\mathbf{K}'_{\Gamma 1}$, but the reader may verify that it is equal to $\frac{1}{8}\mathbf{X}'_{\Gamma}$. From (4.31) we see that the alias matrix of the design Γ relative to $\boldsymbol{\beta}_1$ and $\boldsymbol{\beta}_2$ is equal to $\mathbf{A}_{\Gamma} = \mathbf{I}_4(\mathbf{X}'_{\Gamma 1}\mathbf{X}_{\Gamma 1})^{-1}\mathbf{X}'_{\Gamma 1}\mathbf{X}_{\Gamma 2} = \frac{1}{8}\mathbf{X}'_{\Gamma 1}\mathbf{X}_{\Gamma 2} = \mathbf{0}$, since the columns of $\mathbf{X}_{\Gamma}$ are the remaining columns of $\mathbf{X}_{\rho}$ and thus $\mathbf{X}'_{\Gamma 1}\mathbf{X}_{\Gamma 2} = \mathbf{0}$. Hence $\hat{\boldsymbol{\beta}}_1$ is an unbiased estimator of $\boldsymbol{\beta}_1$ even though $\boldsymbol{\beta}_2$ may not be equal to $\mathbf{0}$. Also note that $\text{MSE}(\hat{\boldsymbol{\beta}}_1) = \text{Cov}(\hat{\boldsymbol{\beta}}_1)$.

In this example the four linear combinations of $\mathbf{Y}_{\Gamma}$ given by $\mathbf{X}'_{\Gamma 2}\mathbf{Y}_{\Gamma}$ may be used to obtain an unbiased estimator of σ^2 if $\boldsymbol{\beta}_2 = \mathbf{0}$. Since $\frac{1}{8}\mathbf{Y}'_{\Gamma}\mathbf{X}_{\Gamma 2}\mathbf{X}'_{\Gamma 2}\mathbf{Y}_{\Gamma}$ has expectation $4\sigma^2$ under Model (4.26) it follows that $\frac{1}{32}\mathbf{Y}'_{\Gamma}\mathbf{X}_{\Gamma 2}\mathbf{X}'_{\Gamma 2}\mathbf{Y}_{\Gamma}$ is an unbiased estimator of σ^2. The reader should verify whether this estimator is the same as that given by expression (3.16).

Example 4.9. For the 2^3 factorial of Example 4.7 consider the design $\Gamma = \{(0,0,0),(0,0,1),(0,1,0),(1,0,0)\}$ with the aim to estimate $\boldsymbol{\beta}'_1 = (\phi_1^0\phi_2^0\phi_3^0, \phi_1^0\phi_2^1\phi_3^1, \phi_1^1\phi_2^0\phi_3^1, \phi_1^1\phi_2^1\phi_3^0)$ under the assumption that σ^2 is known and $\boldsymbol{\beta}'_2 = (\phi_1^0\phi_2^0\phi_3^1, \phi_1^0\phi_2^1\phi_3^0, \phi_1^1\phi_2^0\phi_3^0, \phi_1^1\phi_2^1\phi_3^1)$ is negligible. The model for Γ and $\boldsymbol{\beta}_1$ is directly obtained from the design matrix $\mathbf{X}_{\rho}$ of the minimal complete factorial arrangement ρ, that is,

$$E\begin{bmatrix} Y_{(0,0,0)} \\ Y_{(0,0,1)} \\ Y_{(0,1,0)} \\ Y_{(1,0,0)} \end{bmatrix} = \begin{bmatrix} 1 & 1 & 1 & 1 \\ 1 & -1 & -1 & 1 \\ 1 & -1 & 1 & -1 \\ 1 & 1 & -1 & -1 \end{bmatrix} \begin{bmatrix} \phi_1^0\phi_2^0\phi_3^0 \\ \phi_1^0\phi_2^1\phi_3^1 \\ \phi_1^1\phi_2^0\phi_3^1 \\ \phi_1^1\phi_2^1\phi_3^0 \end{bmatrix}.$$

Since $\mathbf{X}'_{\Gamma 1}\mathbf{X}_{\Gamma 1} = 4\mathbf{I}_4$ the BLUE and covariance in (4.28) and (4.29) take on simplified forms, that is,

$$\hat{\boldsymbol{\beta}}_1 = \tfrac{1}{4}\mathbf{X}'_{\Gamma 1}\mathbf{Y}_{\Gamma}$$

$$= \tfrac{1}{4}\begin{bmatrix} 1 & 1 & 1 & 1 \\ 1 & -1 & -1 & 1 \\ 1 & -1 & 1 & -1 \\ 1 & 1 & -1 & -1 \end{bmatrix} \begin{bmatrix} Y_{(0,0,0)} \\ Y_{(0,0,1)} \\ Y_{(0,1,0)} \\ Y_{(1,0,0)} \end{bmatrix}$$

and

$$\text{Cov}(\hat{\boldsymbol{\beta}}_1) = \frac{\sigma^2}{4}\mathbf{I}_4.$$

If the assumption of $\boldsymbol{\beta}_2 = \mathbf{0}$ is not valid, then the bias of the estimator $\hat{\boldsymbol{\beta}}_1$

according to (4.30) is equal to

$$\begin{aligned}
\mathbf{A}_\Gamma\boldsymbol{\beta}_2 &= \mathbf{I}_4(\mathbf{X}'_{\Gamma 1}\mathbf{X}_{\Gamma 1})^{-1}\mathbf{X}'_{\Gamma 1}\mathbf{X}_{\Gamma 2}\boldsymbol{\beta}_2 \\
&= \tfrac{1}{4}\mathbf{X}'_{\Gamma 1}\mathbf{X}_{\Gamma 2}\boldsymbol{\beta}_2 \\
&= \tfrac{1}{4}\begin{bmatrix} 1 & 1 & 1 & 1 \\ 1 & -1 & -1 & 1 \\ 1 & -1 & 1 & -1 \\ 1 & 1 & -1 & -1 \end{bmatrix}\begin{bmatrix} -1 & -1 & -1 & -1 \\ 1 & -1 & -1 & 1 \\ -1 & 1 & -1 & 1 \\ -1 & -1 & 1 & 1 \end{bmatrix}\boldsymbol{\beta}_2 \\
&= \tfrac{1}{2}\begin{bmatrix} -1 & -1 & -1 & 1 \\ -1 & -1 & 1 & -1 \\ -1 & 1 & -1 & -1 \\ 1 & -1 & -1 & -1 \end{bmatrix}\boldsymbol{\beta}_2.
\end{aligned}$$

4.4 SELECTED ADDITIONAL READING

Starred references are recommended for a first reading.

1. Aitken, A. C. (1933a). On the graduation of data by the orthogonal polynomials of least squares. *Proc. R. Soc. Edinburgh*, **53**, 54–78.
2. Aitken, A. C. (1933b). On fitting polynomials to weighted data by least squares. *Proc. R. Soc. Edinburgh*, **54**, 1–11.
3. Allan, F. E. (1930). The general form of the orthogonal polynomials for simple series with proofs of their simple properties. *Proc. R. Soc. Edinburgh*, **50**, 310–320.

*4. Anderson, R. L., and Houseman, E. E. (1942). Tables of orthogonal polynomial values extended to $N = 104$. *Iowa Agric. Exp. Stat. Res. Bull.* No. 297, 593–672.

5. Anderson, T. W. (1962). The choice of the degree of a polynomial regression as a multiple decision problem. *Ann. Math. Stat.*, **33**, 255–265.
6. Birge, R. T., and Weinberg, J. W. (1947). Least squares fitting of data by means of polynomials. *Rev. Mod. Phys.* **19**, 298–360.
7. Daniel, C., and Wood, F. S. (1971). *Fitting Equations to Data*. Wiley, New York.
8. Fisher, R. A. (1952). *Statistical Methods for Research Workers*, 12th ed. Oliver and Boyd, Edinburgh.
9. Fisher, R. A., and Yates, F. (1953). *Statistical Tables for Biological, Agricultural and Medical Research*, 4th ed. Hafner, New York.
10. Forsyth, G. E. (1957). Generation and use of orthogonal polynomials for data-fitting with a digital computer. *J. Soc. Indust. Appl. Math.*, **5**, 74–88.
11. Graybill, F. A. (1961). *An Introduction to Linear Statistical Models*, Vol. 1. McGraw-Hill, New York.

*12. Hager, H., and Antle, C. (1968). The choice of the degree of a polynomial model. *J. R. Stat. Soc.*, **30**, 469–471.

13. Hartley, H. O. (1951). The fitting of polynomials to equidistant data with missing values. *Biometrika*, **38**, 410–413.

14. Jackson, D. (1941). *Fourier series and orthogonal polynomials* (Carus Monograph No. 6). Mathematical Association of America, Oberlin, Ohio.

15. Kendall, M. G., and Stuart, A. (1966). *The Advanced Theory of Statistics*, Vol. 3 of *Design and Analysis and Time Series*. Hafner, New York.

*16. Nelder, J. A. (1966). Inverse polynomials, a useful group of multi-factor response functions. *Biometrics*, **22**, 128–141.

17. Pearson, E. S., and Hartley, H.O. (1966). *Biometrika Tables for Statisticians*, 3rd ed., Vol. 1. Cambridge University Press, London.

*18. Robson, D. S. (1959). A simple method for constructing orthogonal polynomials when the independent variable is unequally spaced. *Biometrics*, **15**, 187–191.

19. Snedecor, G. W., and Cochran, W. G. (1967). *Statistical Methods*, 6th ed. Iowa State University Press, Ames.

CHAPTER 5

Constraints and Criteria in Selecting Factorial Designs

In this chapter we discuss the problem of selecting a factorial design from what we consider to be the experimenter's point of view. This is achieved by considering a class of designs for estimating a set of specifed linear functions of a subvector of the total parametric vector which is first constrained by the imposition of a fixed budget and then by suitable criteria introduced for the selection of an optimal factorial design. In subsequent chapters some of these criteria are applied for the selection of a design.

5.1 THE PROBLEM OF SELECTING A FACTORIAL DESIGN

In formulating the problem of selecting a factorial design we assume that the following has been established:

(i) The factors $F_1, F_2, \ldots, F_t$ along with the factor-space G have been explicitly defined.
(ii) A linear model has been postulated for the observation vector $\mathbf{Y}_\rho$ corresponding to the minimal complete factorial arrangement ρ.
(iii) The $N \times 1$ parametric vector $\boldsymbol{\beta}_\rho$ is partitioned as $\boldsymbol{\beta}'_\rho = (\boldsymbol{\beta}'_1 \vdots \boldsymbol{\beta}'_2)$, where a specified set of linear functions of the $p \times 1$ vector $\boldsymbol{\beta}_1$ are to be estimated by a design Γ under the assumption that the $(N-p) \times 1$ vector $\boldsymbol{\beta}_2$ is negligible, $1 \leq p \leq N$.
(iv) The assumption that $\boldsymbol{\beta}_2$ is negligible is made by the experimenter based on his knowledge of the model of the phenomenon under consideration.

For discussion purposes let us limit ourselves to the case where the model in (ii) is a polynomial model, keeping in mind that a similar development also carries through for the general linear model. In selecting a design Γ to estimate the specified vector $\mathbf{L}_1'\boldsymbol{\beta}_1$ of linear parametric functions, it is realistic to impose the condition that the design is capable of providing an unbiased estimator of $\mathbf{L}_1'\boldsymbol{\beta}_1$, that is, $\mathbf{L}_1'\boldsymbol{\beta}_1$ is estimable.

Definition 5.1. The class of all designs, denoted by $\Delta(\mathbf{L}_1)$, such that $\mathbf{L}_1'\boldsymbol{\beta}_1$ is estimable by every design in $\Delta(\mathbf{L}_1)$ is called the class of *unbiased designs*.

In most situations, if not all, the experimenter is not free to choose the treatments or design points in an arbitrary fashion. There may be economical, social, political, environmental, and/or other constraints that confine the experimenter to a certain class of designs in $\Delta(\mathbf{L}_1)$. If $\mathbf{C}$ denotes the set of constraints imposed on the designs in $\Delta(\mathbf{L}_1)$, then $\Delta(\mathbf{L}_1, C)$ is the resulting constrained class of unbiased designs. Often the experimenter will need to introduce some reasonable and meaningful quantity associated with a $\Gamma \in \Delta(\mathbf{L}_1, C)$, say $Q(\Gamma)$; he or she will then need to select a Γ^* in $\Delta(\mathbf{L}_1, C)$ such that $Q(\Gamma^*)$ is minimum. In mathematical terms the quantity $Q(\Gamma)$ is called an *objective function*.

In summary, the problem can be formally stated as follows: Given the factors $F_1, F_2, \ldots, F_t$, the factor-space G, $E[\mathbf{Y}_\rho] = \mathbf{X}_\rho\boldsymbol{\beta}_\rho$, $\boldsymbol{\beta}_\rho' = (\boldsymbol{\beta}_1' \vdots \boldsymbol{\beta}_2' = \mathbf{0})$, a class of constraints C, and a quantity $Q(\Gamma)$ associated with every design Γ, then find a Γ^* such that:

(i) $\mathbf{L}_1'\boldsymbol{\beta}_1$ is estimable, that is, $\Gamma^* \in \Delta(\mathbf{L}_1)$.
(ii) Γ^* satisfies all the constraints, that is, $\Gamma^* \in \Delta(\mathbf{L}_1, C)$.
(iii) $Q(\Gamma^*) \leq Q(\Gamma)$ for every Γ in $\Delta(\mathbf{L}_1, C)$.

The resulting design Γ^* in the class $\Delta(\mathbf{L}_1, C)$ is called an *optimal design* with respect to Q.

5.2 CONSTRAINTS IN SELECTING A FACTORIAL DESIGN

As before, let $\Delta(\mathbf{L}_1)$ be the set of all unbiased designs for estimating $\mathbf{L}_1'\boldsymbol{\beta}_1$. Typically in any experiment the experimenter will be confronted with many considerations such as physical, economical, environmental, and political that constrain him to a subclass of designs in $\Delta(\mathbf{L}_1)$. Let C denote the set of all *constraints* and let $\Delta(\mathbf{L}_1, C) = \{\Gamma_1, \Gamma_2, \ldots, \Gamma_s\}$ be the constrained class of unbiased designs. Now, three cases can occur: (i) $\Delta(\mathbf{L}_1, C)$ is empty, in which case the experimenter must modify the class of constraints C and/or

$\mathbf{L}_1$, (ii) $\Delta(\mathbf{L}_1, C)$ contains a single design, so that it will be the selected design, and (iii) $\Delta(\mathbf{L}_1, C)$ has cardinality greater than one, so that it might be better to impose one or more statistical criteria in terms of objective functions for selecting a design. This latter aspect is treated in some detail in Section 5.3.

Not all constraints (for instance, social and political) are mathematically treatable, since they may not easily be stated in a quantitative form. However, the economic aspects of designs can be put in a suitable mathematical form as will be explained later.

In carrying out an experiment the experimenter may have a *budget* assigned to cover the total cost of experimentation. Suppose that the budget equals B dollars. Taking account of the two types of costs, namely the *fixed* (or overhead) cost B_0 and the *variable cost* B_v, we may write

$$B = B_0 + B_v. \tag{5.1}$$

For each design Γ in $\Delta(\mathbf{L}_1)$, we may calculate this variable cost and indicate it by $B_v^*(\Gamma)$. Those designs in $\Delta(\mathbf{L}_1)$ for which $B_v^*(\Gamma) \leq B_v$ form the constrained class $\Delta(\mathbf{L}_1, B)$ of designs contrained by the budget B. If a treatment g in Γ is denoted by $(x_1, x_2, \ldots, x_t)$, where x_j is the x_jth level of the jth factor, then the cost $B_v^*(\Gamma)$ may be written as

$$\begin{aligned} B_v^*(\Gamma) &= \textit{purchase cost} + \textit{application cost} + \textit{maintenance cost} \\ &\quad + \textit{response cost} \\ &= \sum_{(x_1 x_2, \ldots, x_t) \in \Gamma} \big[P(x_1, x_2, \ldots, x_t) + A(x_1, x_2, \ldots, x_t) \\ &\quad + M(x_1, x_2, \ldots, x_t) + R(x_1, x_2, \ldots, x_t) \big]. \end{aligned} \tag{5.2}$$

The costs in Equation (5.2) may be decomposed into sums of costs associated with the levels appearing in a treatment. For example, if it makes sense we can write

$$P(x_1, x_2, \ldots, x_t) = \sum_j P_j(x_j), \tag{5.3}$$

where $P_j(x_j)$ is the purchase cost of the x_jth level of the jth factor. The approach taken previously and adopted here is known as the *fixed budget approach*, in contrast to a *nonfixed budget approach*. In the latter case we start out with a class of designs constrained by estimability of $\mathbf{L}_1'\boldsymbol{\beta}_1$, and possibly other constraints, and then select a design that minimizes the budget.

5.3 CRITERIA FOR SELECTING AN OPTIMAL FACTORIAL DESIGN

In the previous section we denoted the class of unbiased designs for $\mathbf{L}_1'\boldsymbol{\beta}_1$ by $\Delta(\mathbf{L}_1)$ and the constrained class by $\Delta(\mathbf{L}_1, C)$.

Definition 5.2. The class of all designs, denoted by $\Delta(\mathbf{L}_1, C)$, such that every design in $\Delta(\mathbf{L}_1, C)$ satisfies the constraints in C, will be called the class of *feasible designs*.

Now, whenever $\Delta(\mathbf{L}_1, C)$ contains more than one design, there will be a choice, so that a suitable criterion is needed for selecting a design. This is usually formulated in terms of a real objective function Q on $\Delta(\mathbf{L}_1, C)$ and Γ is chosen to minimize it. If several Γ achieve the minimum, a second objective function may be used to choose among them and so on.

Definition 5.3. A design that minimizes $Q(\Gamma)$ over the class of feasible designs $\Delta(\mathbf{L}_1, C)$ is called a *Q-optimal design*.

We now introduce eight functions $Q_1, Q_2, \ldots, Q_8$ as examples of some objective functions that are useful in classifying designs:

(i) Quantities that reflect variances and covariances of $\widehat{L_1'\boldsymbol{\beta}_1}$, ($Q_1$ to Q_6).
(ii) Quantities that reflect the combinatorial nature of the design, (Q_5, Q_8).
(iii) Quantities that measure departure from the initial assumptions on $\boldsymbol{\beta}_p$, (Q_6, Q_7, Q_8).

These eight objective functions are developed in the next two sections.

5.4 SELECTING OPTIMAL DESIGNS WITH RESPECT TO CRITERIA BASED ON THE SPECTRUM OF THE INFORMATION MATRIX

Restricting our attention to the class of feasible designs, let $\lambda_j(\Gamma)$ be the jth nonzero eigenvalue of the *information matrix* $[\mathbf{L}_1'(\mathbf{X}_1'\mathbf{X}_1)^-\mathbf{L}_1]^-$ of the least squares estimator $\widehat{\mathbf{L}_1'\boldsymbol{\beta}_1}$ using design Γ, $j = 1, 2, \ldots, s$ (= rank of $\mathbf{L}$) and $\lambda_j(\Gamma) \geq \lambda_{j+1}(\Gamma)$. The following Q_i are useful objective functions under (i) above:

$$Q_1(\Gamma) = \frac{1}{\prod_{j=1}^{s} \lambda_j(\Gamma)}, \tag{5.4}$$

which is the inverse of the determinant of the information matrix or generalized variance. An optimal design obtained under Q_1 is usually called *determinant optimal* or in short *D-optimal*;

(5.5)
$$Q_2(\Gamma) = \sum_{j=1}^{s} \frac{1}{\lambda_j(\Gamma)},$$

which is proportional to the average variance of the BLUE of the linear parametric functions in $\mathbf{L}_1'\boldsymbol{\beta}_1$. Here an optimal design is referred to in the literature as *average variance optimal* or simply *A-optimal*;

(5.6)
$$Q_3(\Gamma) = \frac{1}{\lambda_s(\Gamma)}.$$

A corresponding optimal design is known as an *eigenvalue optimal* or *E-optimal* design;

(5.7)
$$Q_4(\Gamma) = \max_{g \in \Gamma} \operatorname{Var}\left[\widehat{E(\mathbf{Y}_g)}\right].$$

An optimal design in this case is called *global optimal* or in short *G-optimal* design.

Definition 5.4. A fractional factorial design is said to be *saturated* relative to the $p \times 1$ vector $\boldsymbol{\beta}_1$ if and only if $m = n = p$.

Note that in a saturated factorial design the objective is to estimate p parameters utilizing p distinct treatment combinations. If the rank of the underlying design matrix is equal to p, then all the parameters in $\boldsymbol{\beta}_1$ are estimable when indeed $\boldsymbol{\beta}_2 = \mathbf{0}$. In a saturated design σ^2 cannot be estimated unbiasedly.

Example 5.1. Consider the 2^3 factorial with the orthogonal model of Chapter 4. Let $\boldsymbol{\beta}_1' = (\phi_1^0\phi_2^0\phi_3^0, \phi_1^0\phi_2^0\phi_3^1, \phi_1^0\phi_2^1\phi_3^0, \phi_1^1\phi_2^0\phi_3^0)$ and make the assumption that the remaining four effects, consisting of the two-and three-factor interactions in $\boldsymbol{\beta}_2$, are negligible. For simplicity also assume that the variance σ^2 is known. Let us constrain the class of possible designs only by the constraint that $n = m = 4$; that is, each design is a saturated design. We have seen earlier that there are precisely $\binom{8}{4} = 70$ possible designs of which 58 are of full rank and 12 are less than full rank. Hence the class of feasible designs $\Delta(\mathbf{L}_1 = \mathbf{I}_4, C)$ consists precisely of 58 saturated designs. By complete enumeration, it may be shown that the design $\Gamma_1 = \{(0, 0, 0),$

$(0,1,1),(1,0,1),(1,1,0)\}$ is Q_1-optimal, Q_2-optimal, Q_3-optimal, and Q_4-optimal. This design is not unique in the class $\Delta(\mathbf{I}_4, C)$.

A few words are in order about the intuitive meaning of the preceding criteria. In principle, statistical decision theory, including design construction is based on a structure that includes specification of a *loss* or *disutility* function W, which in the present setting expresses the cost to the experimenter (or society) as a function of the chosen design Γ, the chosen value of the estimate of $\mathbf{L}_1'\boldsymbol{\beta}_1$, and the *true state of nature* (distribution of $\mathbf{Y}_\Gamma$). The estimate is a chance variable that depends on the value taken by $\mathbf{Y}_\Gamma$ and on the estimator of the $\mathbf{L}_1'\boldsymbol{\beta}_1$ used. For a given design and estimator, the expectation of this loss is called the *risk function*, a function of the state of nature. No design and estimator will yield a uniformly smallest risk function, so some real functional Q^* of this risk function is minimized instead. The choice of W and Q^* is often discussed in terms of axioms of rational behavior that will not be treated here. For simplicity let us suppose here that n is fixed, that all factorial designs cost the same amount, and that we are using the least squares estimator of $\mathbf{L}_1'\boldsymbol{\beta}_1$ computed under the assumption that $\boldsymbol{\beta}_2 = 0$. The Q_i listed previously then reflect various functionals of expected loss due to misestimation by this estimator. The functionals Q_1, Q_2 and Q_3 are, respectively, proportional to the determinant, trace, and maximum eigenvalue of the covariance matrix of $\widehat{\mathbf{L}_1'\boldsymbol{\beta}_1}$. Each of these has intuitively the right "shape." Roughly speaking, they are small for designs that tend to estimate $\mathbf{L}_1'\boldsymbol{\beta}_1$ with small error.

It is unlikely in practice that an experimenter will know his Q exactly. Fortunately, there is usually a kind of insensitivity to using a slightly incorrect Q. If Q' is the one he "should" use and Q'' is a slightly different one, then a Q''-optimal design Γ'' will have $Q'(\Gamma'')$ close to the minimum of Q'. This being so, in the absence of exact knowledge of this Q', design theorists often compute Q''-optimal designs for Q'', which makes computation simple; Q_1, Q_2 and Q_3 are of that nature and there are computational algorithms associated with each of them. A slight variation of Q_2 is

$$Q^{(\mathbf{H})}(\Gamma) = \frac{\operatorname{trace}\left[\mathbf{H}\operatorname{Cov}(\widehat{\mathbf{L}_1'\boldsymbol{\beta}_1})\right]}{\sigma^2}, \tag{5.8}$$

where $\mathbf{H}$ is a specified nonnegative definite matrix. It can be shown that any design that is Q-optimal for a Q that depends only on the matrix $\operatorname{Cov}(\widehat{\mathbf{L}_1'\boldsymbol{\beta}_1})/\sigma^2$ in a reasonable way is $Q^{(\mathbf{H})}$-optimal for some $\mathbf{H}$. Thus the class $Q^{(\mathbf{H})}$-optimal designs for all $\mathbf{H}$ is of considerable interest. Computationally, the determination of a $Q^{(\mathbf{H})}$-optimal design can be reduced to that of a Q_2-optimal [$= Q^{(\mathrm{I})}$-optimal] design for a slightly different model. In addition, certain intuitive features of the various criteria may determine the choice among them. For example, Q_1-optimality yields set-estimators of small volume and good performance, in a commonly employed sense, for testing a hypothesis such as "$\mathbf{L}_1'\boldsymbol{\beta}_1 = \mathbf{0}$."

If one is interested perhaps not just in $\mathbf{L}_1'\boldsymbol{\beta}_1$, where $\mathbf{L}_1'$ was chosen merely to give a convenient basis for the whole space of linear functions one is really interested in, then one might consider σ^{-2} Var(estimator of $\mathbf{C}'\mathbf{L}_1'\boldsymbol{\beta}_1$) $= \mathbf{C}' \operatorname{Cov}(\widehat{\mathbf{L}_1'\boldsymbol{\beta}_1})\mathbf{C}/\sigma^2$ for an arbitrary vector $\mathbf{C}$. Normalizing by restricting $\mathbf{C}$ to $\mathbf{C}'\mathbf{C} = 1$ and averaging over that sphere, one is led to consideration of Q_2-optimality. Similarly, if instead of averaging we consider

$$\max_{\mathbf{C}'\mathbf{C}=1} \operatorname{Var}(\text{estimator of } \mathbf{C}'\mathbf{L}_1'\boldsymbol{\beta}_1) \tag{5.9}$$

we are led to Q_3-optimality. Q_4-optimality obviously refers to the accuracy of our estimate of the entire "response-surface" $E[Y_g]$.

If one changes the levels in G_i by a linear transformation, Q_2- and Q_3-optimality are not invariant. For example, if $t = 1$ and one changes the scale of measurement by replacing the levels $\{z_1, z_2, \ldots, z_v\}$ by the set $\{bz_1, bz_2, \ldots, bz_v\}$, then for $\boldsymbol{\beta}_1' = (\theta_0, \theta_1, \theta_2)$ under the polynomial model, we have

$$\sum_{i=0}^{2} \theta_i z^i = \sum_{i=0}^{2} \frac{\theta_i}{b^i}(bz)^i = \sum_{i=0}^{2} \bar{\theta}_i (bz)^i, \tag{5.10}$$

say, and $\bar{\theta}_i = (\theta_i / b^i)$ are the "new regression coefficients." The Q_2-optimal design that minimizes $\Sigma_{i=0}^2 \operatorname{Var}(\hat{\theta}_i)$ does not minimize $\Sigma_{i=0}^2 \operatorname{Var}(\hat{\bar{\theta}}_i)$, but rather minimizes $\Sigma_{i=0}^2 b^{2i} \operatorname{Var}(\hat{\bar{\theta}}_i)$. Thus for equally spaced levels $(z_1, z_2, \ldots, z_v) = (b, 2b, \ldots, vb)$ one would obtain a different design for each b if one were really interested in Q_2-optimality but a single design if one were interested in $\Sigma_{i=0}^2 b^{2i} \operatorname{Var}(\hat{\bar{\theta}}_i)$.

Example 5.2. Consider the 2^3 factorial setting of Example 5.1 with the levels of each factor being -1 and 1 rather than 0 and 1. For any observation, the same model can now be written as

$$E\left[Y_{(x_1, x_2, x_3)}\right] = \alpha_0 + \alpha_1 x_1 + \alpha_2 x_2 + \alpha_3 x_3. \tag{5.11}$$

If the feasible class is now indicated by $\Delta^*(\mathbf{I}_4, C)$ then a Q_2-optimal design is $\Gamma^* = \{(-1,-1,-1), (-1,1,1), (1,-1,1), (1,1,-1)\}$. If the levels of the ith factor are changed from -1 and 1 to $-b_i$ and b_i then according to

(5.10) the model in (5.11) may be rewritten as

$$(5.12) \qquad E\left[Y_{(x_1,x_2,x_3)}\right] = \bar{\alpha}_0 + \bar{\alpha}_1(b_1x_1) + \bar{\alpha}_2(b_2x_2) + \bar{\alpha}_3(b_3x_3),$$

where $\bar{\alpha}_0 = \alpha_0$ and $\bar{\alpha}_i = \alpha_i/b_i$. Now, $\Sigma_{i=0}^3 \mathrm{Var}(\hat{\bar{\alpha}}_i) = \Sigma_{i=0}^3 \mathrm{Var}(\hat{\alpha}_i/b_i) = \Sigma_{i=0}^3 b_i^{-2}\mathrm{Var}(\hat{\alpha}_i)$. It follows that Γ^* is not Q_2-optimal, unless $b_i = 1$ for all i, since it does not minimize $\Sigma_{i=0}^3 b_i^{-2}\mathrm{Var}(\hat{\alpha}_i)$. Note however that Γ^* is optimal relative to the objective function $\Sigma_{i=0}^3 b_i^2 \mathrm{Var}(\hat{\bar{\alpha}}_i)$ since this expression is equal to $\Sigma_{i=0}^3 \mathrm{Var}(\hat{\alpha}_i)$.

5.5 SELECTING OPTIMAL DESIGNS BASED ON OTHER CRITERIA

Let $\mathbf{L}_1'\boldsymbol{\beta}_1$ be a vector of s normalized linear functions of $\boldsymbol{\beta}_1$ and let l_{1j}' be the jth row of $\mathbf{L}_1'$. Also let $a_j(\Gamma)\sigma^2$ be the variance of the BLUE $\widehat{l_{1j}'\boldsymbol{\beta}_1}$. We will need the following definition in the development:

Definition 5.5. A fractional factorial design Γ is said to be *variance balanced* with respect to $\mathbf{L}_1'\boldsymbol{\beta}_1$ if $a_j(\Gamma)\sigma^2$ is independent of j, and it is said to be *covariance balanced* with respect to $\mathbf{L}_1'\boldsymbol{\beta}_1$ if $\mathrm{Cov}(\widehat{l_{1j}'\boldsymbol{\beta}_1}, \widehat{l_{1j'}'\boldsymbol{\beta}_1})$ is independent of j and j', where $j \neq j'$.

The first part of this definition simply says that the diagonal elements of the covariance matrix $\mathbf{L}_1'(\mathbf{X}_\Gamma'\mathbf{X}_\Gamma)^-\mathbf{L}_1$ are all equal. If the experimenter is interested in variance balanced designs, then

$$(5.13) \quad Q_5(\Gamma) = \sum_{j=1}^{s} \left[\left(a_j(\Gamma) - \bar{a}(\Gamma)\right)\right]^2, \qquad \text{for} \quad \bar{a}(\Gamma) = \sum_{j=1}^{s} \frac{a_j(\Gamma)}{s}$$

is a quantity that measures departure from variance balancedness. Hence we have another objective function or criterion for selecting a design in the class $\Delta(\mathbf{L}_1, C)$. The reader may define a similar quantity to reflect departure from covariance balancedness. Of course, it is possible to express these measures and other orthogonally invariant ones in terms of the characteristic roots of the information matrix of $\widehat{\mathbf{L}_1'\boldsymbol{\beta}_1}$. A shortcoming of Q_5 (or its extension) is that $Q_5(\Gamma)$ can be zero for a design with relatively large $\bar{a}(\Gamma)$.

In the literature the reader may encounter the following definition of a balanced design for the 2^t or 3^t factorial: a factorial design is said to be balanced relative to the subvector $\boldsymbol{\beta}_1$ if the covariance matrix of $\hat{\boldsymbol{\beta}}_1$ is invariant under permutation of factor symbols. It is doubtful that this concept can be generalized in a meaningful way to the general $k_1 \times k_2 \times \ldots \times k_t$ factorial design. However, it is clear that a factorial design in this sense is balanced for certain choices of $\mathbf{L}_1$ according to Definition 5.5.

Example 5.3. Consider again saturated main effect designs of the 2^3 factorial as outlined in Example 5.1. The Q_1-, Q_2-, Q_3- and Q_4-optimal design $\Gamma_1 = \{(0,0,0),(0,1,1),(1,0,1),(1,1,0)\}$ is variance and covariance balanced since $\text{Cov}(\hat{\boldsymbol{\beta}}_1) = (\sigma^2/4)\mathbf{I}_4$. On the other hand, the design $\Gamma_2 = \{(0,0,0),(1,0,0),(1,0,1),(1,1,0)\}$ is neither variance nor covariance balanced; the information and covariance matrices are, respectively,

$$\mathbf{X}'_{\Gamma_2 1}\mathbf{X}_{\Gamma_2 1} = \begin{bmatrix} 4 & -2 & -2 & 2 \\ -2 & 4 & 0 & 0 \\ 2 & 0 & 4 & 0 \\ 2 & 0 & 0 & 4 \end{bmatrix},$$

$$\left(\mathbf{X}'_{\Gamma_2 1}\mathbf{X}_{\Gamma_2 1}\right)^{-1} = \tfrac{1}{4}\begin{bmatrix} 4 & 2 & 2 & -2 \\ 2 & 2 & 1 & -1 \\ 2 & 1 & 2 & -1 \\ -2 & -1 & -1 & 2 \end{bmatrix}.$$

Hence the amount of departure from variance balancedness according to (5.13) is obtained by calculating $\bar{a}(\Gamma_2) = \frac{1}{4}[\frac{1}{4}(4+2+2+2)] = \frac{5}{8}$ and then $Q_5(\Gamma) = (1-\frac{5}{8})^2 + (\frac{1}{2}-\frac{5}{8})^2 + (\frac{1}{2}-\frac{5}{8})^2 + (\frac{1}{2}-\frac{5}{8})^2 = \frac{3}{16}$.

The reader is invited to calculate the amount of departure for covariance balancedness for this design.

If the assumption that $\boldsymbol{\beta}_2 = \mathbf{0}$ is violated then we know from (4.32) that the proper measure of accuracy for the estimator $\widehat{\mathbf{L}'_1\boldsymbol{\beta}_1}$ is the mean square error

$$\text{(5.14)} \qquad \text{MSE}(\widehat{\mathbf{L}'_1\boldsymbol{\beta}_1}) = \text{Cov}(\widehat{\mathbf{L}'_1\boldsymbol{\beta}_1}) + \mathbf{A}_\Gamma\boldsymbol{\beta}_2\boldsymbol{\beta}'_2\mathbf{A}'_\Gamma = \mathbf{V}(\Gamma) + \mathbf{B}(\Gamma).$$

One may now introduce a measure to reflect a magnitude of $\text{MSE}(\widehat{\mathbf{L}'_1\boldsymbol{\beta}_1})$ or, in general, a convex combination of $\mathbf{V}(\Gamma)$ and $\mathbf{B}(\Gamma)$. However, this is not an easy problem, because $\text{MSE}(\widehat{\mathbf{L}'_1\boldsymbol{\beta}_1})$ is a function of the unknown parameters σ^2 and $\boldsymbol{\beta}_2$, and thus one would have to deal with an objective function such as

$$\text{(5.15)} \qquad Q_6 = \max_{\boldsymbol{\beta}_2,\, \sigma \in \Lambda} \text{trace}\left[\mathbf{V}(\Gamma) + \mathbf{B}(\Gamma)\right],$$

where Λ is some specified set of parameter values. In certain situations where the experimenter can say something about the relative magnitudes of σ^2 and $\boldsymbol{\beta}'_2\boldsymbol{\beta}_2$ and the measure is the trace of $\text{MSE}(\widehat{\mathbf{L}'_1\boldsymbol{\beta}_1})$, some progress is possible. For example, if $\sigma^2/\boldsymbol{\beta}'_2\boldsymbol{\beta}_2$ is "large" (the usual case in the "philosophy" of this section, that is, of primarily worrying about $\boldsymbol{\beta}_2 = 0$), then the quantity to be minimized is approximately the trace of $\mathbf{V}(\Gamma)$. On the other hand, if $\sigma^2/\boldsymbol{\beta}'_2\boldsymbol{\beta}_2$ is "small," then the trace of $\mathbf{B}(\Gamma)$ should be minimized.

Note that these are approximate statements based on *a priori* knowledge concerning σ^2 and $\boldsymbol{\beta}_2'\boldsymbol{\beta}_2$. These difficulties can be partially overcome or circumvented if the experimenter limits his concern to $\mathbf{V}(\Gamma)$ and to $\mathbf{B}(\Gamma)$ separately. This means that two quantities should be introduced for measuring the magnitudes of $\mathbf{V}(\Gamma)$ and $\mathbf{B}(\Gamma)$. Quantities such as Q_1, Q_2, and Q_3 as introduced previously, can be associated with $\mathbf{V}(\Gamma)$. The trace and similar quantities can be associated with the bias $\mathbf{B}(\Gamma)$.

Example 5.4. For Example 5.1 and the Q_1-, Q_2-, Q_3- and Q_4-optimal design $\Gamma_1 = \{(0,0,1),(0,1,1),(1,0,1),(1,1,0)\}$, the mean square error is obtained as

$$\begin{aligned}
\mathrm{MSE}(\Gamma_1) &= \mathbf{V}(\Gamma_1) + \mathbf{B}(\Gamma_1) \\
&= \frac{\sigma^2}{4}\mathbf{I}_4 + \mathbf{A}_{\Gamma_1}\boldsymbol{\beta}_2\boldsymbol{\beta}'_2\mathbf{A}'_{\Gamma_1} \\
&= \frac{\sigma^2}{4}\mathbf{I}_4 + \mathbf{L}'_1\left(\mathbf{X}'_{\Gamma_1 1}\mathbf{X}_{\Gamma_1 1}\right)^{-1}\mathbf{X}'_{\Gamma_1 1}\mathbf{X}_{\Gamma_1 2}\boldsymbol{\beta}_2\boldsymbol{\beta}'_2\mathbf{X}'_{\Gamma_1 2}\mathbf{X}_{\Gamma_1 1} \\
&\quad \times \left(\mathbf{X}'_{\Gamma_1 1}\mathbf{X}_{\Gamma_1 1}\right)^{-1}\mathbf{L}_1 \\
&= \frac{\sigma^2}{4}\mathbf{I}_4 + \frac{1}{16}\mathbf{X}'_{\Gamma_1 1}\mathbf{X}_{\Gamma_1 2}\boldsymbol{\beta}_2\boldsymbol{\beta}'_2\mathbf{X}'_{\Gamma_1 2}\mathbf{X}_{\Gamma_1 1}.
\end{aligned}$$

Since

$$\begin{aligned}
\mathbf{X}'_{\Gamma_1 1}\mathbf{X}_{\Gamma_1 2} &= \begin{bmatrix} 1 & 1 & 1 & 1 \\ -1 & 1 & 1 & -1 \\ -1 & 1 & -1 & 1 \\ -1 & -1 & 1 & 1 \end{bmatrix}\begin{bmatrix} 1 & 1 & 1 & -1 \\ 1 & -1 & -1 & -1 \\ -1 & 1 & -1 & -1 \\ -1 & -1 & 1 & -1 \end{bmatrix} \\
&= \begin{bmatrix} 0 & 0 & 0 & -4 \\ 0 & 0 & -4 & 0 \\ 0 & -4 & 0 & 0 \\ -4 & 0 & 0 & 0 \end{bmatrix},
\end{aligned}$$

and

$$\boldsymbol{\beta}'_2 = \left(\phi_1^0\phi_2^1\phi_3^1, \phi_1^1\phi_2^0\phi_3^1, \phi_1^1\phi_2^1\phi_3^0, \phi_1^1\phi_2^1\phi_3^1\right),$$

it follows that

$$\mathrm{MSE}(\Gamma_1) = \frac{\sigma^2}{4}\mathbf{I}_4 + \begin{bmatrix} 0 & 0 & 0 & -1 \\ 0 & 0 & -1 & 0 \\ 0 & -1 & 0 & 0 \\ -1 & 0 & 0 & 0 \end{bmatrix}\boldsymbol{\beta}_2\boldsymbol{\beta}'_2\begin{bmatrix} 0 & 0 & 0 & -1 \\ 0 & 0 & -1 & 0 \\ 0 & -1 & 0 & 0 \\ -1 & 0 & 0 & 0 \end{bmatrix}$$

$$= \frac{\sigma^2}{4}\mathbf{I}_4 + \begin{bmatrix} (\phi_1^1\phi_2^1\phi_3^1)^2 & (\phi_1^1\phi_2^1\phi_3^1)(\phi_1^1\phi_2^1\phi_3^0) & (\phi_1^1\phi_2^1\phi_3^1)(\phi_1^1\phi_2^0\phi_3^1) & (\phi_1^1\phi_2^1\phi_3^1)(\phi_1^0\phi_2^1\phi_3^1) \\ (\phi_1^1\phi_2^1\phi_3^0)(\phi_1^1\phi_2^1\phi_3^1) & (\phi_1^1\phi_2^1\phi_3^0)^2 & (\phi_1^1\phi_2^1\phi_3^0)(\phi_1^1\phi_2^0\phi_3^1) & (\phi_1^1\phi_2^1\phi_3^0)(\phi_1^0\phi_2^1\phi_3^1) \\ (\phi_1^1\phi_2^0\phi_3^1)(\phi_1^1\phi_2^1\phi_3^1) & (\phi_1^1\phi_2^0\phi_3^1)(\phi_1^1\phi_2^1\phi_3^0) & (\phi_1^1\phi_2^0\phi_3^1)^2 & (\phi_1^1\phi_2^0\phi_3^1)(\phi_1^0\phi_2^1\phi_3^1) \\ (\phi_1^0\phi_2^1\phi_3^1)(\phi_1^1\phi_2^1\phi_3^1) & (\phi_1^0\phi_2^1\phi_3^1)(\phi_1^1\phi_2^1\phi_3^0) & (\phi_1^0\phi_2^1\phi_3^1)(\phi_1^1\phi_2^0\phi_3^1) & (\phi_1^0\phi_2^1\phi_3^1)^2 \end{bmatrix}.$$

The objective function in (5.15) for Γ_1 depends on the trace of $\text{MSE}(\hat{\boldsymbol{\beta}}_1)$, which is equal to $(\sigma^2 + (\phi_1^0\phi_2^1\phi_3^1)^2 + (\phi_1^1\phi_2^0\phi_3^1)^2 + (\phi_1^1\phi_2^1\phi_3^0)^2 + (\phi_1^1\phi_2^1\phi_3^1)^2$. When the set of values Λ is specified for σ^2 and the four parameters in $\boldsymbol{\beta}_2$, then one may evaluate Q_6 for all the 58 full rank designs and determine a Q_6-optimal design, which may or may not be equal to Γ_1. Note however that when $\sigma^2/\boldsymbol{\beta}_2'\boldsymbol{\beta}_2$ is "large," then trace of $\mathbf{V}(\Gamma)$ should be minimized and we know already that Γ_1 is then optimal. If $\sigma^2/\boldsymbol{\beta}_2'\boldsymbol{\beta}_2$ is "small," then the reader is invited to find a design (with or without the aid of a computer) that minimizes $B(\Gamma)$ for given values of the parameters in $\boldsymbol{\beta}_2$.

A somewhat different approach utilizes the expected value of $\widehat{\mathbf{L}_1'\boldsymbol{\beta}_1}$ rather than $\text{MSE}(\widehat{\mathbf{L}_1'\boldsymbol{\beta}_1})$, which from (4.30) is equal to

(5.16) $$E[\widehat{\mathbf{L}_1'\boldsymbol{\beta}_1}] = \mathbf{L}_1'\boldsymbol{\beta}_1 + \mathbf{A}_\Gamma\boldsymbol{\beta}_2,$$

where

(5.17) $$\mathbf{A}_\Gamma = \mathbf{L}_1'(\mathbf{X}_{\Gamma 1}'\mathbf{X}_{\Gamma 1})^- \mathbf{X}_{\Gamma 1}'\mathbf{X}_{\Gamma 2}.$$

Of the various objective functions that can be introduced, the appealing ones are those that take into account all the entries of $\mathbf{A}_\Gamma$. The following are of this nature and are also *norms* of $\mathbf{A}_\Gamma$ in the mathematical sense:

(5.18) $$\|\mathbf{A}_\Gamma\|_1 = \left(\sum_i \sum_j a_{ij}^2(\Gamma)\right)^{1/2},$$

(5.19) $$\|\mathbf{A}_\Gamma\|_2 = \max_i \sum_j |a_{ij}(\Gamma)|,$$

(5.20) $$\|\mathbf{A}_\Gamma\|_3 = \max_{i,j} |a_{ij}(\Gamma)|,$$

(5.21) $$\|\mathbf{A}_\Gamma\|_4 = \sum_i \sum_j |a_{ij}(\Gamma)|,$$

where $|a_{ij}(\Gamma)|$ indicates the absolute value of $a_{ij}(\Gamma)$ and $\|\cdot\|$ is the tradional norm notation. There are no nontrivial relations between these measures. The first measure $\|\mathbf{A}_\Gamma\|_1$ enjoys some desirable properties that the others do not possess:

(i) $\|\mathbf{A}_\Gamma\|_1$ *is orthogonally invariant*, that is, $\|\mathbf{P}_1\mathbf{A}_\Gamma\|_1 = \|\mathbf{A}_\Gamma\mathbf{P}_2\|_1 = \|\mathbf{A}_\Gamma\|_1$ if $\mathbf{P}_1$ and $\mathbf{P}_2$ are orthogonal matrices.

(ii) $\|\mathbf{A}_\Gamma\|_1 = (\text{trace } \mathbf{A}_\Gamma'\mathbf{A}_\Gamma)^{1/2}$, which implies that $\|\mathbf{A}_\Gamma\|_1$ is the positive square root of the sum of the eigenvalues of $\mathbf{A}_\Gamma'\mathbf{A}_\Gamma$. In particular, if $\mathbf{A}_\Gamma$ is a square matrix, then $\|\mathbf{A}_\Gamma\|_1 = \text{trace } (\mathbf{A}_\Gamma'\mathbf{A}_\Gamma)^{1/2} = \text{trace } (\mathbf{A}_\Gamma\mathbf{A}_\Gamma')^{1/2} = [\Sigma\lambda_i^2(\Gamma)]^{1/2}$, where the λ_i are the eigenvalues of $\mathbf{A}_\Gamma$.

If one is interested in $\text{MSE}(\widehat{\mathbf{L}_1'\boldsymbol{\beta}_1})$, then only $\|\mathbf{A}_\Gamma\|_1$ is relevant and $\|\mathbf{A}_\Gamma\|_2$, $\|\mathbf{A}_\Gamma\|_3$ and $\|\mathbf{A}_\Gamma\|_4$ would not be used; if $\mathbf{A}_\Gamma\boldsymbol{\beta}_2$ is to be considered in a different light, then all these measures might be studied.

As indicated earlier $\mathbf{A}_\Gamma$ is called the *alias matrix* of design Γ relative to $\mathbf{L}_1$, $\boldsymbol{\beta}_1$, and $\boldsymbol{\beta}_2$. It may also be referred to as the *contamination matrix* of design Γ. Because of properties (i) and (ii), we take $\|\mathbf{A}_\Gamma\|_1$ to be our aliasing measure and we formally introduce the following quantity for selecting an *alias optimal* design:

$$Q_7(\Gamma) = \|\mathbf{A}_\Gamma\|_1. \tag{5.22}$$

To complete this section we introduce the following concept:

Definition 5.6. A design Γ is said to be *alias balanced* if $[\Sigma a_{hj}^2(\Gamma)]^{1/2}$ is a constant for all h, where $\mathbf{A}_\Gamma = [a_{hj}(\Gamma)]$.

This definition leads us to the following measure for selecting an optimal design $\Gamma \in \Delta(\mathbf{L}_1, C)$ with respect to departure from alias balancedness:

$$Q_8(\Gamma) = \sum_h \left[b_h(\Gamma) - \bar{b}(\Gamma)\right]^2 \tag{5.23}$$

where

$$\left[\sum_j a_{hj}^2(\Gamma)\right]^{1/2} = b_h(\Gamma), \qquad \bar{b}(\Gamma) = \frac{\sum_h b_h(\Gamma)}{s}.$$

Note that $Q_8(\Gamma)$ carries the same type of shortcomings as $Q_5(\Gamma)$.

We next illustrate the calculation of $Q_7(\Gamma)$ and $Q_8(\Gamma)$ for a given design. The reader is invited to do the same for other designs.

Example 5.5. In Example 5.4 the alias matrix of design $\Gamma_1 = \{(0,0,0), (0,1,1), (1,0,1), (1,1,0)\}$ was equal to

$$\mathbf{A}_\Gamma = \mathbf{I}_4\left(\mathbf{X}_{\Gamma_1 1}'\mathbf{X}_{\Gamma_1 1}\right)^{-1}\mathbf{X}_{\Gamma_1 1}'\mathbf{X}_{\Gamma_1 2} = \tfrac{1}{4}\mathbf{X}_{\Gamma_1 1}'\mathbf{X}_{\Gamma_1 2} = \begin{bmatrix} 0 & 0 & 0 & -1 \\ 0 & 0 & -1 & 0 \\ 0 & -1 & 0 & 0 \\ -1 & 0 & 0 & 0 \end{bmatrix}.$$

Hence $Q_7(\Gamma_1) = \|\mathbf{A}_\Gamma\|_1 = [\Sigma_{i=1}^4 \Sigma_{j=1}^4 a_{ij}^2(\Gamma_1)]^{1/2} = 4^{1/2} = 2$. To evaluate $Q_8(\Gamma_1)$, we first obtain $b_h(\Gamma_1) = 1$, $h = 1,2,3,4$ and $\bar{b}(\Gamma_1) = 4/4 = 1$. Hence $Q_8(\Gamma_1) = 1 - 1 = 0$; that is, design Γ_1 is alias balanced.

The reader is invited to verify that for design $\Gamma_2 = \{(0,0,0), (1,0,0), (1,0,1), (1,1,0)\}$ the values of Q_7 and Q_8 are $Q_7(\Gamma_2) = 4.0$ and $Q_8(\Gamma_2) = 0.80$. Hence we may conclude that design Γ_1 is "alias better" than the alias unbalanced design Γ_2.

5.6 APPROACHES, APPROXIMATIONS, AND APPLICATIONS IN SELECTING OPTIMAL DESIGNS

It is important to be clear on three different aspects in the selection of a design.

(i) *Nature of Domain G of Controlled Variable in the Application Under Consideration by the Experimenter.* One can have a continuum in the t-dimensional Euclidean space or a finite (or countably infinite) set, just as in the "continuous" and in the "discrete" case in probability theory. If the factor space G is infinite (which has been ignored in the development), regularity properties (like continuity) in the natural topology of the regression functions $\boldsymbol{\theta}'\mathbf{f}(g)$ are assumed out of realism or the desire to get anywhere in computing designs. As in other optimization problems, characterization of an extremum is often easier over a continuum than over a large discrete space, but this has nothing to do with whether or not the experimenter is actually faced with a discrete G or a continuum.

(ii) *Approaches.* One can try to choose a design criterion (a) having to do with utility or loss or (b) motivated by simplicity in equal spacing, simple matrices to invert, or other pleasing regularity conditions that do not guarantee optimality in the sense of (a) without further proof in special settings where such "appealing" designs may indeed be optimum.

(iii) *Approximations.* It may be difficult to compute a design satisfying (ii.a) or even (ii.b) (for the latter, there are settings where a design with equal variance of "elementary estimators" is difficult to characterize). So, as elsewhere in mathematics, one sometimes solves instead a closely related problem, which is more tractable, and from that solution obtains an *almost* optimum solution to the original problem. This could take many forms; for example, in terms of (i), a large finite space G might be replaced by a continuum containing it and a solution over the continuum might then be implemented by using a "nearby" element of the original finite space (which will not necessarily be the optimum over the finite space). This last is *not* what is usually meant in the design literature by the *approximate theory*. Rather what is meant, *which applies equally to either case of (i)*, is the solution one obtains if "fractional observations" are allowed (the *exact theory* referring to solving the original problem with integers for replication numbers). It turns out that this approximate theory problem is often easier to solve and can then be implemented by finding a "nearby" integer-valued set of replication numbers, which will yield a design that is often close to the optimum for the exact problem.

5.7 THE DUAL PROBLEM OF PARAMETER SELECTION

In Chapter 3, we mentioned the dual problem of selecting a subset of parameters to be estimated. The criteria developed in this chapter may also be applied to this problem by selecting the appropriate matrices. Besides these, other criteria have been developed for finding the "best" subset of parameters. Similarly, other criteria have also been introduced in the selection of a response surface design. For further reading, the reader is referred to the appropriate references in the list below.

5.8 SELECTED ADDITIONAL READING

Starred references are recommended for a first reading.

*1. Ash, A., and Hedayat, A. (1978). An introduction to design optimality with an overview of the literature. *Commun. Stat.* **A7**, 1295–1325.

2. Atwood, C. L. (1969). Optimal and efficient designs of experiments. *Ann. Math. Stat.*, **40**, 1570–1602.

3. Banerjee, K. S. (1974). On *D*-optimality of Yates' original example in the Yates-Hotelling weighing problem. *Commun. Stat.*, **3**, 185–190.

4. Bodewig, E. (1959). *Matrix Calculus*. North-Holland, Amsterdam.

*5. Box, G. E. P., and Draper, N. R. (1959). A basis for the selection of a response surface design. *J. Am. Stat. Assoc.*, **54**, 622–654.

6. Box, G. E. P., and Draper, N. R. (1975). Robust designs. *Biometrika*, **62**, 347–352.

7. Box, M. J. (1971). An experimental design criterion for precise estimation of a subset of the parameters in a nonlinear model. *Biometrika*, **58**, 149–153.

*8. Box, M. J., and Draper, N. R. (1971). Factorial designs, the $|X'X|$ criterion and some related matters. *Technometrics*, **13**, 731–742.

9. Boyce, D. E., Farhi, A., and Weischedel, R. (1974). *Optimal Subset Selection Multiple Regression, Interdependence and Optimal Network Algorithms*. Springer-Verlag, New York.

10. Draper, N. R., and Lawrence, W. E. (1965). Designs which minimize model inadequacies: cuboidal regions of interest. *Biometrika*, **52**, 111–118.

11. Eccleston, J., and Hedayat, A. (1974). On the theory of connected designs: characterization and optimality. *Ann. Stat.*, **2**, 1238–1255.

12. Eccleston, J., and Russell, K. (1975). Connectedness and orthogonality in multifactor designs. *Biometrika*, **62**, 341–345.

13. Ehrenfeld, S., and Zacks, S. (1963). Optimal strategies in factorial experiments. *Ann. Math. Stat.*, **34**, 780–791.

14. Elfving, G. (1959). Design of Linear Experiments. *Cramér Festschrift*, U. Grenander, Ed. Wiley, New York, pp. 58–74.

15. Federer, W. T., Paik, U. B., Raktoe, B. L., and Werner, L. H. (1973). Some combinatorial problems and results in fractional replication. In *A Survey of Combinatorial Theory*, J. N. Srivastava, F. Harary, C. R. Rao, G.-C. Rota, and S. S. Shrikhande, Eds. North-Holland, Amsterdam, pp. 151–161.

16. Fedorov, V. V. (1972). *Theory of Optimal Experiments*, translated and edited by W. J. Studden and E. M. Klimko. Academic, New York and London.

17. Hadamard, J. (1893). Resolution d'une question relative aux determinants. *Bull. Sci. Math., Astr.* **1**(2), 240–246.

*18. Hedayat, A. (1980). Study of optimality criteria in design of experiments. To appear in *Proceedings of May 1980 Symposium on Statistics and Related Topics*, Carleton University, Ottawa, Canada.

19. Hedayat, A., and Federer, W. T. (1974). Pairwise and variance balanced incomplete block designs. *Ann. Inst. Stat. Math.*, **26**, 331–338.

*20. Hedayat, A., Raktoe, B. L., and Federer, W. T. (1974). On a measure of aliasing due to fitting an incomplete model. *Ann. Stat.*, **12**, 650–660.

21. Herzberg, A. M., and Andrews, D. F. (1976). Some considerations in the optimal design of experiments in non-optimal situations. *J. R. Stat. Soc.*, **B38**, 284–289.

22. Hill, W. J., Hunter, W. G., and Wichern, D. W. (1968). A joint design criterion for the dual problem of model discrimination and parameter estimation. *Technometrics*, **10**, 145–160.

23. Hocking, R. R. (1972). Criteria for selection of a subset regression: which one should be used? *Technometrics*, **14**, 967–970.

24. Hoel, P. G. (1965). Optimum designs for polynomial extrapolation. *Ann. Math. Stat.*, **36**, 1483–1493.

*25. Karson, M. J., Manson, A. R., and Hader, R. J. (1969). Minimum bias estimation and experimental design for response surfaces. *Technometrics*, **11**, 461–475.

26. Kiefer, J. (1958). On the nonrandomized optimality and randomized nonoptimality of symmetrical designs. *Ann. Math. Stat.*, **29**, 675–699.

*27. Kiefer, J. (1959). Optimum experimental designs. *J. R. Stat. Soc.*, **B21**, 272–304.

28. Kiefer, J. (1974). General equivalence theory for optimum designs (approximate theory). *Ann. Math. Stat.*, **2**, 849–879.

*29. Kiefer, J. (1975). Optimal design: variation in structure and performance under change of criterion. *Biometrika*, **62**, 277–288.

30. Kishen, K., and Tyagi, B. N. (1963). Partially balanced asymmetrical factorial designs. *Contributions to Statistics*, C. R. Rao, Ed. (Presented to Professor P. Mahalanobis on his 70th birthday) Pergamon, New York and Statistical Publishing Company, Calcutta, pp. 147–158.

31. Kshirsagar, A. M. (1966). Balanced factorial designs. *J. R. Stat. Soc.*, **B28**, 559–567.

*32. Mood, A. M. (1946). On Hotelling's weighing problem. *Ann. Math. Stat.*, **17**, 432–446.

*33. Plackett, R. L., and Burman, J. P. (1946). The design of optimum multifactorial experiments. *Biometrika*, **33**, 305–325.

*34. Raktoe, B. L. (1974). On classes of equi-information factorial arrangements. *Proceedings of the Eighth International Biometrics Conference*, Editura Academiei, Republicii Socialiste România.

35. Raktoe, B. L., and Federer, W. T. (1973). Balanced optimal saturated main effect plans of the 2^n factorial and their relation to (v, k, λ) configurations. *Ann. Stat.*, **1**, 924–932.

36. Ryser, H. J. (1963). *Combinatorial Mathematics*, (Carus Monograph No. 14). Mathematical Association of America, Oberlin, and Wiley, New York.

*37. Shah, B. V. (1958). On balancing in factorial experiments. *Ann. Math. Stat.*, **29**, 766–779.

*38. Shah, K. R. (1960). Optimality criteria for incomplete block designs. *Ann. Math. Stat.*, **31**, 791–794.

39. Shirakura, T. (1976). A note on the norm of alias matrices in fractional replication. *Aust. J. Stat.*, **18**, 158–160.

40. Soller, M., and Genizi, A. (1967). Optimal experimental designs for realized heritability estimates. *Biometrics*, **23**, 361–365.

41. Srivastava, J. N. (1975). Designs for searching non-negligible effects. In *A Survey of Statistical Design and Linear Models*, J. N. Srivastava, Ed. North-Holland, Amsterdam, pp. 507–519.

*42. Srivastava, J. N., and Anderson, D. A. (1974). A comparison of the determinant, maximum root and trace optimality criteria. *Commun. Stat.*, **3**, 933–940.

43. Srivastava, J. N., Raktoe, B. L., and Pesotan, H. (1976). On invariance and randomization in fractional replication. *Ann. Stat.*, **4**, 423–430.

44. Studden, W. J. (1968). Optimal designs on Tchebycheff points. *Ann. Math. Stat.*, **39**, 1435–1447.

45. Tasaka, M. (1968). Optimum design for polynomial approximation. *Bull. Math. Stat.*, **13**, 25–29.

46. Thomas, M. A., and Myers, R. H. (1973). Optimal designs for the inverse regression method of calibration. *Commun. Stat.*, **2**, 419–432.

47. Thompson, W. O. (1973). Secondary criteria in the selection of minimum bias designs in two variables. *Technometrics*, **15**, 319–328.

*48. Wald, A. (1943). On the efficient design of statistical investigations. *Ann. Math. Stat.*, **14**, 134–140.

49. Weeks, D. L., and Williams, D. R. (1964). A note on the determination of connectedness in a *N*-way cross classification. *Technometrics*, **6**, 319–324. (Errata, **7**, 281.)

50. Wynn, H. P. (1970). The sequential generation of *D*-optimum experimental designs. *Ann. Math. Stat.*, **41**, 1655–1664.

CHAPTER 6

Characterization of Unbiased Designs

In this chapter the concept of minimal unbiased designs for linear parametric functions is introduced. We provide a characterization of these designs for the case where no assumption is made on the total parametric vector and for the case where some elements of this vector are assumed to be zero. These minimal unbiased designs then lead to a class of unbiased designs for the linear parametric function. The development is carried through using the orthogonal polynomial setup, but as remarked in Chapter 5 the results may be extended to other settings.

6.1 THE PROBLEM OF CHARACTERIZING UNBIASED DESIGNS

In Chapter 5 we introduced the class of unbiased designs $\Delta(\mathbf{L}_1)$ relative to the vector of parametric functions $\mathbf{L}_1'\boldsymbol{\beta}_1$. Let $\boldsymbol{\beta}_1$ be total parametric vector $\boldsymbol{\beta}_\rho$, then the first problem in the study and use of factorial design should be the characterization of $\Delta(\mathbf{L})$ with respect to the given $\mathbf{L}'\boldsymbol{\beta}_\rho$. Let Γ be a design in $\Delta(\mathbf{L})$ and let $\mathbf{X}_\Gamma$ be the design matrix associated with Γ. The available theory in linear estimation, which has been summarized in Chapter 3, states that $\mathbf{L}'\boldsymbol{\beta}_\rho$ is estimable if and only if $\mathbf{L}'$ is in the row space of $\mathbf{X}_\Gamma$. Clearly, this tells us little of "immediate use" about which treatments should be in $\Delta(\mathbf{L})$. Some researchers in linear models do the following: they pick a design such that $\boldsymbol{\beta}_\rho$ is estimable, which in turn guarantees estimability of $\mathbf{L}'\boldsymbol{\beta}_\rho$. This means that Γ must be at least a minimal complete factorial design. Of course, all these designs are contained in $\Delta(\mathbf{L})$ and they do not exhaust $\Delta(\mathbf{L})$, if $\mathbf{L}'$ is not the identity matrix. For example, if $\mathbf{L}'$ is $1 \times N$ then $\Delta(\mathbf{L})$ can contain designs of any number of distinct treatment combinations from 1 to N inclusive. The lower bound is clearly achieved whenever $\mathbf{L}'$ is a multiple of a row of $\mathbf{X}_\rho$.

Example 6.1. Consider the 2^3 factorial, as introduced in Example 4.7 and used subsequently in Chapter 5. For convenience we reproduce the model for the minimal complete factorial design ρ, that is,

$$E\begin{bmatrix} Y_{(0,0,0)} \\ Y_{(0,0,1)} \\ Y_{(0,1,0)} \\ Y_{(0,1,1)} \\ Y_{(1,0,0)} \\ Y_{(1,0,1)} \\ Y_{(1,1,0)} \\ Y_{(1,1,1)} \end{bmatrix} = \begin{bmatrix} 1 & -1 & -1 & 1 & -1 & 1 & 1 & -1 \\ 1 & 1 & -1 & -1 & -1 & -1 & 1 & 1 \\ 1 & -1 & 1 & -1 & -1 & 1 & -1 & 1 \\ 1 & 1 & 1 & 1 & -1 & -1 & -1 & -1 \\ 1 & -1 & -1 & 1 & 1 & -1 & -1 & 1 \\ 1 & 1 & -1 & -1 & 1 & 1 & -1 & -1 \\ 1 & -1 & 1 & -1 & 1 & -1 & 1 & -1 \\ 1 & 1 & 1 & 1 & 1 & 1 & 1 & 1 \end{bmatrix} \begin{bmatrix} \phi_1^0\phi_2^0\phi_3^0 \\ \phi_1^0\phi_2^0\phi_3^1 \\ \phi_1^0\phi_2^1\phi_3^0 \\ \phi_1^0\phi_2^1\phi_3^1 \\ \phi_1^1\phi_2^0\phi_3^0 \\ \phi_1^1\phi_2^0\phi_3^1 \\ \phi_1^1\phi_2^1\phi_3^0 \\ \phi_1^1\phi_2^1\phi_3^1 \end{bmatrix}.$$

Then if

$$\mathbf{L}' = [1 \quad -1 \quad -1 \quad 1 \quad -1 \quad 1 \quad 1 \quad -1],$$

the single observation $Y_{(0,0,0)}$ would suffice to estimate $\mathbf{L}'\boldsymbol{\beta}_\rho$, since $E[Y_{(0,0,0)}] = \mathbf{L}'\boldsymbol{\beta}_\rho$. However, if

$$\mathbf{L}' = [0 \; 1 \; 0 \; 0 \; 0 \; 0 \; 0 \; 0],$$

it can be shown that it would be necessary to have all eight observations from the 2^3 factorial in order to estimate $\mathbf{L}'\boldsymbol{\beta}_\rho = \phi_1^0\phi_2^0\phi_3^1$.

Consider now an arbitrary $s \times 1$ vector of linear functions $\mathbf{L}'\boldsymbol{\beta}_\rho$. A design containing treatments corresponding to rows of $\mathbf{X}_\rho$ having nonzero coefficients in the linear combinations clearly is unbiased. In other words, if $\mathbf{l}_i'$ is the ith row of $\mathbf{L}'$ of the form

$$\mathbf{l}_i' = \sum_{j=1}^{N} \alpha_{ij}[\mathbf{R}_j(\rho)], \tag{6.1}$$

where $\mathbf{R}_j(\rho)$ is the jth row of $\mathbf{X}_\rho$, and if Γ_i is a design consisting of those treatments corresponding to the $\mathbf{R}_j(\rho)$'s in $\mathbf{l}_i'$ having nonzero α_{ij}, then the design containing the set-theoretic union of the Γ_i is an unbiased design.

Example 6.2. Consider the 2^2 factorial and the following nonlexicographically ordered model for the minimal complete factorial design ρ:

$$E\begin{bmatrix} Y_{(0,0)} \\ Y_{(1,0)} \\ Y_{(0,1)} \\ Y_{(1,1)} \end{bmatrix} = \frac{1}{2}\begin{bmatrix} 1 & -1 & -1 & 1 \\ 1 & 1 & -1 & -1 \\ 1 & -1 & 1 & -1 \\ 1 & 1 & 1 & 1 \end{bmatrix} \begin{bmatrix} \phi_1^0\phi_2^0 \\ \phi_1^1\phi_2^0 \\ \phi_1^0\phi_2^1 \\ \phi_1^1\phi_2^1 \end{bmatrix}.$$

Let

$$\mathbf{L}' = \begin{bmatrix} 1 & 0 & 0 & 1 \\ \frac{1}{2} & \frac{1}{2} & -\frac{1}{2} & -\frac{1}{2} \end{bmatrix},$$

and suppose that the experimenter is interested in estimating $\mathbf{L}'\boldsymbol{\beta}_\rho$. The traditional linear model theory says that one needs a design containing the minimal complete factorial arrangement ρ, that is, a design containing the four treatments $(0,0),(1,0),(0,1)$, and $(1,1)$, since then $\boldsymbol{\beta}_\rho$ is estimable and hence $\mathbf{L}'\boldsymbol{\beta}_\rho$ is estimable. However, clearly a design containing the treatments $(0,0),(1,0)$, and $(1,1)$ is unbiased and this has fewer treatments. This follows from the fact that

$$\mathbf{L}' = \begin{bmatrix} \mathbf{l}'_1 \\ \mathbf{l}'_2 \end{bmatrix} \begin{bmatrix} 1 & 0 & 0 & 1 \\ 0 & 1 & 0 & 0 \end{bmatrix} \begin{bmatrix} \mathbf{R}_1(\rho) \\ \mathbf{R}_2(\rho) \\ \mathbf{R}_3(\rho) \\ \mathbf{R}_4(\rho) \end{bmatrix},$$

so that $\Gamma_1 = \{(0,0),(1,1)\}$, $\Gamma_2 = \{(1,0)\}$, and $\Gamma_1 \cup \Gamma_2 = \{(0,0),(1,0),(1,1)\}$. Hence any design containing $\Gamma_1 \cup \Gamma_2$ is an unbiased design and from such a design the best linear unbiased estimator is obtained from expressions (3.14) or (4.23).

We should observe that the general problem of characterizing unbiased designs in a useful way is not solved. Next we give results in some special cases. Before doing this, we define the concept of a minimal unbiased design for a vector of linear parametric functions.

Definition 6.1. An unbiased design in $\Delta(\mathbf{L})$ for $\mathbf{L}'_1\boldsymbol{\beta}_1$ is said to be a *minimal unbiased design* if the number of treatments in the design is minimal.

A considerable amount of work has appeared in the literature on saturated designs (see Definition 5.4), which allow estimation of a specified $p \times 1$ subvector $\boldsymbol{\beta}_1$ from p observations. These type of designs are in fact the smallest designs allowing estimation of the parameters in $\boldsymbol{\beta}_1$. The concept of minimal unbiased designs is necessary in order to describe the class $\Delta(\mathbf{L})$ of unbiased designs for estimating $\mathbf{L}'\boldsymbol{\beta}_\rho$ for arbitrary $\mathbf{L}'$.

6.2 CHARACTERIZATION OF MINIMAL UNBIASED DESIGNS FOR $\mathbf{L}'\boldsymbol{\beta}_\rho$ WITH NO ASSUMPTIONS ON $\boldsymbol{\beta}_\rho$

Let $\mathbf{L}'$ be an $s \times N$ matrix and suppose that the experimenter is interested in estimating $\mathbf{L}'\boldsymbol{\beta}_\rho$. The following algorithm generates a unique minimal unbiased design for $\mathbf{L}'\boldsymbol{\beta}_\rho$ under the orthonormal model. Since $\mathbf{L}'$ is in the row space of $\mathbf{X}_\rho$ we may write

$$\mathbf{L}' = \mathbf{C}'\mathbf{X}_\rho, \tag{6.2}$$

where $\mathbf{C}'$ is an $s \times N$ matrix of coefficients. Since $\mathbf{X}_\rho'\mathbf{X}_\rho = \mathbf{I}_N$, the unique solution for $\mathbf{C}'$ is

$$\mathbf{C}' = \mathbf{L}'\mathbf{X}_\rho' \tag{6.3}$$

or $\mathbf{C} = \mathbf{X}_\rho\mathbf{L}$. Hence the unique minimal design is given by those treatments i for which the ith column of $\mathbf{C}'$ (or equivalently ith row of $\mathbf{C}$) is not all zeros. Thus any design containing this minimal design will also be an unbiased design. In the full rank (i.e., rank of $\mathbf{X}_\rho = N$) nonorthogonal setting one proceeds in the same way, replacing $\mathbf{X}_\rho'$ by $\mathbf{X}_\rho^{-1}$.

Example 6.3. In Example 6.1 we claimed that the estimation of $\mathbf{L}'\boldsymbol{\beta}_\rho = [0\ 1\ 0\ 0\ 0\ 0\ 0\ 0]\boldsymbol{\beta}_\rho = \phi_1^0\phi_2^0\phi_3^1$ required all eight treatment combinations of the 2^3 factorial. We now demonstrate this by utilizing the preceding development. Since

$$\begin{aligned}
\mathbf{C}' &= [0\ 1\ 0\ 0\ 0\ 0\ 0\ 0]\mathbf{X}_\rho^{-1} \\
&= [0\ 1\ 0\ 0\ 0\ 0\ 0\ 0]\tfrac{1}{8}\mathbf{X}_\rho' \\
&= \tfrac{1}{8}[0\ 1\ 0\ 0\ 0\ 0\ 0\ 0] \\
&\quad \times \begin{bmatrix}
1 & 1 & 1 & 1 & 1 & 1 & 1 & 1 \\
-1 & 1 & -1 & 1 & -1 & 1 & -1 & 1 \\
-1 & -1 & 1 & 1 & -1 & -1 & 1 & 1 \\
1 & -1 & -1 & 1 & 1 & -1 & -1 & 1 \\
-1 & -1 & -1 & -1 & 1 & 1 & 1 & 1 \\
1 & -1 & 1 & -1 & -1 & 1 & -1 & 1 \\
1 & 1 & -1 & -1 & -1 & -1 & 1 & 1 \\
-1 & 1 & 1 & -1 & 1 & -1 & -1 & 1
\end{bmatrix} \\
&= \tfrac{1}{8}[-1\quad 1\quad -1\quad 1\quad -1\quad 1\quad -1\quad 1],
\end{aligned}$$

and it does not have zero entries, it follows that all eight treatment combinations form a minimal design for estimating $\phi_1^0\phi_2^0\phi_3^1$.

Example 6.4. In Example 6.2 we showed that the design $\Gamma = \{(0,0), (1,0),(1,1)\}$ was an unbiased design for

$$\mathbf{L}'\boldsymbol{\beta}_\rho = \begin{bmatrix} 1 & 0 & 0 & 1 \\ \frac{1}{2} & \frac{1}{2} & -\frac{1}{2} & -\frac{1}{2} \end{bmatrix}\boldsymbol{\beta}_\rho$$

in the 2^2 factorial setup. Is the design a minimal unbiased design? Using the algorithm of this section we see that

$$\mathbf{C}' = \begin{bmatrix} 1 & 0 & 0 & 1 \\ \frac{1}{2} & \frac{1}{2} & -\frac{1}{2} & -\frac{1}{2} \end{bmatrix} \mathbf{X}'_\rho$$

$$= \begin{bmatrix} 1 & 0 & 0 & 1 \\ \frac{1}{2} & \frac{1}{2} & -\frac{1}{2} & -\frac{1}{2} \end{bmatrix} \frac{1}{2} \begin{bmatrix} 1 & 1 & 1 & 1 \\ -1 & 1 & -1 & 1 \\ -1 & -1 & 1 & 1 \\ 1 & -1 & -1 & 1 \end{bmatrix} = \begin{bmatrix} 1 & 0 & 0 & 1 \\ 0 & 1 & 0 & 0 \end{bmatrix}.$$

Hence the first, second, and fourth treatment combinations constitute a minimal unbiased design, and this is precisely equal to Γ.

Example 6.5. Consider the equispaced 3×3 factorial such that the coded levels of the factors are $\{0, 1, 2\}$. Then under the orthogonal polynomial model in nonlexicographic form, we obtain

$$E\begin{bmatrix} Y_{(0\ 0)} \\ Y_{(1\ 0)} \\ Y_{(2\ 0)} \\ Y_{(0\ 1)} \\ Y_{(1\ 1)} \\ Y_{(2\ 1)} \\ Y_{(0\ 2)} \\ Y_{(1\ 2)} \\ Y_{(2\ 2)} \end{bmatrix} = \begin{bmatrix}
\frac{1}{3} & \frac{-1}{\sqrt{6}} & \frac{1}{3\sqrt{2}} & \frac{-1}{\sqrt{6}} & \frac{1}{3\sqrt{2}} & \frac{1}{2} & \frac{-1}{2\sqrt{3}} & \frac{-1}{2\sqrt{3}} & \frac{1}{6} \\
\frac{1}{3} & 0 & \frac{-\sqrt{2}}{3} & \frac{-1}{\sqrt{6}} & \frac{1}{3\sqrt{2}} & 0 & 0 & \frac{1}{\sqrt{3}} & \frac{-1}{3} \\
\frac{1}{3} & \frac{1}{\sqrt{6}} & \frac{1}{3\sqrt{2}} & \frac{-1}{\sqrt{6}} & \frac{1}{3\sqrt{2}} & \frac{-1}{2} & \frac{1}{2\sqrt{3}} & \frac{-1}{2\sqrt{3}} & \frac{1}{6} \\
\frac{1}{3} & \frac{-1}{\sqrt{6}} & \frac{1}{3\sqrt{2}} & 0 & \frac{-\sqrt{2}}{3} & 0 & \frac{1}{\sqrt{3}} & 0 & \frac{-1}{3} \\
\frac{1}{3} & 0 & \frac{-\sqrt{2}}{3} & 0 & \frac{-\sqrt{2}}{3} & 0 & 0 & 0 & \frac{2}{3} \\
\frac{1}{3} & \frac{1}{\sqrt{6}} & \frac{1}{3\sqrt{2}} & 0 & \frac{-\sqrt{2}}{3} & 0 & \frac{-1}{\sqrt{3}} & 0 & \frac{-1}{3} \\
\frac{1}{3} & \frac{-1}{\sqrt{6}} & \frac{1}{3\sqrt{2}} & \frac{1}{\sqrt{6}} & \frac{1}{3\sqrt{2}} & \frac{-1}{2} & \frac{-1}{2\sqrt{3}} & \frac{1}{2\sqrt{3}} & \frac{1}{6} \\
\frac{1}{3} & 0 & \frac{-\sqrt{2}}{3} & \frac{1}{\sqrt{6}} & \frac{1}{3\sqrt{2}} & 0 & 0 & \frac{-1}{\sqrt{3}} & \frac{-1}{3} \\
\frac{1}{3} & \frac{1}{\sqrt{6}} & \frac{1}{3\sqrt{2}} & \frac{1}{\sqrt{6}} & \frac{1}{3\sqrt{2}} & \frac{1}{2} & \frac{1}{2\sqrt{3}} & \frac{1}{2\sqrt{3}} & \frac{1}{6}
\end{bmatrix} \begin{bmatrix} \phi_1^0\phi_2^0 \\ \phi_1^1\phi_2^0 \\ \phi_1^2\phi_2^0 \\ \phi_1^0\phi_2^1 \\ \phi_1^0\phi_2^2 \\ \phi_1^1\phi_2^1 \\ \phi_1^1\phi_2^2 \\ \phi_1^2\phi_2^1 \\ \phi_1^2\phi_2^2 \end{bmatrix}$$

Let

$$\mathbf{L}' = \begin{bmatrix} 0 & 0 & 0 & 1 & 0 & 0 & 0 & 0 & 0 \\ 0 & 0 & 0 & 0 & 0 & 1 & 0 & 0 & 0 \\ 0 & 0 & 0 & 0 & 0 & 0 & 0 & 1 & 0 \end{bmatrix},$$

which states that parameters $\phi_1^0\phi_2^1$, $\phi_1^1\phi_2^1$, and $\phi_1^2\phi_2^1$ are to be estimated. Then it follows that the minimal design for $\mathbf{L}'\boldsymbol{\beta}_\rho$ is determined by the nonzero columns of $\mathbf{C}'$, that is,

$$\mathbf{C}' = \mathbf{L}'\mathbf{X}'_\rho = \begin{bmatrix} \frac{-1}{\sqrt{6}} & \frac{-1}{\sqrt{6}} & \frac{-1}{\sqrt{6}} & 0 & 0 & 0 & \frac{1}{\sqrt{6}} & \frac{1}{\sqrt{6}} & \frac{1}{\sqrt{6}} \\ \frac{1}{2} & 0 & \frac{-1}{2} & 0 & 0 & 0 & \frac{-1}{2} & 0 & \frac{1}{2} \\ \frac{-1}{2\sqrt{3}} & \frac{1}{\sqrt{3}} & \frac{-1}{2\sqrt{3}} & 0 & 0 & 0 & \frac{1}{2\sqrt{3}} & \frac{-1}{\sqrt{3}} & \frac{1}{2\sqrt{3}} \end{bmatrix}.$$

Hence the unique minimal design is

$$\Gamma = \{(0,0),(1,0),(2,0),(0,2),(1,2),(2,2)\}.$$

The development of this section only touches the general problem of characterizing the class of minimal unbiased designs for estimating $\mathbf{L}'\boldsymbol{\beta}_\rho$. The reader is invited to explore the case where $\mathbf{X}_\rho$ is of less than full rank.

Example 6.6. Consider the 3×3 factorial such that the coded levels of the factors are $\{0,\ 1,\ 3\}$ and $\{0,\ 2,\ 3\}$, respectively. Then under the orthogonal polynomial model the design matrix $\mathbf{X}_\rho$ is equal to

$$\frac{1}{3}\begin{bmatrix} 1 & \frac{-4}{\sqrt{42}} & \frac{2}{\sqrt{14}} & \frac{-5}{\sqrt{42}} & \frac{1}{\sqrt{14}} & \frac{20}{42} & \frac{-4}{\sqrt{588}} & \frac{-10}{\sqrt{588}} & \frac{2}{14} \\ 1 & \frac{-1}{\sqrt{42}} & \frac{-3}{\sqrt{14}} & \frac{-5}{\sqrt{42}} & \frac{1}{\sqrt{14}} & \frac{5}{42} & \frac{-1}{\sqrt{588}} & \frac{15}{\sqrt{588}} & \frac{-3}{14} \\ 1 & \frac{5}{\sqrt{42}} & \frac{1}{\sqrt{14}} & \frac{-5}{\sqrt{42}} & \frac{1}{\sqrt{14}} & \frac{-25}{42} & \frac{5}{\sqrt{588}} & \frac{-5}{\sqrt{588}} & \frac{1}{14} \\ 1 & \frac{-4}{\sqrt{42}} & \frac{2}{\sqrt{14}} & \frac{1}{\sqrt{42}} & \frac{-3}{\sqrt{14}} & \frac{4}{42} & \frac{12}{\sqrt{588}} & \frac{2}{\sqrt{588}} & \frac{-6}{14} \\ 1 & \frac{-1}{\sqrt{42}} & \frac{-3}{\sqrt{14}} & \frac{1}{\sqrt{42}} & \frac{-3}{\sqrt{14}} & \frac{-1}{42} & \frac{3}{\sqrt{588}} & \frac{-3}{\sqrt{588}} & \frac{9}{14} \\ 1 & \frac{5}{\sqrt{42}} & \frac{1}{\sqrt{14}} & \frac{1}{\sqrt{42}} & \frac{-3}{\sqrt{14}} & \frac{5}{42} & \frac{-15}{\sqrt{588}} & \frac{1}{\sqrt{588}} & \frac{-3}{14} \\ 1 & \frac{-4}{\sqrt{42}} & \frac{2}{\sqrt{14}} & \frac{4}{\sqrt{14}} & \frac{2}{\sqrt{14}} & \frac{-16}{42} & \frac{-8}{\sqrt{588}} & \frac{8}{\sqrt{588}} & \frac{4}{14} \\ 1 & \frac{-1}{\sqrt{42}} & \frac{-3}{\sqrt{14}} & \frac{4}{\sqrt{42}} & \frac{2}{\sqrt{14}} & \frac{-4}{42} & \frac{-2}{\sqrt{588}} & \frac{-12}{\sqrt{588}} & \frac{-6}{14} \\ 1 & \frac{5}{\sqrt{42}} & \frac{1}{\sqrt{14}} & \frac{4}{\sqrt{42}} & \frac{2}{\sqrt{14}} & \frac{20}{42} & \frac{10}{\sqrt{588}} & \frac{4}{\sqrt{588}} & \frac{2}{14} \end{bmatrix}.$$

Let

$$\mathbf{L}' = \begin{bmatrix} 0 & 0 & 0 & 1 & 0 & 0 & 0 & 0 & 0 \\ 0 & 0 & 0 & 0 & 0 & 1 & 0 & 0 & 0 \\ 0 & 0 & 0 & 0 & 0 & 0 & 0 & 1 & 0 \end{bmatrix}$$

as in Example 6.5; then it follows that the unique minimal design for $\mathbf{L}'\boldsymbol{\beta}_\rho$ is determined by the nonzero columns of $\mathbf{C}'$, which is equal to

$$\mathbf{C}' = \mathbf{L}'\mathbf{X}'_\rho$$

$$= \tfrac{1}{3}\begin{bmatrix} \frac{-5}{\sqrt{42}} & \frac{-5}{\sqrt{42}} & \frac{-5}{\sqrt{42}} & \frac{1}{\sqrt{42}} & \frac{1}{\sqrt{42}} & \frac{1}{\sqrt{42}} & \frac{4}{\sqrt{42}} & \frac{4}{\sqrt{42}} & \frac{4}{\sqrt{42}} \\ \frac{20}{42} & \frac{5}{42} & \frac{-25}{42} & \frac{-4}{42} & \frac{-1}{42} & \frac{5}{42} & \frac{-16}{42} & \frac{-4}{42} & \frac{20}{42} \\ \frac{-10}{\sqrt{588}} & \frac{15}{\sqrt{588}} & \frac{-5}{\sqrt{588}} & \frac{2}{\sqrt{588}} & \frac{-3}{\sqrt{588}} & \frac{1}{\sqrt{588}} & \frac{8}{\sqrt{588}} & \frac{-12}{\sqrt{588}} & \frac{4}{\sqrt{588}} \end{bmatrix}.$$

Hence the unique minimal design is the MFD, which consists of the entire set of nine combinations $\{(00),(02),(03),(10),(12),(13),(30),(32),(33)\}$. It should be noted that the equal spacing of levels in Example 6.5 allowed the use of six treatments to estimate $\mathbf{L}'\boldsymbol{\beta}_\rho$ whereas the unequal spacing in this example required that all nine observations be utilized. Other linear functions may result in a different minimal design.

6.3 CHARACTERIZATION OF MINIMAL UNBIASED DESIGNS FOR $\mathbf{L}_1'\boldsymbol{\beta}_1$ UNDER THE ASSUMPTION THAT $\boldsymbol{\beta}_2 = \mathbf{0}$

We assume that the full parametric vector $\boldsymbol{\beta}_\rho$ is partitioned as $\boldsymbol{\beta}_\rho' = (\boldsymbol{\beta}_1' : \boldsymbol{\beta}_2' = \mathbf{0})$, where $\boldsymbol{\beta}_1$ is a $p \times 1$ vector of parameters. The problem is now to find a minimal unbiased design for $\mathbf{L}_1'\boldsymbol{\beta}_1$. Thc case $p = N$ has already been discussed in Section 6.2 and in this section we assume that $p < N$. Recall that the model for the MCFD in this case is

$$E[\mathbf{Y}_\rho] = \mathbf{X}_{\rho 1}\boldsymbol{\beta}_1, \tag{6.4}$$

where $\mathbf{X}_{\rho 1}$ is an $N \times p$ matrix. For $\mathbf{L}_1'$ to be in the row space of $\mathbf{X}_{\rho 1}$ there must exist a matrix $\mathbf{C}_1'$ such that

$$\mathbf{L}_1' = \mathbf{C}_1'\mathbf{X}_{\rho 1}. \tag{6.5}$$

Clearly a solution for $\mathbf{C}_1'$ is given by

$$\mathbf{C}_1' = \mathbf{L}_1'\mathbf{X}_{\rho 1}', \tag{6.6}$$

since $\mathbf{X}'_{\rho 1}\mathbf{X}_{\rho 1} = \mathbf{I}_P$. Hence an unbiased design for $\mathbf{L}'_1\boldsymbol{\beta}_1$ is given by those treatments i for which the ith column of $\mathbf{C}'_1$ is not all zeros. Such a design is not necessarily unique or minimal as the following example indicates.

Example 6.7. Consider the 2^2 factorial of Example 6.2 and let $\boldsymbol{\beta}'_1 = (\phi_1^0\phi_2^0, \phi_1^1\phi_2^0)$. The induced model relative to $\boldsymbol{\beta}'_1$ is given by

$$E\begin{bmatrix} Y_{(0,0)} \\ Y_{(0,1)} \\ Y_{(1,0)} \\ Y_{(1,1)} \end{bmatrix} = \tfrac{1}{2}\begin{bmatrix} 1 & -1 \\ 1 & -1 \\ 1 & 1 \\ 1 & 1 \end{bmatrix}\begin{bmatrix} \phi_1^0\phi_2^0 \\ \phi_1^1\phi_2^0 \end{bmatrix} = \mathbf{X}_{\rho 1}\boldsymbol{\beta}_1.$$

If $\mathbf{L}'_1 = \frac{1}{2}(1 \ \ -1)$, then a solution for $\mathbf{C}'_1$ is given by

$$\begin{aligned} \mathbf{C}'_1 = \mathbf{L}'_1\mathbf{X}'_{\rho 1} &= \tfrac{1}{2}(1 \ \ -1)\begin{bmatrix} 1 & 1 & 1 & 1 \\ -1 & -1 & 1 & 1 \end{bmatrix} \\ &= (1 \ 1 \ 0 \ 0). \end{aligned}$$

Thus an unbiased design determined by $\mathbf{C}'_1$ is given by $\Gamma = \{(0,0),(0,1)\}$. However, the designs $\Gamma_1 = \{(0,0)\}$ and $\Gamma_2 = \{(0,1)\}$ are two minimal unbiased designs for this problem. This clearly follows from the nonuniqueness of $\mathbf{C}_1$, which in turn reflects the dependency of the rows of $\mathbf{X}_{\rho 1}$. For these minimal designs the reader may verify that the solutions for the coefficient matrices are, respectively,

$$\mathbf{C}_1 = \begin{bmatrix} 1 \\ 0 \\ 0 \\ 0 \end{bmatrix}, \qquad \mathbf{C}_2 = \begin{bmatrix} 0 \\ 1 \\ 0 \\ 0 \end{bmatrix}.$$

It follows that the problem of determining minimal unbiased designs in this setting is solved by those solutions to $\mathbf{C}_1$ in the equation $\mathbf{L}'_1 = \mathbf{C}'_1\mathbf{X}_{\rho 1}$ for which $\mathbf{C}'_1$ has maximum number of columns with all elements equal to zero.

If $\mathbf{L}'_1 = \mathbf{I}_P$, then from (6.6) a solution for $\mathbf{C}'_1$ is

$$\mathbf{C}'_1 = \mathbf{I}_P\mathbf{X}'_{\rho 1} = \mathbf{X}'_{\rho 1}$$

and an unbiased design then is given by those treatments i for which the corresponding columns in $\mathbf{X}'_{\rho 1}$ are nonzero. Such a design need not be minimal, as can easily be verified with Example 6.6. In this setting a minimal design consists of exactly p treatment combinations and it is desirable to develop algorithms for finding such minimal designs by just inspecting a p-subset of treatment combinations from among the possible $C(N, p)$ possible subsets. This problem is known in the literature as the *nonsingularity or connectedness problem* relative to $\boldsymbol{\beta}_1$.

In Chapter 7 we discuss in more detail the partitioning of $\boldsymbol{\beta}_\rho$ and the resulting minimal unbiased design problem for the case wherein the experimenter is interested in elements of $\boldsymbol{\beta}_\rho$ themselves rather than in linear functions of these parameters.

6.4 ESTIMATION OF σ^2

Thus far we have ignored the problem of estimating the variance σ^2. Since σ^2 will usually be unknown and hence needs to be estimated, a design should be selected that allows the unbiased estimation of σ^2. Suppose that only the p parameters in $\boldsymbol{\beta}_1$ are to be estimated and $\boldsymbol{\beta}_2 = 0$. The estimation of σ^2 may be accomplished in two ways, either by replicating, that is, by obtaining repetitions of treatments, or by taking observations on additional treatments or both. In either case we know from Chapter 3 that σ^2 is estimated unbiasedly by $\hat{\sigma}^2 = [(\mathbf{Y}_\Gamma - \mathbf{X}_{\Gamma 1}\hat{\boldsymbol{\beta}}_1)'(\mathbf{Y}_\Gamma - \mathbf{X}_{\Gamma 1}\hat{\boldsymbol{\beta}}_1)]/(n-p)$.

From what has been stated earlier it follows that the use of p observations on p distinct treatments to estimate the parameters in $\boldsymbol{\beta}_1$ results in a *saturated design*, which is automatically minimal. The use of repetitions or additions of treatments results in a design that is neither saturated nor minimal.

Example 6.8. Suppose that $\boldsymbol{\beta}_1' = (\phi_1^0\phi_2^0, \phi_1^1\phi_2^0, \phi_1^2\phi_2^0, \phi_1^0\phi_2^1, \phi_1^0\phi_2^2)$ for the equispaced 3^2 factorial. Then a minimal and saturated unbiased design for estimating $\mathbf{L}_1'\boldsymbol{\beta}_1 = \boldsymbol{\beta}_1$ is

$$\Gamma = \{(0,0), (1,0), (2,0), (0,1), (0,2)\}.$$

An unsaturated and nonminimal unbiased design is given by

$$\Gamma^* = \{(0,0), (1,0), (1,0), (2,0), (0,1), (0,1), (0,2)\}.$$

The design Γ does not allow estimation of σ^2 since $n-p=0$; however, Γ^* provides an unbiased estimator of σ^2, because $n-p=2$. The reader may verify that the unsaturated nonminimal unbiased design

$$\Gamma^{**} = \{(0,0), (1,0), (2,0), (0,1), (0,2), (2,1), (1,2)\}$$

also provides an unbiased estimator of σ^2. Obviously, repetitions and additional treatments may be combined to produce, for example, the design

$$\Gamma^{***} = \{(0,0),(1,0),(1,0),(2,0),(0,1),(0,1),(0,2),(1,2),(2,1)\}.$$

Remark. The results of Section 6.3 can be extended to a more general partitioning of $\boldsymbol{\beta}_\rho$ (e.g., see Chapter 7).

6.5 SELECTED ADDITIONAL READING

1. Federer, W. T., Hedayat, A., and Raktoe, B. L. (1975). Minimal unbiased designs for linear parametric functions. In *A Survey of Statistical Design and Linear Models*, J. N. Srivastava, Ed. North-Holland, Amsterdam, pp. 145–153.
2. Fraser, W., and Raktoe, B. L. (1976). A theorem on invariance of estimability of linear parametric functions in linear models. *Commun. Stat.*, **A5**, 977–983.
3. Graybill, F. A. (1976). *Theory and Application of the Linear Model*. Duxbury Press, North Scituate, Mass.
4. Srivastava, J. N. (1975). Designs for searching non-negligible effects. In *A Survey of Statistical Design and Linear Models*, J. N. Srivastava, Ed. North-Holland, Amsterdam, pp. 507–519.

CHAPTER 7

Resolution and Confounding in Factorial Designs

In this chapter we introduce a general partitioning of the full parametric vector $\boldsymbol{\beta}_\rho$ in order to estimate some or all of its components with or without assuming that the remaining parameters are negligible. The partitioning leads to four exhaustive cases, which contain the classical designs of arbitrary *resolutions* as special cases. In addition we point out the problems of searching for minimal unbiased designs in each of the four cases. The concept of a *confounded design* is introduced and illustrated. A further study of these concepts is carried out in Chapters 9–11. The concept of *randomized factorial designs* is introduced in Chapter 9. This concept and corresponding results are related to Case 3 of Section 7.1. In Chapter 10 we study resolution III designs in some detail, and Chapter 11 deals with designs of resolution IV and V.

7.1 THE PARTITIONING OF $\boldsymbol{\beta}_\rho$ FROM THE EXPERIMENTER'S VIEWPOINT

In some factorial experiments, the experimenter will be interested in estimating all the elements of $\boldsymbol{\beta}_\rho$ if he has no advance knowledge concerning any of them. In other situations, the experimenter knows some of the elements of $\boldsymbol{\beta}_\rho$ and he or she is interested in estimating some specified subset of the remaining ones. More precisely, these two situations lead to the following formulation.

Without loss of generality the total parametric vector $\boldsymbol{\beta}_\rho$ can be partitioned as

$$\boldsymbol{\beta}_\rho' = \left(\boldsymbol{\beta}_1' \;\vdots\; \boldsymbol{\beta}_2' \;\vdots\; \boldsymbol{\beta}_3'\right) \tag{7.1}$$

where $\boldsymbol{\beta}_1$ is an $N_1 \times 1$ vector to be estimated, $\boldsymbol{\beta}_2$ is an $N_2 \times 1$ vector not of interest and not assumed to be known, and $\boldsymbol{\beta}_3$ is an $N_3 \times 1$ vector of parameters assumed to be known (which without loss of generality can be taken to be zero), such that $1 \leq N_1 \leq N$, $0 \leq N_2 \leq N-1$, and $0 \leq N_3 = N - N_1 - N_2 \leq N-1$. Explicitly, we then have the cases:

(1) $N_1 = N$, $N_2 = N_3 = 0$.
(2) $N_2 = 0$, $N_3 \neq 0$.
(3) $N_2 \neq 0$, $N_3 \neq 0$.
(4) $N_2 \neq 0$, $N_3 = 0$.

The cases in Sections 6.2 and 6.3 are Cases 1 and 2, respectively. Also, the reader should not confuse the above-mentioned $\boldsymbol{\beta}_2$ with the $\boldsymbol{\beta}_2$ in previous chapters. Recalling the definition of a factorial effect of degree or order k, $1 \leq k \leq t$, from Chapter 4, we may now connect the concept of "*resolution*" with the partitioning (7.1).

Definition 7.1. A factorial design is said to be of *resolution R* if all factorial effects up to order k are estimable, where k is the greatest integer *less* than $R/2$, under the assumption that all factorial effects of order $R-k$ and higher are zero, with the additional convention that if R is odd then the general mean is also estimable and if R is even the general mean is not of interest for estimation.

The designs of resolution R have been traditionally divided into two types, that is:

(i) $R = 2r$, known as designs of *even resolution*.
(ii) $R = 2r + 1$, known as designs of *odd resolution*.

Thus an even resolution design is such that all factorial effects of order less than or equal to $r-1$ are estimable under the assumption that all factorial effects of order $r+1$ or more are negligible. In this case the general mean and factorial effects of order r are not to be estimated and are not assumed to be equal to zero. On the other hand, an odd resolution design is a design that allows unbiased estimation of the general mean and all factorial effects of order less than or equal to r assuming that all factorial effects of order $r+1$ or higher are zero.

It follows that an even resolution design is a special case of 3 and an odd resolution design is a special case of 2.

Remark. The reader may question the reason for including the general mean for estimation in odd resolution designs but not in even resolution designs. The literature is vague on this point but we may justify this convention as follows. (i) In odd resolution designs we estimate all parameters not assumed to be zero. Thus, if we can also estimate the general mean, we will have a predictive model. However, in even resolution designs we do not estimate some of the nonzero parameters because they are not of interest to us. Therefore, if we estimate the general mean we cannot build a predictive model anyway. (ii) If we include the general mean as a part of the parameters not to be estimated we will end up with some less cumbersome mathematics in even resolution designs.

Finally, we should point out that in many even resolution designs if we include a few additional, judiciously selected treatments, we can also estimate the general mean.

Let us now study each of the four cases 1 to 4 separately.

Case 1. Since we have to estimate all N components of $\boldsymbol{\beta}_\rho$, it is clear that the minimal unbiased design is the minimal complete factorial design ρ. Hence any design containing ρ will be an unbiased design for $\boldsymbol{\beta}_\rho$. If Γ is such a design, then the BLUE of $\boldsymbol{\beta}_\rho$ using design Γ is

$$(7.2)\qquad \hat{\boldsymbol{\beta}}_\rho = (\mathbf{X}_\Gamma' \mathbf{X}_\Gamma)^{-1} \mathbf{X}_\Gamma' \mathbf{Y}_\Gamma,$$

where the rows of $\mathbf{X}_\Gamma$ are the rows of $\mathbf{X}_\rho$ taking repetitions into account.

Case 2. The results in this case follow immediately from Section 6.3. The basic model for the minimal complete factorial design ρ is

$$(7.3)\qquad E[\mathbf{Y}_\rho] = \mathbf{X}_\rho \boldsymbol{\beta}_\rho = \begin{bmatrix} \mathbf{X}_{\rho 1} & \vdots & \mathbf{X}_{\rho 3} \end{bmatrix} \begin{bmatrix} \boldsymbol{\beta}_1 \\ \cdots \\ \boldsymbol{\beta}_3 \end{bmatrix} = \mathbf{X}_{\rho 1} \boldsymbol{\beta}_1,$$

since $\boldsymbol{\beta}_3$ is equal to zero. It follows that the search for a minimal unbiased design reduces to finding N_1 independent rows of $\mathbf{X}_{\rho 1}$, which in general leads to many designs. Any design Γ containing a minimal unbiased design is then an unbiased design for $\boldsymbol{\beta}_1$. The BLUE of $\boldsymbol{\beta}_1$ using design Γ is obtained as

$$(7.4)\qquad \hat{\boldsymbol{\beta}}_1 = (\mathbf{X}_{\Gamma 1}' \mathbf{X}_{\Gamma 1})^{-1} \mathbf{X}_{\Gamma 1}' \mathbf{Y}_\Gamma.$$

Exhibiting the whole class of minimal unbiased designs is at present unresolved for a general setting. In some particular cases, such as designs of odd resolution (e.g., resolutions III and V), some work has been done and is further discussed in Chapters 10–12.

Example 7.1. Consider the 2^3 factorial with levels zero and one for each factor and the model given in Example 3.12, that is,

$$E\begin{bmatrix} Y_{(0,0,0)} \\ Y_{(1,0,0)} \\ Y_{(0,1,0)} \\ Y_{(0,0,1)} \\ Y_{(1,1,0)} \\ Y_{(1,0,1)} \\ Y_{(0,1,1)} \\ Y_{(1,1,1)} \end{bmatrix} = \begin{bmatrix} 1 & -1 & -1 & -1 & 1 & 1 & 1 & -1 \\ 1 & 1 & -1 & -1 & -1 & -1 & 1 & 1 \\ 1 & -1 & 1 & -1 & -1 & 1 & -1 & 1 \\ 1 & -1 & -1 & 1 & 1 & -1 & -1 & 1 \\ 1 & 1 & 1 & -1 & 1 & -1 & -1 & -1 \\ 1 & 1 & -1 & 1 & -1 & 1 & -1 & -1 \\ 1 & -1 & 1 & 1 & -1 & -1 & 1 & -1 \\ 1 & 1 & 1 & 1 & 1 & 1 & 1 & 1 \end{bmatrix} \begin{bmatrix} \phi_1^0\phi_2^0\phi_3^0 \\ \phi_1^1\phi_2^0\phi_3^0 \\ \phi_1^0\phi_2^1\phi_3^0 \\ \phi_1^0\phi_2^0\phi_3^1 \\ \phi_1^1\phi_2^1\phi_3^0 \\ \phi_1^1\phi_2^0\phi_3^1 \\ \phi_1^0\phi_2^1\phi_3^1 \\ \phi_1^1\phi_2^1\phi_3^1 \end{bmatrix} = \mathbf{X}_\rho \boldsymbol{\beta}_\rho.$$

Under Case 1 all eight components of $\boldsymbol{\beta}_\rho$ are to be estimated, since $N_1 = 8$ and $N_2 = N_3 = 0$. Any design Γ containing the preceding eight treatment combinations would suffice. If the minimal complete factorial arrangement ρ itself is used, then the BLUE is $\hat{\boldsymbol{\beta}}_\rho = \frac{1}{8}\mathbf{X}'_\rho \mathbf{Y}_\rho$. Now, let $\boldsymbol{\beta}'_1 = (\phi_1^1\phi_2^0\phi_3^0, \phi_1^1\phi_2^1\phi_3^1)$ and let $\boldsymbol{\beta}_3$ contain the remaining six parameters. It can be easily verified that the designs $\{(0,0,0),(1,0,1)\}$ and $\{(1,1,0),(1,1,1)\}$ are minimal unbiased designs. The reader may exhibit many more. Next, consider designs of odd resolutions under Case 2. The possible values of R for the 2^3 factorial are resolutions I, III, V, and VII. If $R = \text{I}$ then $r = 0$, so that there are no factorial effects to be estimated and the only parameter in the model is the mean $\phi_1^0\phi_2^0\phi_3^0$. In this case there are eight minimal designs of resolution I for estimating the mean. Any design consisting of one or more (not necessarily distinct) treatment combinations will lead to estimation of the mean. If $R = \text{III}$ then $r = 1$, so that $\boldsymbol{\beta}_1$ comprises the three main effects $\phi_1^1\phi_2^0\phi_3^0$, $\phi_1^0\phi_2^1\phi_3^0$, $\phi_1^0\phi_2^0\phi_3^1$ and $\boldsymbol{\beta}_3$ is the vector of two-and three- factor interactions. If the mean $\phi_1^0\phi_2^0\phi_3^0$ is not negligible (which is the usual assumption), then it is usually taken as an element of $\boldsymbol{\beta}_1$. For this reason, the vector $\boldsymbol{\beta}_1$ in resolution III designs of the general 2^t factorial is always taken to consist of the mean and the t main effects. From Example 3.4, we may deduce that for the 2^3 factorial, there are precisely 58 minimal resolution III designs for estimating $\boldsymbol{\beta}'_1 = (\phi_1^0\phi_2^0\phi_3^0, \phi_1^1\phi_2^0\phi_3^0, \phi_1^0\phi_2^1\phi_3^0,$

$\phi_1^0\phi_2^0\phi_3^1$). One of these is $\{(0,0,0),(0,1,1),(1,0,1),(1,1,0)\}$, which was shown in Example 5.1 to be a (Q_1,Q_2,Q_3,Q_4)-optimal design. As an exercise, the reader may try to find (algebraically, geometrically, or using the computer) the number (or bounds on the number) of minimal resolution III designs of the 2^4 factorial. Now, let $R=\text{V}$, then $r=2$ so that three main effects and three two-factor interactions are to be estimated. Including the mean there will be seven parameters in $\boldsymbol{\beta}_1$, and $\boldsymbol{\beta}_3$ consists of the singleton $\phi_1^1\phi_2^1\phi_3^1$. A minimal design to estimate $\boldsymbol{\beta}_1$ must consist of at least seven distinct treatment combinations. From the fact that any 7×7 submatrix of the columnwise orthogonal 8×8 matrix $\mathbf{X}_\rho$ is nonsingular, it follows that there are exactly $\text{C}(8,7)=8$ minimal resolution V designs for the 2^3 factorial. Any design Γ containing one of these designs is then automatically a resolution V design. Finally, take $R=\text{VII}$, so that $r=3$. Then we must estimate all the seven factorial effects. Including the mean in $\boldsymbol{\beta}_1$, we observe that $N_1=8$ and $N_3=0$, so that we are in Case 1, which has already been discussed.

Case 3. The basic model for this case is conveniently written as

$$(7.5)\qquad E[\mathbf{Y}_\rho]=\mathbf{X}_\rho\boldsymbol{\beta}_\rho=\begin{bmatrix}\mathbf{X}_{\rho 1} & \vdots & \mathbf{X}_{\rho 2} & \vdots & \mathbf{X}_{\rho 3}\end{bmatrix}\begin{bmatrix}\boldsymbol{\beta}_1\\ \cdots\\ \boldsymbol{\beta}_2\\ \cdots\\ \boldsymbol{\beta}_3\end{bmatrix}$$

$$=\begin{bmatrix}\mathbf{X}_1 & \vdots & \mathbf{X}_2\end{bmatrix}\begin{bmatrix}\boldsymbol{\beta}_1\\ \cdots\\ \boldsymbol{\beta}_2\end{bmatrix},$$

since $\boldsymbol{\beta}_3=\mathbf{0}$. For any arbitrary design Γ, the model obtained from (7.5) is

$$(7.6)\qquad E[\mathbf{Y}_\Gamma]=\begin{bmatrix}\mathbf{X}_{\Gamma 1} & \vdots & \mathbf{X}_{\Gamma 2}\end{bmatrix}\begin{bmatrix}\boldsymbol{\beta}_1\\ \cdots\\ \boldsymbol{\beta}_2\end{bmatrix}.$$

Applying the least squares procedure to (7.6), we obtain the normal equation

$$(7.7)\qquad \begin{bmatrix}\mathbf{X}'_{\Gamma 1}\mathbf{X}_{\Gamma 1} & \vdots & \mathbf{X}'_{\Gamma 1}\mathbf{X}_{\Gamma 2}\\ \cdots\cdots & \vdots & \cdots\cdots\\ \mathbf{X}'_{\Gamma 2}\mathbf{X}_{\Gamma 1} & \vdots & \mathbf{X}'_{\Gamma 2}\mathbf{X}_{\Gamma 2}\end{bmatrix}\begin{bmatrix}\boldsymbol{\beta}_1\\ \cdots\\ \boldsymbol{\beta}_2\end{bmatrix}=\begin{bmatrix}\mathbf{X}'_{\Gamma 1}\\ \cdots\\ \mathbf{X}'_{\Gamma 2}\end{bmatrix}\mathbf{Y}_\Gamma.$$

Expressing $\boldsymbol{\beta}_2$ in the second equation of (7.7) in terms of $\boldsymbol{\beta}_1$ and substituting the result in the first equation of (7.7) an equation involving $\boldsymbol{\beta}_1$ only is obtained; that is,

$$(7.8)\quad \left[\mathbf{X}'_{\Gamma 1}\mathbf{X}_{\Gamma 1}-\mathbf{X}'_{\Gamma 1}\mathbf{X}_{\Gamma 2}(\mathbf{X}'_{\Gamma 2}\mathbf{X}_{\Gamma 2})^{-}\mathbf{X}'_{\Gamma 2}\mathbf{X}_{\Gamma 1}\right]\boldsymbol{\beta}_1 = \left[\mathbf{X}'_{\Gamma 1}-\mathbf{X}'_{\Gamma 1}\mathbf{X}_{\Gamma 2}(\mathbf{X}'_{\Gamma 2}\mathbf{X}_{\Gamma 2})^{-}\mathbf{X}'_{\Gamma 2}\right]\mathbf{Y}_{\Gamma},$$

where $\mathbf{A}^-$ denotes a generalized inverse of the matrix $\mathbf{A}$. Hence it follows that a necessary and sufficient condition for Γ to be an unbiased design is that the rank of $[\mathbf{X}'_{\Gamma 1}\mathbf{X}_{\Gamma 1}-\mathbf{X}'_{\Gamma 1}\mathbf{X}_{\Gamma 2}(\mathbf{X}'_{\Gamma 2}\mathbf{X}_{\Gamma 2})^{-}\mathbf{X}'_{\Gamma 2}\mathbf{X}_{\Gamma 1}]$ is equal to N_1. Therefore, the search of a minimal unbiased design involves searching a minimal set of rows of $[\mathbf{X}_{\rho 1}\vdots\mathbf{X}_{\rho 2}]$ such that the stated rank condition is satisfied. Note that if $\mathbf{X}'_{\Gamma 1}\mathbf{X}_{\Gamma 2}=\mathbf{0}$, then the preceding necessary and sufficient condition for Γ to be an unbiased design becomes the condition that the rank of $\mathbf{X}'_{\Gamma 1}\mathbf{X}_{\Gamma 1}$ is equal to N_1. This follows directly from equation (7.8).

The general rank condition involves considerable calculation. An equivalent and possibly more amenable necessary and sufficient condition for Γ to be an unbiased design is that the rank of $\mathbf{X}_1$ is N_1 and no column of $\mathbf{X}_1$ is in the column space of $\mathbf{X}_2$ (or no column of $\mathbf{X}_2$ is in the column space of $\mathbf{X}_1$). The proof of this is as follows. First, without loss of generality, suppose that the first column of $\mathbf{X}_1$ is in the column space of $\mathbf{X}_2$. That is, if $\mathbf{X}_{11},\mathbf{X}_{12},\ldots,\mathbf{X}_{1N_1}$ are the columns of $\mathbf{X}_1$ and $\mathbf{X}_{21},\mathbf{X}_{22},\ldots,\mathbf{X}_{2N_2}$ are the columns of $\mathbf{X}_2$, then $\mathbf{X}_{11}=\Sigma_{i=1}^{N_2}c_i\mathbf{X}_{2i}$, where not all the c_i are 0. If Γ is an unbiased design, that is, if $\boldsymbol{\beta}'_1=(\beta_{11},\beta_{12},\ldots,\beta_{1N_1})$ is estimable, then there exists a $1\times n$ vector $\mathbf{q}'$ such that $E[\mathbf{q}'\mathrm{Y}_\Gamma]=\beta_{11}$. Now,

$$E[\mathbf{q}'Y_\Gamma]=\mathbf{q}'\left[\mathbf{X}_{\Gamma 1}\ \vdots\ \mathbf{X}_{\Gamma 2}\right]\begin{bmatrix}\boldsymbol{\beta}_1\\ \cdots\\ \boldsymbol{\beta}_2\end{bmatrix}$$

$$=\mathbf{q}'\left[\mathbf{X}_{11}\beta_{11}+\mathbf{X}_{12}\beta_{12}+\ldots+\mathbf{X}_{1N_1}\beta_{1N_1}+\mathbf{X}_{21}\beta_{21}+\mathbf{X}_{22}\beta_{22}+\ldots+\mathbf{X}_{2N_2}\beta_{2N_2}\right].$$

Hence $\mathbf{q}'\mathbf{X}_{11}=1$ and $\mathbf{q}'\mathbf{X}_{12}\beta_{12}=\ldots=\mathbf{q}'\mathbf{X}_{1N_1}\beta_{1N_1}=\mathbf{q}'\mathbf{X}_{21}\beta_{21}=\ldots=\mathbf{q}'\mathbf{X}_{2N_2}\beta_{2N_2}=0$ so that $\mathbf{q}'\mathbf{X}_{12}=\ldots=\mathbf{q}'\mathbf{X}_{1N_1}=\mathbf{q}'\mathbf{X}_{21}=\ldots=\mathbf{q}'\mathbf{X}_{2N_2}=0$. However, $\mathbf{q}'\mathbf{X}_{11}=\mathbf{q}'\sum_{i=1}^{N_2}c_i\mathbf{X}_{2i}=0$, which leads to a contradiction; hence the necessity is proved. The sufficiency part of the proof is left as an exercise for the reader. The BLUE of $\boldsymbol{\beta}_1$ for an unbiased design in Case 3 is given by

$$\hat{\boldsymbol{\beta}}_1 = \left[\mathbf{X}'_{\Gamma 1}\mathbf{X}_{\Gamma 1} - \mathbf{X}'_{\Gamma 1}\mathbf{X}_{\Gamma 2}(\mathbf{X}'_{\Gamma 2}\mathbf{X}_{\Gamma 2})^{-}\mathbf{X}'_{\Gamma 2}\mathbf{X}_{\Gamma 1}\right]^{-1} \cdot \left[\mathbf{X}'_{\Gamma 1} - \mathbf{X}'_{\Gamma 1}\mathbf{X}_{\Gamma 2}(\mathbf{X}'_{\Gamma 2}\mathbf{X}_{\Gamma 2})^{-}\mathbf{X}'_{\Gamma 2}\right]\mathbf{Y}_{\Gamma}. \tag{7.9}$$

The problem of finding unbiased designs (not of the trivial type, i.e., those that contain ρ) is an unsolved problem at present. In some cases, such as resolution IV, certain classes of unbiased designs have been constructed, and we present some of these in Chapters 11 and 12.

Example 7.2. Consider the 2^3 factorial of the previous example, and let $\boldsymbol{\beta}'_1 = (\phi_1^1\phi_2^0\phi_3^0, \phi_1^0\phi_2^1\phi_3^0, \phi_1^0\phi_2^0\phi_3^1)$, $\boldsymbol{\beta}'_2 = (\phi_1^0\phi_2^0\phi_3^0, \phi_1^1\phi_2^1\phi_3^0, \phi_1^1\phi_2^0\phi_3^1, \phi_1^0\phi_2^1\phi_3^1)$, and $\boldsymbol{\beta}_3 = (\phi_1^1\phi_2^1\phi_3^1)$. From what has been stated earlier it follows that any design consisting of $m=7$ treatment combinations, for example, $\Gamma^* = \{(0,0,0),(1,0,0),(0,1,0),(0,0,1),(1,1,0),(1,0,1),(0,1,1)\}$, is a resolution IV design for estimating $\boldsymbol{\beta}_1$. This also follows directly from the fact that for the 2^3 factorial such designs are of resolution V relative to estimation of $\boldsymbol{\beta}_1$ and $\boldsymbol{\beta}_2$. These designs consisting of $m=7$ treatments are, however, not minimal. It can be deduced that minimal resolution IV designs for the 2^3 factorial consist of $m=6$ treatment combinations. Such a design is $\Gamma = \{(1,0,0),(0,1,0),(0,0,1),(0,1,1),(1,0,1),(1,1,0)\}$, for which the matrices $\mathbf{X}_{\Gamma 1}$ and $\mathbf{X}_{\Gamma 2}$ are

$$\mathbf{X}_{\Gamma 1} = \begin{bmatrix} 1 & -1 & -1 \\ -1 & 1 & -1 \\ -1 & -1 & 1 \\ -1 & 1 & 1 \\ 1 & -1 & 1 \\ 1 & 1 & -1 \end{bmatrix}, \quad \mathbf{X}_{\Gamma 2} = \begin{bmatrix} 1 & -1 & -1 & 1 & 1 \\ 1 & -1 & 1 & -1 & 1 \\ 1 & 1 & -1 & -1 & 1 \\ 1 & -1 & -1 & 1 & -1 \\ 1 & -1 & 1 & -1 & -1 \\ 1 & 1 & -1 & -1 & -1 \end{bmatrix}.$$

The reader may verify that $\mathbf{X}'_{\Gamma 1}\mathbf{X}_{\Gamma 2} = \mathbf{0}$.

Case 4. Since in this case $N_3 = 0$, the basic model is equal to

$$E[\mathbf{Y}_\rho] = [\mathbf{X}_\rho\boldsymbol{\beta}_\rho] = \left[\mathbf{X}_{\rho 1} \;\vdots\; \mathbf{X}_{\rho 2}\right]\begin{bmatrix} \boldsymbol{\beta}_1 \\ \cdots \\ \boldsymbol{\beta}_2 \end{bmatrix} \tag{7.10}$$

so that for any design Γ the induced model is

$$E[\mathbf{Y}_\Gamma] = \begin{bmatrix} \mathbf{X}_{\Gamma 1} & \vdots & \mathbf{X}_{\Gamma 2} \end{bmatrix} \begin{bmatrix} \boldsymbol{\beta}_1 \\ \cdots \\ \boldsymbol{\beta}_2 \end{bmatrix}. \tag{7.11}$$

Since $N_1 + N_2 = N$, the only difference between Cases 3 and 4 is that now $\boldsymbol{\beta}_1$ and $\boldsymbol{\beta}_2$ exhaust $\boldsymbol{\beta}_p$. Thus an analysis similar to that of Case 3 can be carried out and a necessary and sufficient condition for Γ to be an unbiased design is that the rank of $[\mathbf{X}'_{\Gamma 1}\mathbf{X}_{\Gamma 1} - \mathbf{X}'_{\Gamma 1}\mathbf{X}_{\Gamma 2}(\mathbf{X}'_{\Gamma 2}\mathbf{X}_{\Gamma 2})^{-}\mathbf{X}'_{\Gamma 2}\mathbf{X}_{\Gamma 1}]$ is N_1. As before, an equivalent necessary and sufficient condition is that the rank of $\mathbf{X}_1$ is N_1 and no column of $\mathbf{X}_{\Gamma 1}$ is in the column space of $\mathbf{X}_{\Gamma 2}$. The BLUE of $\boldsymbol{\beta}_1$ for a design of Case 4 is, as in (7.9), given by

$$\hat{\boldsymbol{\beta}}_1 = \left[\mathbf{X}'_{\Gamma 1}\mathbf{X}_{\Gamma 1} - \mathbf{X}'_{\Gamma 1}\mathbf{X}_{\Gamma 2}(\mathbf{X}'_{\Gamma 2}\mathbf{X}_{\Gamma 2})^{-}\mathbf{X}'_{\Gamma 2}\mathbf{X}_{\Gamma 1}\right]^{-1} \cdot \left[\mathbf{X}'_{\Gamma 1} - \mathbf{X}'_{\Gamma 1}\mathbf{X}_{\Gamma 2}(\mathbf{X}'_{\Gamma 2}\mathbf{X}_{\Gamma 2})^{-}\mathbf{X}'_{\Gamma 2}\right]\mathbf{Y}_\Gamma. \tag{7.12}$$

Example 7.3. Consider the 3×3 equispaced factorial with the nonnormalized orthogonal polynomial model. Let $\boldsymbol{\beta}'_1 = (\phi_1^1\phi_2^0, \phi_1^0\phi_2^1, \phi_1^1\phi_2^2, \phi_1^2\phi_2^1)$ so that $\boldsymbol{\beta}'_2 = (\phi_1^0\phi_2^0, \phi_1^1\phi_2^1, \phi_1^2\phi_2^0, \phi_1^0\phi_2^2, \phi_1^2\phi_2^2)$. A trivial unbiased design for $\boldsymbol{\beta}_1$ is any factorial design containing the minimal complete factorial design. For $m = 8$, an unbiased design is $\Gamma = \{(0,0),(1,0),(2,0),(0,2),(1,2),(2,2),(0,1),(2,1)\}$. The matrices $\mathbf{X}_{\Gamma 1}$ and $\mathbf{X}_{\Gamma 2}$ are, respectively,

$$\mathbf{X}_{\Gamma 1} = \begin{bmatrix} -1 & -1 & -1 & -1 \\ 0 & -1 & 0 & 2 \\ 1 & -1 & 1 & -1 \\ -1 & 1 & -1 & 1 \\ 0 & 1 & 0 & -2 \\ 1 & 1 & 1 & 1 \\ -1 & 0 & 2 & 0 \\ 1 & 0 & -2 & 0 \end{bmatrix},$$

$$\mathbf{X}_{\Gamma 2} = \begin{bmatrix} 1 & 1 & 1 & 1 & 1 \\ 1 & 0 & -2 & 1 & -2 \\ 1 & -1 & 1 & 1 & 1 \\ 1 & -1 & 1 & 1 & 1 \\ 1 & 0 & -2 & 1 & -2 \\ 1 & 1 & 1 & 1 & 1 \\ 1 & 0 & 1 & -2 & -2 \\ 1 & 0 & 1 & -2 & -2 \end{bmatrix},$$

and

$$\mathbf{X}'_{\Gamma 1}\mathbf{X}_{\Gamma 1}=\begin{bmatrix} 6 & 0 & 0 & 0 \\ 0 & 6 & 0 & 0 \\ 0 & 0 & 12 & 0 \\ 0 & 0 & 0 & 12 \end{bmatrix}.$$

It may be verified that $\mathbf{X}'_{\Gamma 1}\mathbf{X}_{\Gamma 2}=\mathbf{0}$. The reader is invited to verify whether Γ is a minimal design.

7.2 CONFOUNDING AND ALIASING

Consider the $k_1 \times k_2 \times \ldots \times k_t$ factorial and the general linear model associated with the minimal complete factorial design ρ, that is, $E[\mathbf{Y}_\rho]=\mathbf{X}_\rho\boldsymbol{\theta}$ and $\text{Cov}(\mathbf{Y}_\rho)=\sigma^2\mathbf{I}_N$, where the rank of $\mathbf{X}_\rho$ is less than or equal to the number of columns of $\mathbf{X}_\rho$ and $\boldsymbol{\theta}$ is a parametric vector arising in a model related to any of the four cases in Section 7.1.

Definition 7.2. Let ψ_1 and ψ_2 be two algebraically independent linear parametric functions of $\boldsymbol{\theta}$. For a given design Γ let $\widehat{\psi_1}$ be any linear unbiased estimator of ψ_1 under the assumption that $\psi_2=0$. If $\psi_2 \neq 0$ and $E[\hat{\psi}_1]=\psi_1 + c\psi_2$, $c \neq 0$, then ψ_1 is said to be *confounded* (or *aliased*) with ψ_2 under Γ, and the design Γ is referred to as a *confounded design*.

We would like to emphasize the model dependency of the concept of confounded designs and also note that a confounded design does not admit an unbiased estimator of ψ_1 if ψ_2 is not equal to zero; that is, a confounded design is a *biased design* for ψ_1 if $\psi_2 \neq 0$. Also, Definition 7.2 can be generalized in a straightforward manner to a set of algebraically independent parametric functions. The relation of confounding is symmetric; that is, if ψ_1 is confounded with ψ_2 under Γ, then ψ_2 is confounded with ψ_1 under Γ.

Example 7.4. Consider the 2×3 factorial with equispaced levels and the nonnormalized orthogonal polynomial model

$$E\begin{bmatrix} Y_{(0,0)} \\ Y_{(0,1)} \\ Y_{(0,2)} \\ Y_{(1,0)} \\ Y_{(1,1)} \\ Y_{(1,2)} \end{bmatrix} = \begin{bmatrix} 1 & -1 & 1 & -1 & 1 & -1 \\ 1 & 0 & -2 & -1 & 0 & 2 \\ 1 & 1 & 1 & -1 & -1 & -1 \\ 1 & -1 & 1 & 1 & -1 & 1 \\ 1 & 0 & -2 & 1 & 0 & -2 \\ 1 & 1 & 1 & 1 & 1 & 1 \end{bmatrix} \begin{bmatrix} \phi_1^0\phi_2^0 \\ \phi_1^0\phi_2^1 \\ \phi_1^0\phi_2^2 \\ \phi_1^1\phi_2^0 \\ \phi_1^1\phi_2^1 \\ \phi_1^1\phi_2^2 \end{bmatrix} = \mathbf{X}_\rho\boldsymbol{\beta}_\rho.$$

The vector $\boldsymbol{\beta}_\rho$ itself constitutes a set of algebraically independent linear functions of $\boldsymbol{\beta}_\rho$, since $\boldsymbol{\beta}_\rho = \mathbf{I}_6\boldsymbol{\beta}_\rho$. Consider the design $\Gamma = \{(0,1)\}$ and the estimator $\widehat{\phi_1^0\phi_2^0} = Y_{(0,1)}$ of the parameter $\phi_1^0\phi_2^0$. This estimator is a linear unbiased estimator of $\phi_1^0\phi_2^0$ if $\phi_1^0\phi_2^2 = \phi_1^1\phi_2^0 = \phi_1^1\phi_2^2 = 0$. If these parameters are not equal to zero, then $E[\widehat{\phi_1^0\phi_2^0}] = \phi_1^0\phi_2^0 - 2\,\phi_1^0\phi_2^2 - \phi_1^1\phi_2^0 + 2\,\phi_1^1\phi_2^2$. The design $\Gamma = \{(0,1)\}$ is then a confounded design, since $\phi_1^0\phi_2^0$ is confounded with $\phi_1^0\phi_2^2$, $\phi_1^1\phi_2^0$ and $\phi_1^1\phi_2^2$. Note that $\phi_1^0\phi_2^0$ is unconfounded with $\phi_1^0\phi_2^1$ and $\phi_1^1\phi_2^1$.

Example 7.5. Consider the resolution III design $\Gamma = \{(0,0,0), (0,1,1),(1,0,1),(1,1,0)\}$ that was introduced in Example 7.1 for the 2^3 factorial. If the assumption that $\boldsymbol{\beta}_3 = \mathbf{0}$ is not valid, then $E[\hat{\boldsymbol{\beta}}_1] = \boldsymbol{\beta}_1 + \mathbf{A}_\Gamma\boldsymbol{\beta}_3$, where $\mathbf{A}_\Gamma$ is the alias matrix, which has already been calculated in Example 5.5. Hence

$$E[\hat{\boldsymbol{\beta}}_1] = \begin{bmatrix} \phi_1^0\phi_2^0\phi_3^0 \\ \phi_1^0\phi_2^0\phi_3^1 \\ \phi_1^0\phi_2^1\phi_3^0 \\ \phi_1^1\phi_2^0\phi_3^0 \end{bmatrix} + \begin{bmatrix} 0 & 0 & 0 & -1 \\ 0 & 0 & -1 & 0 \\ 0 & -1 & 0 & 0 \\ -1 & 0 & 0 & 0 \end{bmatrix} \begin{bmatrix} \phi_1^0\phi_2^1\phi_3^1 \\ \phi_1^1\phi_2^0\phi_3^1 \\ \phi_1^1\phi_2^1\phi_3^0 \\ \phi_1^1\phi_2^1\phi_3^1 \end{bmatrix}.$$

It follows that when $\boldsymbol{\beta}_3 \neq \mathbf{0}$, then $\phi_1^0\phi_2^0\phi_3^0$ is confounded with $\phi_1^1\phi_2^1\phi_3^1$, $\phi_1^0\phi_2^0\phi_3^1$ is confounded with $\phi_1^1\phi_2^1\phi_3^0$, $\phi_1^0\phi_2^1\phi_3^0$ is confounded with $\phi_1^1\phi_2^0\phi_3^1$, and $\phi_1^1\phi_2^0\phi_3^0$ is confounded with $\phi_1^0\phi_2^1\phi_3^1$. The design Γ is thus a confounded design if $\boldsymbol{\beta}_3 \neq \mathbf{0}$. It is, of course, an unconfounded design if $\boldsymbol{\beta}_3$ is indeed zero.

The reader is invited to provide examples of confounded designs for Cases 3 and 4.

Remark. In many experiments, units are grouped into relatively homogeneous groups called *incomplete blocks*. If treatment combinations of the $k_1 \times k_2 \times \ldots \times k_t$ factorial are allocated to the experimental units in such a way that effects associated with the factor "incomplete blocks" are confounded with a set of factorial effects, then such an experimental design or plan is commonly referred to in the literature as "*confounded incomplete block design*" or as a "*confounded factorial design*". The incomplete blocks may be further grouped into *complete replicates*, that is, all the treatments in the incomplete blocks within one replicate exhaust the minimal complete factorial ρ. If a given set of factorial effects is confounded with incomplete block effects in *all* the replicates, then the design is called a *completely confounded incomplete block design*; otherwise, it is called a *partially confounded incomplete block design*. Frequently, the $k_1 \times k_2 \times \ldots \times k_t$ factorial is associated with a set of $v = \prod k_i$ treatments, which do not possess such a structure. Such a setup is referred to as a *pseudo-or quasi-factorial* and the resulting confounded incomplete block designs are known as *pseudo-factorial* or *lattice designs*. We will not delve into the relationship between the preceding designs and fractional replication. For the interested reader, we provide several references in the list at the end of this chapter. However, to illustrate the confounding aspects in an incomplete block design we provide the following example.

Example 7.6. Consider an incomplete block design Γ for the 2^3 factorial in two blocks of four plots each:

Block 0	Block 1
000	001
011	010
101	100
110	111

Here "blocks" is a fourth factor with two levels so that the resulting factorial is now a 2^4 factorial. Put *all* factorial effects of the first three factors and the mean in $\boldsymbol{\beta}_1$ and all factorial effects involving the fourth factor in $\boldsymbol{\beta}_3$. Then the usual model when $\boldsymbol{\beta}_3$ is assumed to be equal to $\mathbf{0}$ is

$$E\begin{bmatrix} Y_{(0,0,0,0)} \\ Y_{(0,1,1,0)} \\ Y_{(1,0,1,0)} \\ Y_{(1,1,0,0)} \\ Y_{(0,0,1,1)} \\ Y_{(0,1,0,1)} \\ Y_{(1,0,0,1)} \\ Y_{(1,1,1,1)} \end{bmatrix} = \begin{bmatrix} 1 & -1 & -1 & -1 & 1 & 1 & 1 & -1 \\ 1 & -1 & 1 & 1 & -1 & -1 & 1 & -1 \\ 1 & 1 & -1 & 1 & -1 & 1 & -1 & -1 \\ 1 & 1 & 1 & -1 & 1 & -1 & -1 & -1 \\ 1 & -1 & -1 & 1 & 1 & -1 & -1 & 1 \\ 1 & -1 & 1 & -1 & -1 & 1 & -1 & 1 \\ 1 & 1 & -1 & -1 & -1 & -1 & 1 & 1 \\ 1 & 1 & 1 & 1 & 1 & 1 & 1 & 1 \end{bmatrix} \begin{bmatrix} \phi_1^0\phi_2^0\phi_3^0\phi_4^0 \\ \phi_1^1\phi_2^0\phi_3^0\phi_4^0 \\ \phi_1^0\phi_2^1\phi_3^0\phi_4^0 \\ \phi_1^0\phi_2^0\phi_3^1\phi_4^0 \\ \phi_1^1\phi_2^1\phi_3^0\phi_4^0 \\ \phi_1^1\phi_2^0\phi_3^1\phi_4^0 \\ \phi_1^0\phi_2^1\phi_3^1\phi_4^0 \\ \phi_1^1\phi_2^1\phi_3^1\phi_4^0 \end{bmatrix}$$

$$= \mathbf{X}_{\Gamma 1}\boldsymbol{\beta}_1.$$

Under the preceding model the BLUE of $\boldsymbol{\beta}_1$ is equal to

$$\hat{\boldsymbol{\beta}}_1 = \tfrac{1}{8}\mathbf{X}'_{\Gamma 1}\mathbf{Y}_{\Gamma}.$$

If, however, in $\boldsymbol{\beta}_3$, the main effect $\phi_1^0\phi_2^0\phi_3^0\phi_4^1$ due to blocks is not equal to zero, but the rest are, then

$$E[\hat{\boldsymbol{\beta}}_1] = \boldsymbol{\beta}_1 + \begin{bmatrix} 0 \\ 0 \\ 0 \\ 0 \\ 0 \\ 0 \\ 0 \\ \phi_1^0\phi_2^0\phi_3^0\phi_4^1 \end{bmatrix}.$$

Hence the incomplete block design Γ is a confounded design relative to $\phi_1^1\phi_2^1\phi_3^1\phi_4^0$ and $\phi_1^0\phi_2^0\phi_3^0\phi_4^1$ since the effect $\phi_1^1\phi_2^1\phi_3^1\phi_4^0$ is confounded with $\phi_1^0\phi_2^0\phi_3^0\phi_4^1$. On the other hand, Γ is an unconfounded design relative to the other effects in $\boldsymbol{\beta}_1$ and $\phi_1^0\phi_2^0\phi_3^0\phi_4^1$. Note that in the literature, Γ is referred to as an incomplete block design with the interaction "$F_1F_2F_3$ confounded with blocks." Finally, an exercise for the reader is to investigate confounding aspects of Γ when other effects in $\boldsymbol{\beta}_3$ are indeed not zero.

Remark. Let Γ be a factorial design of arbitrary resolution, and let the mean $\phi_1^0\phi_2^0\ldots\phi_t^0$ be included in $\boldsymbol{\beta}_1$ and written as its first element. Let $\widehat{\boldsymbol{\beta}_1}$ be the BLUE of $\boldsymbol{\beta}_1$ under the assumption that $\boldsymbol{\beta}_3 = \mathbf{0}$. If $\boldsymbol{\beta}_3 \neq \mathbf{0}$, then the first element of the vector $E[\hat{\boldsymbol{\beta}}_1]$ is referred to as the *generalized defining relationship*, and the whole vector is called the *aliasing structure* of Γ relative to $\boldsymbol{\beta}_1$, $\boldsymbol{\beta}_2$, and $\boldsymbol{\beta}_3$. Example 7.5 provides an illustration of these concepts. For further reading, the following relevant references are recommended.

7.3 SELECTED ADDITIONAL READING

Starred references are recommended for a first reading.

*1. Addelman, S. (1972). Recent developments in the design of factorial experiments. *J. Am. Stat. Assoc.*, **67**, 103–111.

2. Anderson, D. A., and Srivastava, J. N. (1972). Resolution IV designs of the $2^m \times 3$ series. *J. R. Stat. Soc.*, **B34**, 377–384.

3. Banerjee, A. K., and Das, M. N. (1969). On a method of construction of confounded asymmetrical factorial designs. *Calcutta Stat. Assoc. Bull.*, **18**, 163–177.

4. Banerjee, K. S., and Federer, W. T. (1966). On estimation and construction in fractional replication. *Ann. Math. Stat.*, **37**, 1033–1039.

*5. Box, G. E. P., and Hunter, J. S. (1961). The 2^{k-p} fractional factorial designs I. *Technometrics*, **3**, 311–351.

*6. Box, G. E. P., and Wilson, K. B. (1951). On the experimental attainment of optimum conditions. *J. R. Stat. Soc.*, **B13**, 1–45.

7. Burton, R. C., and Connor, W. S. (1957). On the identity relationship for fractional replicates of the 2^n series. *Ann. Math. Stat.*, **28**, 762–767.

8. DeGray, R. J. (1968). Design for interactions. *Technometrics*, **10**, 389–391.

9. Draper, N. R., and Mitchell, T. J. (1968). Construction of the set of 256-run designs of resolution ≥ 5 and the set of even 512-run designs of resolution ≥ 6 with special reference to the unique saturated designs. *Ann. Math. Stat.*, **39**, 246–255.

10. Federer, W. T., and Raktoe, B. L. (1965). General theory of prime-power lattice designs: lattice rectangles for $v = s^m$ treatments in s^r rows and s^c columns for $r + c = m$, $r \neq c$, and $v < 1000$. *J. Am. Stat. Assoc.*, **60**, 891–904.

11. Fisher, R. A. (1945). A system of confounding for factors with more than two alternatives giving completely orthogonal cubes and higher powers. *Ann. Eugen.*, **12**, 283–290.

12. Gulati, B. R. (1972). Construction of two-level saturated symmetrical factorial designs of resolution VI. *Ann. Math. Stat.*, **43**, 1652–1663.

13. Healy, M. J. R., and Gower, J. C. (1961). Aliasing in partially confounded factorial experiments. *Biometrika*, **48**, 218–220.

*14. Kempthorne, 0. (1947). A simple approach to confounding and fractional replication in factorial experiments. *Biometrika*, **34**, 255–272.

15. Kishen, K., and Srivastava, J. N. (1959). Mathematical theory of confounding in asymmetrical and symmetrical factorial designs. *J. Indian Soc. Agric. Stat.*, **11**, 73–110.

16. Margolin, B. H. (1967). Systematic methods for analyzing 2^n3^m factorial experiments with applications. *Technometrics*, **9**, 245–259.

*17. Margolin, B. H. (1968). Orthogonal main-effect 2^n3^m designs and two-factor interaction aliasing. *Technometrics*, **10**, 559–573.

18. Nair, K. R., and Rao, C. R. (1948). Confounding in asymmetrical factorial experiments. *J. R. Stat. Soc.*, **B10**, 109–131.

19. Pesotan, H., and Raktoe, B. L. (1975). Remarks on extensions of the Anderson-Federer (0,1)-matrix procedure for main effects to effects of a higher degree. *Commun. Stat.*, **4**, 797–811.

20. Pesotan, H., Raktoe, B. L., and Federer, W. T. (1975). On complexes of Abelian groups with applications to fractional factorial designs. *Ann. Inst. Stat. Math.*, **27**, 117–142.

*21. Raktoe, B. L. (1976). On alias matrices and generalized defining relationships of equi-information factorial arrangements. *J. R. Stat. Soc.*, **B38**, 279–283.

22. Raktoe, B. L., and Federer, W. T. (1970). A theorem on saturated plans and their complements. *Ann. Math. Stat.*, **41**, 2184–2185.

23. Raktoe, B. L., Pesotan, H., and Federer, W. T. (1980). On aliasing and generalized defining relationships of factorial arrangements. *Can. J. Stat.*, **8**, 65–77.

24. Raktoe, B. L., Rayner, A. A., and Chalton, D. O. (1978). On construction of confounded mixed factorial and lattice designs. *Aust. J. Stat.*, **20**, 209–218.

25. Srivastava, J. N. (1968). Analysis and construction of fractional and confounded factorial designs with emphasis on the asymmetrical case. In *Contributions in Statistics and Agricultural Science*, Indian Society of Agricultural Statistics. (Felicitation volume presented to Dr. V. G. Panse).

26. Webb, S. R. (1968a). Non-orthogonal designs of even resolutions. *Technometrics*, **10**, 291–299.

27. Webb, S. R. (1968b). Saturated sequential factorial designs. *Technometrics*, **10**, 535–550.

28. Williams, E. J. (1949). Confounding and fractional replication in factorial experiments. *Aust. J. Agric. Sci.*, **15**, 145–153.

*29. Yates, F. (1933). The principles of orthogonality and confounding in replicated experiments. *J. Agric. Sci.*, **23**, 108–145.

*30. Youden, W. G. (1961). Partial confounding in fractional replication. *Technometrics*, **3**, 353–358.

CHAPTER 8

On Orthogonality and Balancedness of Factorial Designs

This chapter is concerned with orthogonal and balanced fractional factorial designs. These types of designs arise when certain structures are imposed on the covariance matrix or the information matrix. The concepts are developed for the general partitioning of the parametric vector $\boldsymbol{\beta}_\rho$ and are then applied in some specific settings.

8.1 PURPOSE OF ORTHOGONALITY AND BALANCEDNESS

Imposing orthogonality and/or balancedness on factorial designs is sometimes justified in several ways:

(i) Many optimal designs in various settings turn out to be orthogonal and/or balanced.

(ii) The problem of finding an optimal design with respect to a given optimality criterion Q (see Chapter 5) among all nontrivial unbiased designs is in general intractable. One way to reduce the size of the class is to impose orthogonality (or balancedness) and search for an optimal design in the reduced class. This makes sense if the class of orthogonal (or balanced) designs is rich enough so that the restriction to this class does not cost too much in terms of the optimality criterion selected.

(iii) Even though at present one can handle very complex analyses with modern electronic computers, the two preceding concepts will help in reducing the number of steps in any kind of computing and in the verification that the optimality criterion is satisfied.

(iv) These two properties are desired by many experimenters, because orthogonal designs provide estimators that are uncorrelated and balanced designs give estimators that are of equal precision.

8.2 CONCEPTS OF ORTHOGONALITY AND BALANCEDNESS

Consider the partitioned parametric vector $\boldsymbol{\beta}_p$ of the previous chapter, that is,

$$\boldsymbol{\beta}_p' = \left(\boldsymbol{\beta}_1' \vdots \boldsymbol{\beta}_2' \vdots \boldsymbol{\beta}_3'\right),$$

where $\boldsymbol{\beta}_1$ is the $N_1 \times 1$ vector to be estimated, $\boldsymbol{\beta}_2$ is the $N_2 \times 1$ vector not to be estimated and not assumed to be known, and $\boldsymbol{\beta}_3$ is the $N_3 \times 1$ vector assumed to be known, which without loss of generality can be taken to be equal to zero.

Definition 8.1. An unbiased design Γ for the parametric vector $\boldsymbol{\beta}_1$ is said to be *orthogonal* if

$$\text{Cov}\left(\hat{\boldsymbol{\beta}}_1\right) = \sigma^2 \mathbf{V}, \tag{8.1}$$

where $\mathbf{V}$ is a diagonal matrix of order $\mathbf{N}_1$.

Note that Definition 8.1 implies that the information matrix $\mathbf{M} = \mathbf{V}^{-1}$ is diagonal. The definition of orthogonality could have been stated in terms of the information matrix. Since $\mathbf{V}$ is a diagonal matrix, it follows that orthogonal designs provide uncorrelated estimators for the elements of $\boldsymbol{\beta}_1$.

The notions of *variance balancedness* and *covariance balancedness* were introduced in Definition 5.5. From this definition, it follows that in our setting a design will be variance balanced if

$$\text{Cov}\left(\hat{\boldsymbol{\beta}}_1\right) = (u\mathbf{I} + \mathbf{W})\sigma^2 \tag{8.2}$$

and covariance balanced if

$$\text{Cov}\left(\hat{\boldsymbol{\beta}}_1\right) = (\mathbf{D} + h\mathbf{J}^*)\sigma^2, \tag{8.3}$$

where u and h are scalars, $\mathbf{W}$ is a matrix with zeros in the diagonal, $\mathbf{D}$ is a diagonal matrix, and $\mathbf{J}^*$ is a matrix with zeros in the diagonal and ones elsewhere.

The concept of variance balancedness can be generalized to the case where $\boldsymbol{\beta}_1$ is partitioned as $\boldsymbol{\beta}_1' = (\boldsymbol{\beta}_{11}' \vdots \boldsymbol{\beta}_{12}' \vdots \cdots \vdots \boldsymbol{\beta}_{1v}')$ so that the covariance matrix of $\hat{\boldsymbol{\beta}}_1$ is of the form

$$\text{Cov}(\hat{\boldsymbol{\beta}}_1) = (\mathbf{Z} + \mathbf{W})\sigma^2 \tag{8.4}$$

with $\mathbf{Z} = u_1\mathbf{I}_{(1)} \oplus u_2\mathbf{I}_{(2)} \oplus \cdots \oplus u_v\mathbf{I}_{(v)}$, $v < N_1$, where $\oplus$ denotes the usual direct sum operation, that is,

$$\mathbf{Z} = \begin{bmatrix} u_1\mathbf{I}_{(1)} & & 0 \\ & u_2\mathbf{I}_{(2)} & \\ 0 & & u_v\mathbf{I}_{(v)} \end{bmatrix}. \tag{8.5}$$

The objective of partitioning $\boldsymbol{\beta}_1$ into subvectors is to come up with a variance balanced design according to (8.4), so that elements in each subvector will have estimators of equal variance.

Similarly, a generalized form of covariance balancedness is given by

$$\begin{aligned} \text{Cov}(\hat{\boldsymbol{\beta}}_{1i}) &= (\mathbf{D}_i + c_{ii}\mathbf{J}_i^*)\sigma^2, \\ \text{Cov}(\hat{\boldsymbol{\beta}}_{1i}, \hat{\boldsymbol{\beta}}_{1j}) &= (c_{ij}\mathbf{J}_{ij})\sigma^2, \qquad i<j \end{aligned} \tag{8.6}$$

where as before $\mathbf{D}_i$ is a diagonal matrix, the c_{ij} are scalars, $\mathbf{J}_i^*$ is a matrix with zeros in the diagonal and ones elsewhere, and $\mathbf{J}_{ij}$ are matrices of all ones.

Example 8.1. Suppose $\boldsymbol{\beta}_1' = (\boldsymbol{\beta}_{11}' \vdots \boldsymbol{\beta}_{12}' \vdots \boldsymbol{\beta}_{13}')$, where $\boldsymbol{\beta}_{11}$ consists of a single parameter, $\boldsymbol{\beta}_{12}$ is a 2×1 vector, and $\boldsymbol{\beta}_{13}$ is a 3×1 vector. If a design Γ is such that

$$\text{Cov}(\hat{\boldsymbol{\beta}}_1) = \left[\begin{array}{c|cc|ccc} u_1 & c_{12} & c_{12} & c_{13} & c_{13} & c_{13} \\ \hline c_{12} & u_2 & c_{22} & c_{23} & c_{23} & c_{23} \\ c_{12} & c_{22} & u_2 & c_{23} & c_{23} & c_{23} \\ \hline c_{13} & c_{23} & c_{23} & u_3 & c_{33} & c_{33} \\ c_{13} & c_{23} & c_{23} & c_{33} & u_3 & c_{33} \\ c_{13} & c_{23} & c_{23} & c_{33} & c_{33} & u_3 \end{array}\right]\sigma^2,$$

then Γ is variance balanced in the general sense, since

$$\mathbf{Z} = \begin{bmatrix} u_1\mathbf{I}_1 & & \mathbf{0} \\ & u_2\mathbf{I}_2 & \\ \mathbf{0} & & u_3\mathbf{I}_3 \end{bmatrix},$$

which satisfies (8.5). The design Γ is also covariance balanced because it satisfies (8.6); that is,

$$\operatorname{Cov}(\hat{\boldsymbol{\beta}}_{11}) = u_1\sigma^2 = (\mathbf{D}_1 + c_{11}\mathbf{J}^*)\sigma^2, \qquad \text{where } c_{11} = 0$$

$$\operatorname{Cov}(\hat{\boldsymbol{\beta}}_{12}) = \begin{bmatrix} u_2 & c_{22} \\ c_{22} & u_2 \end{bmatrix}\sigma^2 = (\mathbf{D}_2 + c_{22}\mathbf{J}^*)\sigma^2,$$

$$\operatorname{Cov}(\hat{\boldsymbol{\beta}}_{13}) = \begin{bmatrix} u_3 & c_{33} & c_{33} \\ c_{33} & u_3 & c_{33} \\ c_{33} & c_{33} & u_3 \end{bmatrix}\sigma^2 = (\mathbf{D}_3 + c_{33}\mathbf{J}^*)\sigma^2,$$

$$\operatorname{Cov}(\hat{\boldsymbol{\beta}}_{11}, \hat{\boldsymbol{\beta}}_{12}) = [c_{12}\, c_{12}]\sigma^2 = (c_{12}\mathbf{J}_{12})\sigma^2,$$

$$\operatorname{Cov}(\hat{\boldsymbol{\beta}}_{11}, \hat{\boldsymbol{\beta}}_{13}) = [c_{13}\, c_{13}\, c_{13}]\sigma^2 = (c_{13}\mathbf{J}_{13})\sigma^2,$$

$$\operatorname{Cov}(\hat{\boldsymbol{\beta}}_{12}, \hat{\boldsymbol{\beta}}_{13}) = \begin{bmatrix} c_{23} & c_{23} & c_{23} \\ c_{23} & c_{23} & c_{23} \end{bmatrix}\sigma^2 = (c_{23}\mathbf{J}_{23})\sigma^2.$$

The partitioning of $\boldsymbol{\beta}_1$ into subvectors arises frequently in practice, since in many situations all parameters are not of equal importance. The experimenter is thus interested in estimating sets of parameters with different precisions. Aside from the variances, the covariances between estimators of parameters in a given subset are also taken to be the same. These considerations lead us to adopt the following definition of a balanced design.

Definition 8.2. A design Γ is said to be *completely balanced* relative to the vector $\boldsymbol{\beta}_1$ if it is variance balanced according to (8.2) and covariance balanced according to (8.3); Γ is *partially balanced* relative to the partitioned vector $\boldsymbol{\beta}_1' = (\boldsymbol{\beta}_{11}' \vdots \boldsymbol{\beta}_{12}' \vdots \dots \vdots \boldsymbol{\beta}_{1v}')$ if it is variance balanced according to (8.4) and covariance balanced according to (8.6).

Note that the covariance matrices of completely balanced and partially balanced designs, respectively, may be rewritten as

$$\operatorname{Cov}(\hat{\boldsymbol{\beta}}_1) = (a\mathbf{I}_{N_1} + b\mathbf{J}_{N_1})\sigma^2 \tag{8.7}$$

and

$$\begin{aligned} \operatorname{Cov}(\hat{\boldsymbol{\beta}}_{1i}) &= (a_i\mathbf{I}_{N_{1i}} + b_i\mathbf{J}_{N_{1i}})\sigma^2 \\ \operatorname{Cov}(\hat{\boldsymbol{\beta}}_{1i}, \hat{\boldsymbol{\beta}}_{1j}) &= (c_{ij}\mathbf{J}_{N_{1i}\times N_{1j}})\sigma^2, \qquad i \neq j, \end{aligned} \tag{8.8}$$

where the identity and **J**-matrices are of the dimensions indicated. Also, a partially balanced design Γ is a completely balanced design relative to each subvector $\boldsymbol{\beta}_{1i}$, $i = 1, 2, \dots, v$, $v < N_1$.

Many results have been obtained concerning unbiased designs that possess either the orthogonality property or the balancedness property or both. Let us now clarify the orthogonality and balancedness concepts in the four settings of Chapter 7.

8.3 ORTHOGONALITY AND BALANCEDNESS IN VARIOUS SETTINGS

Case 1. $N_2 = N_3 = 0$. Assuming the normalized orthogonal model $E[\mathbf{Y}_\rho] = \mathbf{X}_\rho \boldsymbol{\beta}_\rho$, where $\mathbf{X}'_\rho \mathbf{X}_\rho = \mathbf{I}_N$, it follows that a design Γ is an unbiased orthogonal design if and only if it contains ρ a fixed number of times. Such a design will be completely balanced, since $\text{Cov}(\hat{\boldsymbol{\beta}}_1) = \text{Cov}(\hat{\boldsymbol{\beta}}_\rho) = r^{-1}\mathbf{I}_N \sigma^2$, where r is the number of times ρ is contained in Γ. It will be seen in the following example that orthogonality and balancedness of a design in Case 1 are affected by unequal repetition of treatment combinations in Γ.

Example 8.2. Take the equispaced 2×3 factorial with the normalized orthogonal polynomial model

$$E\begin{bmatrix} Y_{(0,0)} \\ Y_{(0,1)} \\ Y_{(0,2)} \\ Y_{(1,0)} \\ Y_{(1,1)} \\ Y_{(1,2)} \end{bmatrix} = \begin{bmatrix} \frac{1}{\sqrt{6}} & \frac{-1}{2} & \frac{1}{\sqrt{12}} & \frac{-1}{\sqrt{6}} & \frac{1}{2} & \frac{-1}{\sqrt{12}} \\ \frac{1}{\sqrt{6}} & 0 & \frac{-2}{\sqrt{12}} & \frac{-1}{\sqrt{6}} & 0 & \frac{2}{\sqrt{12}} \\ \frac{1}{\sqrt{6}} & \frac{1}{2} & \frac{1}{\sqrt{12}} & \frac{-1}{\sqrt{6}} & \frac{-1}{2} & \frac{-1}{\sqrt{12}} \\ \frac{1}{\sqrt{6}} & \frac{-1}{2} & \frac{1}{\sqrt{12}} & \frac{1}{\sqrt{6}} & \frac{-1}{2} & \frac{1}{\sqrt{12}} \\ \frac{1}{\sqrt{6}} & 0 & \frac{-2}{\sqrt{12}} & \frac{1}{\sqrt{6}} & 0 & \frac{-2}{\sqrt{12}} \\ \frac{1}{\sqrt{6}} & \frac{1}{2} & \frac{1}{\sqrt{12}} & \frac{1}{\sqrt{6}} & \frac{1}{2} & \frac{1}{\sqrt{12}} \end{bmatrix} \begin{bmatrix} \phi_1^0\phi_2^0 \\ \phi_1^0\phi_2^1 \\ \phi_1^0\phi_2^2 \\ \phi_1^1\phi_2^0 \\ \phi_1^1\phi_2^1 \\ \phi_1^1\phi_2^2 \end{bmatrix} = \mathbf{X}_\rho \boldsymbol{\beta}_\rho .$$

If Γ contains r copies of ρ, then it is easily verified that $\mathbf{X}'_\Gamma \mathbf{X}_\Gamma = r\mathbf{I}_6$. Hence Γ is orthogonal and completely balanced. In equation (8.7), $a = r^{-1}$ and $b = 0$. If the nonnormalized orthogonal version of the above is adopted, then $\mathbf{X}'_\Gamma \mathbf{X}_\Gamma = \text{diagonal}\,(6,4,12,6,4,12)$, so that Γ will be orthogonal, but not completely balanced. If we had initially partitioned $\boldsymbol{\beta}'_1 = \boldsymbol{\beta}'_\rho$ as $\boldsymbol{\beta}'_{11} =$

$(\phi_1^0\phi_2^0, \phi_1^1\phi_2^0)$, $\boldsymbol{\beta}_{12}' = (\phi_1^0\phi_2^1, \phi_1^1\phi_2^1)$, and $\boldsymbol{\beta}_{13}' = (\phi_1^0\phi_2^2, \phi_1^1\phi_2^2)$, then under the nonnormalized orthogonal model Γ will be partially balanced. It is left to the reader to find the quantities in equation (8.8). Finally, if either in the normalized or nonnormalized orthogonal model treatment (0,0) in Γ is repeated twice and the rest of the treatments of ρ appear once in Γ, then Γ is neither orthogonal nor completely balanced for $\boldsymbol{\beta}_\rho$. This may be verified by calculating the information matrix $\mathbf{X}_\Gamma'\mathbf{X}_\Gamma$. The disappearance of orthogonality and complete balancedness is a consequence of unequal replication of the treatment combinations in Γ.

Case 2. $N_2 = 0$, $N_3 \neq 0$. Any unbiased orthogonal design Γ should satisfy in addition to unbiasedness, the condition

$$(\mathbf{X}_{\Gamma 1}'\mathbf{X}_{\Gamma 1})^{-1} = \mathbf{V}, \tag{8.9}$$

where $\mathbf{V}$ is a diagonal matrix. An unbiased completely balanced design satisfies, besides unbiasedness, the condition

$$(\mathbf{X}_{\Gamma 1}'\mathbf{X}_{\Gamma 1})^{-1} = a\mathbf{I}_{N_1} + b\mathbf{J}_{N_1}. \tag{8.10}$$

Example 8.3. Consider the 2^3 factorial with levels 0 and 1 for each factor and the model of Example 7.1. Let $\boldsymbol{\beta}_1$ be the vector consisting of the mean and the main effects, that is, $\boldsymbol{\beta}_1' = (\phi_1^0\phi_2^0\phi_3^0, \phi_1^1\phi_2^0\phi_3^0, \phi_1^0\phi_2^1\phi_3^0, \phi_1^0\phi_2^0\phi_3^1)$. The design $\Gamma = \{(1,0,0),(0,1,0),(0,0,1),(1,1,1)\}$ is an orthogonal and completely balanced design with $\mathbf{X}_{\Gamma 1}'\mathbf{X}_{\Gamma 1} = 4\mathbf{I}_4$. On the other hand, the design $\Gamma^* = \{(0,0,0),(1,0,0),(0,1,0),(0,0,1)\}$ is not orthogonal and not completely balanced. If the mean is taken by itself as a subvector and the main effects form the other subvector, then Γ^* is a partially balanced design with information matrix equal to

$$\mathbf{X}_{\Gamma^*1}'\mathbf{X}_{\Gamma^*1} = \begin{bmatrix} 4 & -2 & -2 & -2 \\ -2 & 4 & 0 & 0 \\ -2 & 0 & 4 & 0 \\ -2 & 0 & 0 & 4 \end{bmatrix} = \begin{bmatrix} 4 & \vdots & -2\mathbf{J} \\ \cdots & & \cdots \\ -2\mathbf{J}' & \vdots & 4\mathbf{I} \end{bmatrix}.$$

Hence the quantities in equation (8.8) are $a_1 = a_2 = 4$, $b_1 = b_2 = 0$, and $c_{12} = -2$.

Case 3. $N_2 \neq 0$, $N_3 \neq 0$. Apart from unbiasedness, the conditions for orthogonality and complete balancedness, respectively, are given by

$$[\mathbf{X}_{\Gamma 1}'\mathbf{X}_{\Gamma 1} - \mathbf{X}_{\Gamma 1}'\mathbf{X}_{\Gamma 2}(\mathbf{X}_{\Gamma 2}'\mathbf{X}_{\Gamma 2})^{-}\mathbf{X}_{\Gamma 2}'\mathbf{X}_{\Gamma 1}]^{-1} = \mathbf{V} \tag{8.11}$$

and

$$\text{(8.12)} \qquad \left[\mathbf{X}'_{\Gamma 1}\mathbf{X}_{\Gamma 1} - \mathbf{X}'_{\Gamma 1}\mathbf{X}_{\Gamma 2}(\mathbf{X}'_{\Gamma 2}\mathbf{X}_{\Gamma 2})^{-}\mathbf{X}'_{\Gamma 2}\mathbf{X}_{\Gamma 1}\right]^{-1} = a\mathbf{I}_{N_1} + b\mathbf{J}_{N_1},$$

where $\mathbf{V}$ is a diagonal matrix.

Example 8.4. Consider the 3×3 equispaced factorial with the levels coded as $0, 1, 2$ and the normalized orthogonal polynomial model of Example 6.5, that is,

$$E[\mathbf{Y}_\rho] = (\mathbf{X}_3 \otimes \mathbf{X}_3)\boldsymbol{\beta}_\rho = \mathbf{X}_\rho\boldsymbol{\beta}_\rho,$$

where

$$\mathbf{X}_3 = \begin{bmatrix} \frac{1}{\sqrt{3}} & \frac{-1}{\sqrt{2}} & \frac{1}{\sqrt{6}} \\ \frac{1}{\sqrt{3}} & 0 & \frac{-2}{\sqrt{6}} \\ \frac{1}{\sqrt{3}} & \frac{1}{\sqrt{2}} & \frac{1}{\sqrt{6}} \end{bmatrix}.$$

Let $\boldsymbol{\beta}'_1 = (\phi_1^1\phi_2^0, \phi_1^2\phi_2^0)$, $\boldsymbol{\beta}'_2 = (\phi_1^0\phi_2^0, \phi_1^0\phi_2^1, \phi_1^0\phi_2^2)$, and $\boldsymbol{\beta}'_3 = (\phi_1^1\phi_2^1, \phi_1^1\phi_2^2, \phi_1^2\phi_2^1, \phi_1^2\phi_2^2)$. Then the design $\Gamma = \{(0,0),(1,0),(2,0)\}$ is a minimal unbiased, orthogonal, and completely balanced design, since

$$\mathbf{X}_{\Gamma 1} = \begin{bmatrix} \frac{-1}{\sqrt{6}} & \frac{1}{3\sqrt{2}} \\ 0 & \frac{-2}{3\sqrt{2}} \\ \frac{1}{\sqrt{6}} & \frac{1}{3\sqrt{2}} \end{bmatrix}, \qquad \mathbf{X}_{\Gamma 2} = \begin{bmatrix} \frac{1}{3} & \frac{-1}{\sqrt{6}} & \frac{1}{3\sqrt{2}} \\ \frac{1}{3} & \frac{-1}{\sqrt{6}} & \frac{1}{3\sqrt{2}} \\ \frac{1}{3} & \frac{-1}{\sqrt{6}} & \frac{1}{3\sqrt{2}} \end{bmatrix},$$

where $\mathbf{X}'_{\Gamma 1}\mathbf{X}_{\Gamma 1} = \frac{1}{3}\mathbf{I}_3$ and $\mathbf{X}'_{\Gamma 1}\mathbf{X}_{\Gamma 2} = \mathbf{0}$.

Case 4. $N_2 \neq 0$, $N_3 = 0$. A design Γ in this final case must satisfy along with unbiasedness, the conditions

$$\text{(8.13)} \qquad \left[\mathbf{X}'_{\Gamma 1}\mathbf{X}_{\Gamma 1} - \mathbf{X}'_{\Gamma 1}\mathbf{X}_{\Gamma 2}(\mathbf{X}'_{\Gamma 2}\mathbf{X}_{\Gamma 2})^{-}\mathbf{X}'_{\Gamma 2}\mathbf{X}_{\Gamma 1}\right]^{-1} = \mathbf{V},$$

$\mathbf{V}$ being an orthogonal matrix, and

$$(8.14) \qquad \left[\mathbf{X}'_{\Gamma 1}\mathbf{X}_{\Gamma 1} - \mathbf{X}'_{\Gamma 1}\mathbf{X}_{\Gamma 2}(\mathbf{X}'_{\Gamma 2}\mathbf{X}_{\Gamma 2})^{-}\mathbf{X}'_{\Gamma 2}\mathbf{X}_{\Gamma 1}\right]^{-1} = a\mathbf{I}_{N_1} + b\mathbf{J}_{N_1}$$

for orthogonality and complete balancedness, respectively.

Note that in both Cases 3 and 4 when $\mathbf{X}'_{\Gamma 1}\mathbf{X}_{\Gamma 2} = \mathbf{0}$, then the conditions simplify to

$$(8.15) \qquad (\mathbf{X}'_{\Gamma 1}\mathbf{X}_{\Gamma 1})^{-1} = \mathbf{V}$$

and

$$(8.16) \qquad (\mathbf{X}'_{\Gamma 1}\mathbf{X}_{\Gamma 1})^{-1} = a\mathbf{I}_{N_1} + b\mathbf{J}_{N_1},$$

respectively.

Example 8.5. The design $\Gamma = \{(0,0), (1,0), (2,0), (0,2), (1,2), (2,2), (0,1), (2,1)\}$ for $\boldsymbol{\beta}'_1 = (\phi_1^1\phi_2^0, \phi_1^0\phi_2^1, \phi_1^1\phi_2^2, \phi_1^2\phi_2^1)$ and $\boldsymbol{\beta}_2$ being the vector of remaining parameters in the 3×3 equispaced factorial has already been shown in Example 7.3 to be an orthogonal design. The reader is invited to investigate its balancedness property.

A special instance of Case 2 is discussed in depth in Chapter 10, where construction and analysis of resolution III designs are considered. The reading list at the end of this chapter contains many examples of all the four cases.

Finally, there is a relationship between *Hadamard matrices* and orthogonal resolution III designs of the 2^n factorial and also between balanced fractional factorial designs and other combinatorial structures, such as balanced incomplete block designs. For the benefit of the reader, several references are provided in the following list. A further exploration of relationships is provided in Chapter 12 and the references given therein.

8.4 SELECTED ADDITIONAL READING

Starred references are recommended for a first reading.

*1. Addelman, S. (1962). Orthogonal main-effect plans for asymmetrical factorial experiments. *Technometrics*, **4**, 21–46.

2. Agrawal, H. (1966). A method of construction of three factor balanced designs. *J. Indian Stat. Assoc.*, **4**, 10–13.

3. Anderson, D. A. (1972). Designs with partial factorial balance. *Ann. Math. Stat.*, **43**, 1333–1341.

4. Anderson, D. A., and Federer, W. T. (1976). Multidimensional balanced designs. *Commun. Stat.*, **A5**, 1193–1204.

5. Atiqullah, M. (1961). On a property of balanced designs. *Biometrika*, **48**, 215–218.

6. Banerjee, K. S. (1950). How balanced incomplete block designs may be made to furnish orthogonal estimates in weighing designs. *Biometrika*, **37**, 50–58.

7. Dykstra, O., Jr. (1966). The orthogonalization of undesigned experiments. *Technometrics*, **8**, 279–290. (Correction, **8**, 731.)

*8. Eccleston, J., and Russell, K. (1975). Connectedness and orthogonality in multifactor designs. *Biometrika*, **62**, 341–345.

9. Efron, B. (1971). Forcing a sequential experiment to be balanced. *Biometrika*, **58**, 403–417.

10. Geramita, A. V., and Seberry, J. (1979). *Orthogonal Designs*. Marcel Dekker, New York.

11. Hadamard, J. (1873). Resolution d'une question relative aux determinants. *Bull. Sci. Math., Astr.* **(1)2,** 240–246.

12. Hall, M., Jr. (1967). *Combinatorial Theory*. Blaisdell, Waltham, Mass.

*13. Hedayat, A., and Federer, W. T. (1974). Pairwise and variance balanced incomplete block designs. *Ann. Inst. Stat. Math.*, **26**, 331–338.

14. Hedayat, A., and Wallis, W. D. (1978). Hadamard matrices and their applications. *Ann. Stat.*, **6**, 1184–1238.

15. John, J. A., and Smith, T. M. F. (1972). Two factorial experiments in non-orthogonal designs. *J. R. Stat. Soc.*, **B43**, 401–409.

16. John, P. W. M. (1969). Some non-orthogonal fractions of 2^n designs. *J. R. Stat. Soc.*, **B31**, 270–275.

17. Kishen, K., and Tyagi, B. N. (1964a). On the construction and analysis of some balanced asymmetrical factorial designs. *Calcutta Stat. Assoc. Bull.*, **13**, 123–149.

18. Kishen, K., and Tyagi, B. N. (1964b). Partially balanced asymmetrical factorial designs. In *Contributions to Statistics*, C. R. Rao, ed. Pergamon, New York and Statistical Publishing Company, Calcutta, pp. 147–158.

*19. Kshirsagar, A. M. (1966). Balanced factorial designs. *J. R. Stat. Soc.*, **B28**, 559–567.

20. Margolin, B. H. (1968). Orthogonal main-effect 2^n3^m designs and two-factor interaction aliasing. *Technometrics*, **10**, 559–573.

*21. Margolin, B. H. (1969). Orthogonal main-effect plans permitting estimation of all two-factor interactions for the 2^n3^m factorial series of design. *Technometrics*, **11**, 747–762.

22. Margolin, B. H. (1972). Non-orthogonal main effect designs for asymmetrical factorial experiments. *J. R. Stat. Soc.*, **B34**, 431–440.

23. Muller, E. R. (1966). Balanced confounding of factorial experiments. *Biometrika*, **53**, 507–524.

24. Paley, R. E. A. C. (1933). On orthogonal matrices. *J. Math. Phys.*, **12**, 311–320.

*25. Preece, D. A. (1977). Orthogonality and designs: a terminological muddle. *Util. Math.*, **12**, 201–223.

*26. Puri, P. D., and Nigam, A. K. (1976). Balanced factorial experiments. I. *Commun. Stat.*, **A5**, 599–620.

27. Raghavarao, D. (1959). Some optimum weighing designs. *Ann. Math. Stat.*, **30**, 295–303.

*28. Raktoe, B. L., and Federer, W. T. (1973). Balanced optimal saturated main effect plans of the 2^n factorial and their relation to (v, k, λ) configurations. *Ann. Stat.*, **1**, 924–932.

29. Saha, G. M., and Mohanty, S. (1970). On non-orthogonal main effect plans for asymmetrical factorials. *Ann. Inst. Stat. Math.*, **22**, 159–169.

30. Shah, B. V. (1958). On balancing in factorial experiments. *Ann. Math. Stat.*, **29**, 766–779. [Corrections, **30**, (1959), 1267.]

31. Shah, B. V. (1959). A note on orthogonality of experimental designs. *Calcutta Stat. Assoc. Bull.*, **8**, 73–80.

*32. Shah, B. V. (1960). Balanced factorial experiments. *Ann. Math. Stat.*, **31**, 502–514.

33. Srivastava, J. N. (1970). Optimal balanced 2^m fractional factorial designs. In *S. N. Roy Memorial Volume*, University of North Carolina, Chapel Hill and Indian Statistical Institute, New Delhi, pp. 689–706.

*34. Srivastava, J. N., and Anderson, D. A. (1970). Optimal fractional factorial plans for main effects orthogonal to two-factor interactions: 2^m series. *J. Am. Stat. Assoc.*, **65**, 828–843.

35. Srivastava, J. N., and Anderson, D. A. (1971). Factorial association schemes with applications to the construction of multidimensional partially balanced designs. *Ann. Math. Stat.*, **42**, 1167–1181.

36. Starks, T. H. (1964). A note on small orthogonal main effect plans for factorial experiments. *Technometrics*, **6**, 220–222.

37. Yamamoto, S., Shirakura, T., and Kuwada, M. (1975). Balanced arrays of strength $2l$ and balanced fractional 2^m factorial designs. *Ann. Inst. Stat. Math.*, **27**, 143–157.

*38. Yates, F. (1933). The principles of orthogonality and confounding in replicated experiments. *J. Agric. Sci.*, **23**, 108–145.

CHAPTER 9

Randomized Factorial Designs and Regular Factorial Designs

This chapter introduces another technique that leads to unbiased estimation of a set of linear parametric functions of $\boldsymbol{\beta}_1$, if the total parametric vector is partitioned as $\boldsymbol{\beta}_\rho' = (\boldsymbol{\beta}_1' \vdots \boldsymbol{\beta}_2')$, where some components of $\boldsymbol{\beta}_2$ may or may not be equal to zero. Note that this partitioning of $\boldsymbol{\beta}_\rho$ combines Cases 3 and 4 of the partitioning (7.1) in Chapter 7. It is shown that under certain conditions on $\boldsymbol{\beta}_1$ and under a particular model, there always exists a class of factorial designs with the properties that: (i) the design matrix associated with $\boldsymbol{\beta}_1$ for each member of this class has rank N_1; (ii) if we select a design from this class with equal probability, then the selected design, referred to as a *randomized factorial design*, can be utilized to estimate $\boldsymbol{\beta}_1$ unbiasedly. The concept of randomized designs and the related estimation aspects of linear parametric functions of $\boldsymbol{\beta}_1$ are formally introduced and treated in this chapter.

Apart from the concept of randomized factorial designs, the notion of *regular factorial designs* of a prime powered factorial is introduced and illustrated. A class of regular factorial designs is obtained to which the theory of randomized factorial designs can then be applied.

9.1 THE GENERAL PROBLEM OF RANDOMIZED FACTORIAL DESIGNS

Let $\Delta = \{\Gamma_1, \Gamma_2, \ldots, \Gamma_s\}$ be a class of designs, each consisting of n treatment combinations. Suppose that an experimenter may run any of the designs, but that $\mathbf{L}_1'\boldsymbol{\beta}_1$ is not estimable under any of these designs under the model

$$(9.1) \qquad E[\mathbf{Y}_{\Gamma_i}] = \mathbf{X}_{\Gamma_i 1}\boldsymbol{\beta}_1 + \mathbf{X}_{\Gamma_i 2}\boldsymbol{\beta}_2, \qquad \operatorname{Cov}(\mathbf{Y}_{\Gamma_i}) = \sigma^2 \mathbf{I}_n.$$

We now show that if the class of available designs satisfies certain conditions and if the experimenter is willing to select his design in a randomized manner, then there is a solution to the preceding problem.

Suppose that the class of available designs Δ has the property that there exists a probability vector $\boldsymbol{\xi}' = (\xi_1, \xi_2, \ldots, \xi_s)$, $\sum_{i=1}^{s} \xi_i = 1$, such that

$$(9.2) \qquad \xi_1 \mathbf{A}_{\Gamma_1} + \xi_2 \mathbf{A}_{\Gamma_2} + \cdots + \xi_s \mathbf{A}_{\Gamma_s} = \mathbf{0},$$

where

$$\mathbf{A}_{\Gamma_i} = \mathbf{L}_1'(\mathbf{X}_{\Gamma_i 1}'\mathbf{X}_{\Gamma_i 1})^{-} \mathbf{X}_{\Gamma_i 1}'\mathbf{X}_{\Gamma_i 2}.$$

Then if the experimenter selects his design according to $\boldsymbol{\xi}$, that is, if Γ_i has probability ξ_i of being selected, there is an unbiased estimator of $\mathbf{L}_1'\boldsymbol{\beta}_1$, that is,

$$(9.3) \qquad \widehat{\mathbf{L}_1'\boldsymbol{\beta}_1} = \mathbf{L}_1'(\mathbf{X}_{\Gamma_i 1}'\mathbf{X}_{\Gamma_i 1})^{-} \mathbf{X}_{\Gamma_i}'\mathbf{Y}_{\Gamma_i}.$$

This is so, since

$$(9.4) \qquad \begin{aligned} E_{\boldsymbol{\xi}}[E(\widehat{\mathbf{L}_1'\boldsymbol{\beta}_1})] &= E_{\boldsymbol{\xi}}[\mathbf{L}_1'\boldsymbol{\beta}_1 + \mathbf{A}_{\Gamma_i}\boldsymbol{\beta}_2] \\ &= \mathbf{L}_1'\boldsymbol{\beta}_1 + \sum_{i=1}^{s} \xi_i \mathbf{A}_{\Gamma_i}\boldsymbol{\beta}_2 \\ &= \mathbf{L}_1'\boldsymbol{\beta}_1 + \mathbf{0} = \mathbf{L}_1'\boldsymbol{\beta}_1. \end{aligned}$$

Note that implicit in this result is the fact that $\mathbf{L}_1'$ must be in the row space of each $\mathbf{X}_{\Gamma_i 1}$.

In general, the search for the existence or nonexistence of the probability vector $\boldsymbol{\xi}$, which satisfies equation (9.2), for a given class of designs is a difficult task. The literature on this topic is mostly oriented towards a uniform probability vector, that is, each $\xi_i = s^{-1}$, and certain classes of designs and types of linear parametric functions. We state formally the following definition.

Definition 9.1. A design, with the objective to estimate $\mathbf{L}_1'\boldsymbol{\beta}_1$, which is selected from the class $\Delta = \{\Gamma_1, \Gamma_2, \ldots, \Gamma_s\}$ such that Γ_i has probability ξ_i of being selected is called a *randomized design* relative to $\boldsymbol{\xi}$. If $\xi_i = s^{-1}$, then it is referred to as a *uniform randomized design*. If $\sum_{i=1}^{s} \xi_i \mathbf{A}_{\Gamma_i} = \mathbf{0}$, then it is called an *unbiased randomized design* relative to $\boldsymbol{\xi}$.

Example 9.1. Consider the 2^2 factorial with the following model for the minimal complete factorial design ρ:

$$E\begin{bmatrix} Y_{(0,0)} \\ Y_{(1,0)} \\ Y_{(0,1)} \\ Y_{(1,1)} \end{bmatrix} = \begin{bmatrix} 1 & -1 & -1 & 1 \\ 1 & 1 & -1 & -1 \\ 1 & -1 & 1 & -1 \\ 1 & 1 & 1 & 1 \end{bmatrix}\begin{bmatrix} \phi_1^0\phi_2^0 \\ \phi_1^1\phi_2^0 \\ \phi_1^0\phi_2^1 \\ \phi_1^1\phi_2^1 \end{bmatrix} = \mathbf{X}_\rho\boldsymbol{\beta}_\rho.$$

Let $\boldsymbol{\beta}_1' = (\phi_1^0\phi_2^0, \phi_1^1\phi_2^0)$ and $\boldsymbol{\beta}_2' = (\phi_1^0\phi_2^1, \phi_1^1\phi_2^1)$ and consider the class of designs $\Delta = \{\Gamma_1, \Gamma_2, \Gamma_3, \Gamma_4\}$ for estimating $\mathbf{L}_1'\boldsymbol{\beta}_1 = \mathbf{I}\boldsymbol{\beta}_1 = \boldsymbol{\beta}_1$, where

$$\Gamma_1 = \{(0,0),(1,0)\},$$

$$\Gamma_2 = \{(1,1),(0,1)\},$$

$$\Gamma_3 = \{(0,0),(1,1)\},$$

$$\Gamma_4 = \{(1,0),(0,1)\}.$$

The relevant matrices required to show that $\xi_1\mathbf{A}_{\Gamma_1} + \xi_2\mathbf{A}_{\Gamma_2} + \xi_3\mathbf{A}_{\Gamma_3} + \xi_4\mathbf{A}_{\Gamma_4}$ is zero for the various $\boldsymbol{\xi}$ are

	$\mathbf{X}_{\Gamma 1}$	$(\mathbf{X}'_{\Gamma 1}\mathbf{X}_{\Gamma 1})^{-1}$	$(\mathbf{X}'_{\Gamma 1}\mathbf{X}_{\Gamma 1})^{-1}\mathbf{X}'_{\Gamma 1}$	$\mathbf{X}_{\Gamma 2}$	$(\mathbf{X}'_{\Gamma 1}\mathbf{X}_{\Gamma 1})^{-1}\mathbf{X}'_{\Gamma 1}\mathbf{X}_{\Gamma 2}$
Γ_1:	$\begin{bmatrix} 1 & -1 \\ 1 & 1 \end{bmatrix}$	$\begin{bmatrix} 2 & 0 \\ 0 & 2 \end{bmatrix}$	$\frac{1}{2}\begin{bmatrix} 1 & 1 \\ -1 & 1 \end{bmatrix}$	$\begin{bmatrix} -1 & 1 \\ -1 & -1 \end{bmatrix}$	$\begin{bmatrix} -1 & 0 \\ 0 & -1 \end{bmatrix}$
Γ_2:	$\begin{bmatrix} 1 & 1 \\ 1 & -1 \end{bmatrix}$	$\begin{bmatrix} 2 & 0 \\ 0 & 2 \end{bmatrix}$	$\frac{1}{2}\begin{bmatrix} 1 & 1 \\ 1 & -1 \end{bmatrix}$	$\begin{bmatrix} 1 & 1 \\ 1 & -1 \end{bmatrix}$	$\begin{bmatrix} 1 & 0 \\ 0 & 1 \end{bmatrix}$
Γ_3:	$\begin{bmatrix} 1 & -1 \\ 1 & 1 \end{bmatrix}$	$\begin{bmatrix} 2 & 0 \\ 0 & 2 \end{bmatrix}$	$\frac{1}{2}\begin{bmatrix} 1 & 1 \\ -1 & 1 \end{bmatrix}$	$\begin{bmatrix} -1 & 1 \\ 1 & 1 \end{bmatrix}$	$\begin{bmatrix} 0 & 1 \\ 1 & 0 \end{bmatrix}$
Γ_4:	$\begin{bmatrix} 1 & 1 \\ 1 & -1 \end{bmatrix}$	$\begin{bmatrix} 2 & 0 \\ 0 & 2 \end{bmatrix}$	$\frac{1}{2}\begin{bmatrix} 1 & 1 \\ 1 & -1 \end{bmatrix}$	$\begin{bmatrix} -1 & -1 \\ 1 & -1 \end{bmatrix}$	$\begin{bmatrix} 0 & -1 \\ -1 & 0 \end{bmatrix}$

There are numerous $\boldsymbol{\xi}$ such that (9.4) is satisfied. As long as $\xi_1 = \xi_2$ and $\xi_3 = \xi_4$, the resulting randomized design will lead to unbiased estimation of $\boldsymbol{\beta}_1$. The reader may observe later that we could have started out with a smaller class, for example, $\{\Gamma_1, \Gamma_2\}$ or $\{\Gamma_3, \Gamma_4\}$. The first class arises essentially when $\xi_3 = \xi_4 = 0$ and the second one, when $\xi_1 = \xi_2 = 0$. If one of these classes is considered by itself then the only unbiased randomized design is a uniform type of design; that is, for the first class by itself $\xi_1 = \xi_2 = \frac{1}{2}$ and for the second class by itself $\xi_3 = \xi_4 = \frac{1}{2}$. It is seen in the next section that the two classes $\{\Gamma_1, \Gamma_2\}$ and $\{\Gamma_3, \Gamma_4\}$ are each generated by level permutations.

For a given vector of linear parametric functions $\mathbf{L}_1'\boldsymbol{\beta}_1$, it is in general a difficult problem to determine a *minimal class* of designs, which provides an unbiased randomized design for $\mathbf{L}_1'\boldsymbol{\beta}_1$. It is clear that the number of designs in such a class must be equal to two. In Example 9.1 the classes $\{\Gamma_1, \Gamma_2\}$ and $\{\Gamma_3, \Gamma_4\}$ are both minimal for unbiased randomized estimation of $\boldsymbol{\beta}_1' = (\phi_1^0\phi_2^0, \phi_1^1\phi_2^0)$.

In the following development we construct a class of designs that for a certain type of $\boldsymbol{\beta}_1$ in the partitioned vector $\boldsymbol{\beta}_\rho' = (\boldsymbol{\beta}_1' \vdots \boldsymbol{\beta}_2')$ and under a certain model, leads to a uniform unbiased randomized design for $\boldsymbol{\beta}_1$.

9.2 ADMISSIBLE PARAMETRIC VECTORS AND PERMUTED DESIGNS

We adopt the following model throughout the following sections:

$$E[\mathbf{Y}_\rho] = \mathbf{X}_\rho\boldsymbol{\beta}_\rho; \qquad \operatorname{Cov}[\mathbf{Y}_\rho] = \sigma^2\mathbf{I}_N, \tag{9.5}$$

where $\mathbf{X}_\rho = \mathbf{X}_1 \otimes \mathbf{X}_2 \otimes \cdots \otimes \mathbf{X}_t$, $\mathbf{X}_i$ being a real orthonormal $k_i \times k_i$ matrix with first column having the same entry so that the sum of all terms in the other columns is equal to zero. Such matrices arise in the orthonormal polynomial model, which was introduced and discussed in Chapter 4. The vector $\boldsymbol{\beta}_\rho$ consists of the mean $\phi_1^0\phi_2^0 \cdots \phi_t^0$, the main effects of the factors and the two, and higher order interaction effects.

Let parametric vector $\boldsymbol{\beta}_\rho$ be partitioned as

$$\boldsymbol{\beta}_\rho' = \left(\boldsymbol{\beta}_1' \vdots \boldsymbol{\beta}_2'\right), \tag{9.6}$$

where $\boldsymbol{\beta}_1$ is an $N_1 \times 1$ vector and $\boldsymbol{\beta}_2$ is an $N_2\ (= N - N_1) \times 1$ vector such that $N_1 > 0$. The vector $\boldsymbol{\beta}_1$ is not assumed to be zero and $\boldsymbol{\beta}_2$ may or may not contain components that are equal to zero.

Let Γ be any design, then its model relative to $\boldsymbol{\beta}_\rho' = (\boldsymbol{\beta}_1' \vdots \boldsymbol{\beta}_2')$ is

$$E[\mathbf{Y}_\Gamma] = \left[\mathbf{X}_{\Gamma 1} \vdots \mathbf{X}_{\Gamma 2}\right]\begin{bmatrix} \boldsymbol{\beta}_1 \\ \cdots \\ \boldsymbol{\beta}_2 \end{bmatrix}, \qquad \operatorname{Cov}(\mathbf{Y}_\Gamma) = \sigma^2\mathbf{I}_n. \tag{9.7}$$

If the rank of $\mathbf{X}_{\Gamma 1}$ is N_1, then we know that the estimator

$$\hat{\boldsymbol{\beta}}_1 = (\mathbf{X}_{\Gamma 1}'\mathbf{X}_{\Gamma 1})^{-1}\mathbf{X}_{\Gamma 1}'\mathbf{Y}_\Gamma \tag{9.8}$$

is a biased estimator of $\boldsymbol{\beta}_1$, which has expectation

$$E\left[\hat{\boldsymbol{\beta}}_1\right] = \boldsymbol{\beta}_1 + (\mathbf{X}_{\Gamma 1}'\mathbf{X}_{\Gamma 1})^{-1}\mathbf{X}_{\Gamma 1}'\mathbf{X}_{\Gamma 2}\boldsymbol{\beta}_2 = \boldsymbol{\beta}_1 + \mathbf{A}_\Gamma\boldsymbol{\beta}_2. \tag{9.9}$$

We will need the following definition in the sequel.

Definition 9.2. A subvector $\boldsymbol{\beta}_1$ is said to be *admissible* if and only if whenever $\phi_1^{i_1}\phi_2^{i_2}\cdots\phi_t^{i_t}$ belongs to $\boldsymbol{\beta}_1$ and $i_j\neq 0$ $(1\le j\le t)$, then $\phi_1^{i_1}\phi_2^{i_2}\cdots\phi_{j-1}^{i_{j-1}}\phi_j^{u}\phi_{j+1}^{i_{j+1}}\cdots\phi_t^{i_t}$ belongs to $\boldsymbol{\beta}_1$ for all $u\neq 0$.

From Definition 9.2 it is clear that admissibility of a subvector $\boldsymbol{\beta}_1$ implies that if a particular main effect for a factor belongs to it, then all main effects for that factor belong to it, and if a particular interaction effect of a set of factors belong to it then all interaction effects of that set belong to it. It also follows that if $\boldsymbol{\beta}_1$ is admissible, then $\boldsymbol{\beta}_2$ in the partitioning (9.7) is also admissible. Before proceeding further let us illustrate the concept of admissibility with an example.

Example 9.2. Consider the 2×3 factorial with the orthonormal polynomial model $E[\mathbf{Y}_\rho]=[\mathbf{X}_1\otimes\mathbf{X}_2]\boldsymbol{\beta}_\rho$, $\text{Cov}(\mathbf{Y}_\rho)=\sigma^2\mathbf{I}_6$, that is,

$$\mathbf{X}_1=\begin{bmatrix}\frac{1}{\sqrt{2}} & \frac{-1}{\sqrt{2}}\\ \frac{1}{\sqrt{2}} & \frac{1}{\sqrt{2}}\end{bmatrix},\qquad \mathbf{X}_2=\begin{bmatrix}\frac{1}{\sqrt{3}} & \frac{-1}{\sqrt{2}} & \frac{1}{\sqrt{6}}\\ \frac{1}{\sqrt{3}} & 0 & \frac{-2}{\sqrt{6}}\\ \frac{1}{\sqrt{3}} & \frac{1}{\sqrt{2}} & \frac{1}{\sqrt{6}}\end{bmatrix},$$

so that

$$E\begin{bmatrix}Y_{(0,0)}\\ Y_{(0,1)}\\ Y_{(0,2)}\\ Y_{(1,0)}\\ Y_{(1,1)}\\ Y_{(1,2)}\end{bmatrix}=\begin{bmatrix}\frac{1}{\sqrt{6}} & \frac{-1}{2} & \frac{1}{\sqrt{12}} & \frac{-1}{\sqrt{6}} & \frac{1}{2} & \frac{-1}{\sqrt{12}}\\ \frac{1}{\sqrt{6}} & 0 & \frac{-2}{\sqrt{12}} & \frac{-1}{\sqrt{6}} & 0 & \frac{2}{\sqrt{12}}\\ \frac{1}{\sqrt{6}} & \frac{1}{2} & \frac{1}{\sqrt{12}} & \frac{-1}{\sqrt{6}} & \frac{-1}{2} & \frac{-1}{\sqrt{12}}\\ \frac{1}{\sqrt{6}} & \frac{-1}{2} & \frac{1}{\sqrt{12}} & \frac{1}{\sqrt{6}} & \frac{-1}{2} & \frac{1}{\sqrt{12}}\\ \frac{1}{\sqrt{6}} & 0 & \frac{-2}{\sqrt{12}} & \frac{1}{\sqrt{6}} & 0 & \frac{-2}{\sqrt{12}}\\ \frac{1}{\sqrt{6}} & \frac{1}{2} & \frac{1}{\sqrt{12}} & \frac{1}{\sqrt{6}} & \frac{1}{2} & \frac{1}{\sqrt{12}}\end{bmatrix}\begin{bmatrix}\phi_1^0\phi_2^0\\ \phi_1^0\phi_2^1\\ \phi_1^0\phi_2^2\\ \phi_1^1\phi_2^0\\ \phi_1^1\phi_2^1\\ \phi_1^1\phi_2^2\end{bmatrix}$$

$$=\mathbf{X}_\rho\boldsymbol{\beta}_\rho.$$

First of all, notice that the mean $\phi_1^0\phi_2^0$ and the whole parametric vector $\boldsymbol{\beta}_p$ are admissible. The subvector $\boldsymbol{\beta}_1^{*\prime}=(\phi_1^1\phi_2^0,\phi_1^1\phi_2^1)$ is not admissible. However, $\boldsymbol{\beta}_1'=(\phi_1^1\phi_2^0,\phi_1^1\phi_2^1,\phi_1^1\phi_2^2)$ is admissible, since all main effects (there is only one!) for the first factor and all interactions (there are two) between the two factors belong to it.

Assume that the levels of the ith factor are always labeled $0,1,2,\ldots,k_i-1$ and denote the resulting set of levels by G_i, $i=1,2,\ldots,t$. Let $\Omega_i=\{\omega_{i1},\omega_{i2},\ldots,\omega_{ik_i!}\}$ be the set of all permutations on G_i and let Ω be the set of all permutations of the form $\Omega=X_{i=1}^t\Omega_i=\{\omega:\omega=(\omega_1,\omega_2,\ldots,\omega_t),\ \omega_i\in\Omega_i,\ i=1,2,\ldots,t\}$. Notice that Ω is simply the set of *level permutations*.

Definition 9.3. For a permutation $\omega\in\Omega$ and a design Γ, we define the *permuted design* $\omega(\Gamma)$ as the collection of treatments obtained by the action of ω on each treatment in Γ.

We may describe the permuted design $\omega(\Gamma)$ explicitly as $\omega(\Gamma)=\{(\omega_1(x_1),\omega_2(x_2),\ldots,\omega_t(x_t)):(x_1,x_2,\ldots,x_t)\in\Gamma\}$. The models for the design Γ and the permuted design $\omega(\Gamma)$ are obtained from equation (9.7) and are equal, respectively, to

$$E[\mathbf{Y}_\Gamma]=\left[\mathbf{X}_{\Gamma1}\vdots\mathbf{X}_{\Gamma2}\right]\begin{bmatrix}\boldsymbol{\beta}_1\\ \cdots\\ \boldsymbol{\beta}_2\end{bmatrix},\qquad \mathrm{Cov}(\mathbf{Y}_\Gamma)=\sigma^2\mathbf{I}_n, \tag{9.10}$$

and

$$E\left[\mathbf{Y}_{\omega(\Gamma)}\right]=\left[\mathbf{X}_{\omega(\Gamma)1}\vdots\mathbf{X}_{\omega(\Gamma)2}\right]\begin{bmatrix}\boldsymbol{\beta}_1\\ \cdots\\ \boldsymbol{\beta}_2\end{bmatrix},\qquad \mathrm{Cov}\left(\mathbf{Y}_{\omega(\Gamma)}\right)=\sigma^2\mathbf{I}_n. \tag{9.11}$$

It may be shown in a straightforward but rather laborious manner that when $\boldsymbol{\beta}_1$ is an admissible vector, then

$$\begin{aligned}\mathbf{X}_{\omega(\Gamma)1}&=\mathbf{X}_{\Gamma1}\mathbf{P}_{\omega1},\\ \mathbf{X}_{\omega(\Gamma)2}&=\mathbf{X}_{\Gamma2}\mathbf{P}_{\omega2},\end{aligned} \tag{9.12}$$

where $\mathbf{P}_{\omega i}$ is an orthogonal matrix of order N_i, $i=1,2$. It follows not only that the rank of $\mathbf{X}_{\omega(\Gamma)i}$ is equal to that of $\mathbf{X}_{\Gamma i}$, but also that the matrices $\mathbf{X}_{\omega(\Gamma)i}'\mathbf{X}_{\omega(\Gamma)i}$ and $\mathbf{X}_{\Gamma i}'\mathbf{X}_{\Gamma i}$ have the same set of characteristic roots, since they are *orthogonally similar*.

Now let Γ be a given initial design and let $\Omega(\Gamma)$ be the collection of designs obtained by the action of the elements of Ω on Γ. Explicitly this collection is equal to

$$\Omega(\Gamma)=\{\omega(\Gamma):\omega\in\Omega\}. \tag{9.13}$$

The following property is enjoyed by the orthogonal matrices

$$(9.14)\qquad \frac{1}{(\Pi k_i!)}\sum_{\omega\in\Omega}\mathbf{P}_{\omega i}=\begin{cases}\mathbf{0} & \text{if } \phi_1^0\phi_2^0\cdots\phi_t^0\neq\boldsymbol{\beta}_i\\ 1 & \text{if } \phi_1^0\phi_2^0\cdots\phi_t^0=\boldsymbol{\beta}_i\end{cases},\quad i=1,2.$$

Example 9.3. For the 2×3 factorial of Example 9.2, we have the following permutations: $\Omega_1=\{\omega_{11},\omega_{12}\}=\left\{\begin{smallmatrix}0\\1\end{smallmatrix}\rightarrow\begin{smallmatrix}0\\1\end{smallmatrix},\ \begin{smallmatrix}0\\1\end{smallmatrix}\rightarrow\begin{smallmatrix}1\\0\end{smallmatrix}\right\}$ and

$$\begin{aligned}\Omega_2&=\{\omega_{21},\omega_{22},\omega_{23},\omega_{24},\omega_{25},\omega_{26}\}\\&=\left\{\begin{matrix}0\\1\\2\end{matrix}\rightarrow\begin{matrix}0\\1\\2\end{matrix},\quad \begin{matrix}0\\1\\2\end{matrix}\rightarrow\begin{matrix}2\\1\\0\end{matrix},\quad \begin{matrix}0\\1\\2\end{matrix}\rightarrow\begin{matrix}1\\0\\2\end{matrix},\quad \begin{matrix}0\\1\\2\end{matrix}\rightarrow\begin{matrix}0\\2\\1\end{matrix},\quad \begin{matrix}0\\1\\2\end{matrix}\rightarrow\begin{matrix}1\\2\\0\end{matrix},\quad \begin{matrix}0\\1\\2\end{matrix}\rightarrow\begin{matrix}2\\0\\1\end{matrix}\right\}.\end{aligned}$$

Let $\Gamma=\{(0,0),(1,0),(1,1),(1,2)\}$ and $\boldsymbol{\beta}_1'=(\phi_1^0\phi_2^0,\phi_1^0\phi_2^1,\phi_1^0\phi_2^2)$, which is clearly an admissible vector. Note that $(\omega_{11},\omega_{21})$ is the identity permutation; that is, $(\omega_{11},\omega_{21})(\Gamma)=\Gamma_1=\{(0,0),(1,0),(1,1),(1,2)\}$. The second permutation results in $(\omega_{11},\omega_{22})(\Gamma)=\Gamma_2=\{(0,2),(1,2),(1,1),(1,0)\}$. The matrices in (9.12) for Γ_1 and Γ_2 are

$$\mathbf{X}_{\Gamma_1 1}=\begin{bmatrix}\frac{1}{\sqrt{6}} & \frac{-1}{2} & \frac{1}{\sqrt{12}}\\ \frac{1}{\sqrt{6}} & \frac{-1}{2} & \frac{1}{\sqrt{12}}\\ \frac{1}{\sqrt{6}} & 0 & \frac{-2}{\sqrt{12}}\\ \frac{1}{\sqrt{6}} & \frac{1}{2} & \frac{1}{\sqrt{12}}\end{bmatrix},\quad \mathbf{X}_{\Gamma_2 1}=\begin{bmatrix}\frac{1}{\sqrt{6}} & \frac{1}{2} & \frac{1}{\sqrt{12}}\\ \frac{1}{\sqrt{6}} & \frac{1}{2} & \frac{1}{\sqrt{12}}\\ \frac{1}{\sqrt{6}} & 0 & \frac{-2}{\sqrt{12}}\\ \frac{1}{\sqrt{6}} & \frac{-1}{2} & \frac{1}{\sqrt{12}}\end{bmatrix},$$

$$\mathbf{X}_{\Gamma_1 2}=\begin{bmatrix}\frac{-1}{\sqrt{6}} & \frac{1}{2} & \frac{-1}{\sqrt{12}}\\ \frac{1}{\sqrt{6}} & \frac{-1}{2} & \frac{1}{\sqrt{12}}\\ \frac{1}{\sqrt{6}} & 0 & \frac{-2}{\sqrt{12}}\\ \frac{1}{\sqrt{6}} & \frac{1}{2} & \frac{1}{\sqrt{12}}\end{bmatrix},\quad \mathbf{X}_{\Gamma_2 2}=\begin{bmatrix}\frac{-1}{\sqrt{6}} & \frac{-1}{2} & \frac{-1}{\sqrt{12}}\\ \frac{1}{\sqrt{6}} & \frac{1}{2} & \frac{1}{\sqrt{12}}\\ \frac{1}{\sqrt{6}} & 0 & \frac{-2}{\sqrt{12}}\\ \frac{1}{\sqrt{6}} & \frac{-1}{2} & \frac{1}{\sqrt{12}}\end{bmatrix},$$

where $\mathbf{X}_{(\omega_{11},\omega_{22})(\Gamma)1}=\mathbf{X}_{\Gamma_2 1}, \mathbf{X}_{(\omega_{11},\omega_{22})(\Gamma)2}=\mathbf{X}_{\Gamma_2 2}$,

$$\mathbf{P}_{(\omega_{11},\omega_{22})1}=\begin{bmatrix}1 & 0 & 0\\ 0 & -1 & 0\\ 0 & 0 & 1\end{bmatrix}, \quad \text{and} \quad \mathbf{P}_{(\omega_{11},\omega_{22})2}=\begin{bmatrix}1 & 0 & 0\\ 0 & -1 & 0\\ 0 & 0 & 1\end{bmatrix}.$$

It can be verified that $\mathbf{X}_{\Gamma_2 1}=\mathbf{X}_{\Gamma_1 1}\mathbf{P}_{(\omega_{11},\omega_{22})1}$ and $\mathbf{X}_{\Gamma_2 2}=\mathbf{X}_{\Gamma_1 2}\mathbf{P}_{(\omega_{11},\omega_{22})2}$.

Example 9.4. Consider the 2^3 factorial with the model $E[\mathbf{Y}_\rho]=[\mathbf{X}_2\otimes\mathbf{X}_2\otimes\mathbf{X}_2]\boldsymbol{\beta}_\rho$ and $\mathrm{Cov}(\mathbf{Y}_\rho)=\sigma^2\mathbf{I}_8$, where

$$\mathbf{X}_2=\begin{bmatrix}1 & -1\\ 1 & 1\end{bmatrix}.$$

and $\boldsymbol{\beta}_\rho'=(\phi_1^0\phi_2^0\phi_3^0, \phi_1^0\phi_2^0\phi_3^1, \phi_1^0\phi_2^1\phi_3^0, \phi_1^0\phi_2^1\phi_3^1, \phi_1^1\phi_2^0\phi_3^0, \phi_1^1\phi_2^0\phi_3^1, \phi_1^1\phi_2^1\phi_3^0, \phi_1^1\phi_2^1\phi_3^1)$. Let the initial design be $\Gamma=\{(0,0,0),(1,0,0),(0,1,0),(0,0,1)\}$ for estimating $\boldsymbol{\beta}_1'=(\phi_1^0\phi_2^0\phi_3^0, \phi_1^0\phi_2^0\phi_3^1, \phi_1^0\phi_2^1\phi_3^0, \phi_1^1\phi_2^0\phi_3^0)$. Since $\Omega=\Omega_1\times\Omega_2\times\Omega_3=\{\omega_{11},\omega_{12}\}\times\{\omega_{21},\omega_{22}\}\times\{\omega_{31},\omega_{32}\}$, where $\Omega_i=\{\begin{smallmatrix}0\\1\end{smallmatrix}\to\begin{smallmatrix}0\\1\end{smallmatrix}, \begin{smallmatrix}0\\1\end{smallmatrix}\to\begin{smallmatrix}1\\0\end{smallmatrix}\}$, we have the following eight designs in $\Omega(\Gamma)$:

$$(\omega_{11},\omega_{21},\omega_{31})(\Gamma)=\Gamma_1=\{(0,0,0),(1,0,0),(0,1,0),(0,0,1)\},$$
$$(\omega_{11},\omega_{21},\omega_{32})(\Gamma)=\Gamma_2=\{(0,0,1),(1,0,1),(0,1,1),(0,0,0)\},$$
$$(\omega_{11},\omega_{22},\omega_{31})(\Gamma)=\Gamma_3=\{(0,1,0),(1,1,0),(0,0,0),(0,1,1)\},$$
$$(\omega_{11},\omega_{22},\omega_{32})(\Gamma)=\Gamma_4=\{(0,1,1),(1,1,1),(0,0,1),(0,1,0)\},$$
$$(\omega_{12},\omega_{21},\omega_{31})(\Gamma)=\Gamma_5=\{(1,0,0),(0,0,0),(1,1,0),(1,0,1)\},$$
$$(\omega_{12},\omega_{21},\omega_{32})(\Gamma)=\Gamma_6=\{(1,0,1),(0,0,1),(1,1,1),(1,0,0)\},$$
$$(\omega_{12},\omega_{22},\omega_{31})(\Gamma)=\Gamma_7=\{(1,1,0),(0,1,0),(1,0,0),(1,1,1)\},$$
$$(\omega_{12},\omega_{22},\omega_{32})(\Gamma)=\Gamma_8=\{(1,1,1),(0,1,1),(1,0,1),(1,1,0)\}.$$

To verify equation (9.14) we set up the following table of matrices:

$$\begin{array}{ccccc}
 & \mathbf{X}_{\Gamma 1} & \mathbf{X}_{\Gamma 2} & \mathbf{P}_{\omega 1} & \mathbf{P}_{\omega 2}\\
\Gamma_1: & \begin{bmatrix}1 & -1 & -1 & -1\\ 1 & -1 & -1 & 1\\ 1 & -1 & 1 & -1\\ 1 & 1 & -1 & -1\end{bmatrix} & \begin{bmatrix}1 & 1 & 1 & -1\\ 1 & -1 & -1 & 1\\ -1 & 1 & -1 & 1\\ -1 & -1 & 1 & 1\end{bmatrix} & \begin{bmatrix}1 & 0 & 0 & 0\\ 0 & 1 & 0 & 0\\ 0 & 0 & 1 & 0\\ 0 & 0 & 0 & 1\end{bmatrix} & \begin{bmatrix}1 & 0 & 0 & 0\\ 0 & 1 & 0 & 0\\ 0 & 0 & 1 & 0\\ 0 & 0 & 0 & 1\end{bmatrix}\\
\Gamma_2: & \begin{bmatrix}1 & 1 & -1 & -1\\ 1 & 1 & -1 & 1\\ 1 & 1 & 1 & -1\\ 1 & -1 & -1 & -1\end{bmatrix} & \begin{bmatrix}-1 & -1 & 1 & 1\\ -1 & 1 & -1 & -1\\ 1 & -1 & -1 & -1\\ 1 & 1 & 1 & -1\end{bmatrix} & \begin{bmatrix}1 & 0 & 0 & 0\\ 0 & -1 & 0 & 0\\ 0 & 0 & 1 & 0\\ 0 & 0 & 0 & 1\end{bmatrix} & \begin{bmatrix}-1 & 0 & 0 & 0\\ 0 & -1 & 0 & 0\\ 0 & 0 & 1 & 0\\ 0 & 0 & 0 & -1\end{bmatrix}
\end{array}$$

$$\Gamma_3: \begin{bmatrix} 1 & -1 & 1 & -1 \\ 1 & -1 & 1 & 1 \\ 1 & -1 & -1 & -1 \\ 1 & 1 & 1 & -1 \end{bmatrix}\begin{bmatrix} -1 & 1 & -1 & 1 \\ -1 & -1 & 1 & -1 \\ 1 & 1 & 1 & -1 \\ 1 & -1 & -1 & -1 \end{bmatrix}\begin{bmatrix} 1 & 0 & 0 & 0 \\ 0 & 1 & 0 & 0 \\ 0 & 0 & -1 & 0 \\ 0 & 0 & 0 & 1 \end{bmatrix}\begin{bmatrix} -1 & 0 & 0 & 0 \\ 0 & 1 & 0 & 0 \\ 0 & 0 & -1 & 0 \\ 0 & 0 & 0 & -1 \end{bmatrix}$$

$$\Gamma_4: \begin{bmatrix} 1 & 1 & 1 & -1 \\ 1 & 1 & 1 & 1 \\ 1 & 1 & -1 & -1 \\ 1 & -1 & 1 & -1 \end{bmatrix}\begin{bmatrix} 1 & -1 & -1 & -1 \\ 1 & 1 & 1 & 1 \\ -1 & -1 & 1 & 1 \\ -1 & 1 & -1 & 1 \end{bmatrix}\begin{bmatrix} 1 & 0 & 0 & 0 \\ 0 & -1 & 0 & 0 \\ 0 & 0 & -1 & 0 \\ 0 & 0 & 0 & 1 \end{bmatrix}\begin{bmatrix} 1 & 0 & 0 & 0 \\ 0 & -1 & 0 & 0 \\ 0 & 0 & -1 & 0 \\ 0 & 0 & 0 & 1 \end{bmatrix}$$

$$\Gamma_5: \begin{bmatrix} 1 & -1 & -1 & 1 \\ 1 & -1 & -1 & -1 \\ 1 & -1 & 1 & 1 \\ 1 & 1 & -1 & 1 \end{bmatrix}\begin{bmatrix} 1 & -1 & -1 & 1 \\ 1 & 1 & 1 & -1 \\ -1 & -1 & 1 & -1 \\ -1 & 1 & -1 & -1 \end{bmatrix}\begin{bmatrix} 1 & 0 & 0 & 0 \\ 0 & 1 & 0 & 0 \\ 0 & 0 & 1 & 0 \\ 0 & 0 & 0 & -1 \end{bmatrix}\begin{bmatrix} 1 & 0 & 0 & 0 \\ 0 & -1 & 0 & 0 \\ 0 & 0 & -1 & 0 \\ 0 & 0 & 0 & -1 \end{bmatrix}$$

$$\Gamma_6: \begin{bmatrix} 1 & 1 & -1 & 1 \\ 1 & 1 & -1 & -1 \\ 1 & 1 & 1 & 1 \\ 1 & -1 & -1 & 1 \end{bmatrix}\begin{bmatrix} -1 & 1 & -1 & -1 \\ -1 & -1 & 1 & 1 \\ 1 & 1 & 1 & 1 \\ 1 & -1 & -1 & 1 \end{bmatrix}\begin{bmatrix} 1 & 0 & 0 & 0 \\ 0 & -1 & 0 & 0 \\ 0 & 0 & 1 & 0 \\ 0 & 0 & 0 & -1 \end{bmatrix}\begin{bmatrix} -1 & 0 & 0 & 0 \\ 0 & 1 & 0 & 0 \\ 0 & 0 & -1 & 0 \\ 0 & 0 & 0 & 1 \end{bmatrix}$$

$$\Gamma_7: \begin{bmatrix} 1 & -1 & 1 & 1 \\ 1 & -1 & 1 & -1 \\ 1 & -1 & -1 & 1 \\ 1 & 1 & 1 & 1 \end{bmatrix}\begin{bmatrix} -1 & -1 & 1 & -1 \\ -1 & 1 & -1 & 1 \\ 1 & -1 & -1 & 1 \\ 1 & 1 & 1 & 1 \end{bmatrix}\begin{bmatrix} 1 & 0 & 0 & 0 \\ 0 & 1 & 0 & 0 \\ 0 & 0 & -1 & 0 \\ 0 & 0 & 0 & -1 \end{bmatrix}\begin{bmatrix} -1 & 0 & 0 & 0 \\ 0 & -1 & 0 & 0 \\ 0 & 0 & 1 & 0 \\ 0 & 0 & 0 & 1 \end{bmatrix}$$

$$\Gamma_8: \begin{bmatrix} 1 & 1 & 1 & 1 \\ 1 & 1 & 1 & -1 \\ 1 & 1 & -1 & 1 \\ 1 & -1 & 1 & 1 \end{bmatrix}\begin{bmatrix} 1 & 1 & 1 & 1 \\ 1 & -1 & -1 & -1 \\ -1 & 1 & -1 & -1 \\ -1 & -1 & 1 & -1 \end{bmatrix}\begin{bmatrix} 1 & 0 & 0 & 0 \\ 0 & -1 & 0 & 0 \\ 0 & 0 & -1 & 0 \\ 0 & 0 & 0 & -1 \end{bmatrix}\begin{bmatrix} 1 & 0 & 0 & 0 \\ 0 & 1 & 0 & 0 \\ 0 & 0 & 1 & 0 \\ 0 & 0 & 0 & -1 \end{bmatrix}.$$

Now, observe that

$$\frac{1}{(\Pi k_i!)} \sum_{\omega \in \Omega} \mathbf{P}_{\omega 1} = \tfrac{1}{8}\begin{bmatrix} 8 & 0 & 0 & 0 \\ 0 & 0 & 0 & 0 \\ 0 & 0 & 0 & 0 \\ 0 & 0 & 0 & 0 \end{bmatrix} = \begin{bmatrix} 1 & 0 & 0 & 0 \\ 0 & 0 & 0 & 0 \\ 0 & 0 & 0 & 0 \\ 0 & 0 & 0 & 0 \end{bmatrix}$$

and

$$\frac{1}{(\Pi k_i!)} \sum_{\omega \in \Omega} \mathbf{P}_{\omega 2} = \tfrac{1}{8}\begin{bmatrix} 0 & 0 & 0 & 0 \\ 0 & 0 & 0 & 0 \\ 0 & 0 & 0 & 0 \\ 0 & 0 & 0 & 0 \end{bmatrix} = \begin{bmatrix} 0 & 0 & 0 & 0 \\ 0 & 0 & 0 & 0 \\ 0 & 0 & 0 & 0 \\ 0 & 0 & 0 & 0 \end{bmatrix}.$$

9.3 UNIFORM RANDOMIZED FACTORIAL DESIGNS AND UNBIASED ESTIMATION OF β_1

Let Γ be a given initial design such that $\mathbf{X}_{\Gamma 1}$ is of full rank and let $\Omega(\Gamma)$ be the collection of designs obtained by the action of the elements of Ω. Note that now each design in $\Omega(\Gamma)$ is also of full rank. If a design in $\Omega(\Gamma)$ is selected at random, that is, with equal probability $(\prod k_i!)^{-1}$, then such a design will be a uniform randomized design according to Definition 9.1. Consider the estimator

$$\hat{\boldsymbol{\beta}}_1' = (\mathbf{X}'_{\omega(\Gamma)1}\mathbf{X}_{\omega(\Gamma)1})^{-1}\mathbf{X}'_{\omega(\Gamma)1}\mathbf{Y}_{\omega(\Gamma)} \tag{9.15}$$

obtained from the randomized design $\omega(\Gamma)$ for the admissible vector $\boldsymbol{\beta}_1$. Then

$$\begin{aligned} E[\hat{\boldsymbol{\beta}}_1] &= \sum_{\omega\in\Omega} (\prod k!)^{-1} E[(\mathbf{X}'_{\omega(\Gamma)1}\mathbf{X}_{\omega(\Gamma)1})^{-1}\mathbf{X}'_{\omega(\Gamma)1}\mathbf{Y}_{\omega(\Gamma)}] \\ &= \boldsymbol{\beta}_1 + (\prod k_i!)^{-1} \sum_{\omega\in\Omega} \mathbf{P}'_{\omega 1}(\mathbf{X}'_{\Gamma 1}\mathbf{X}_{\Gamma 1})^{-1}\mathbf{X}'_{\Gamma 1}\mathbf{X}_{\Gamma 2}\mathbf{P}_{\omega 2}\boldsymbol{\beta}_2 \\ &= \boldsymbol{\beta}_1 + (\prod k_i!)^{-1} \sum_{\omega\in\Omega} \mathbf{A}_{\omega(\Gamma)}\boldsymbol{\beta}_2. \end{aligned} \tag{9.16}$$

It can be shown that $\sum_{\omega\in\Omega}\mathbf{A}_{\omega(\Gamma)} = \mathbf{0}$, so that for a randomized design $\hat{\boldsymbol{\beta}}_1$ is an unbiased estimator of $\boldsymbol{\beta}_1$. Hence under $\xi_i = (\prod k_i!)^{-1}$ the class $\Omega(\Gamma)$ provides a class of unbiased uniform randomized designs.

Example 9.5. Consider the 2^3 factorial of Example 9.4 with the same $\boldsymbol{\beta}_1$ and $\boldsymbol{\beta}_2$, and the nonrandomized D-optimal design $\Gamma = \{(0,0,0), (1,1,0), (1,0,1), (0,1,1)\}$ under the assumption that $\boldsymbol{\beta}_2 = \mathbf{0}$. Under Ω we obtain exactly two distinct designs, $\Gamma_1 = \Gamma$ and $\Gamma_2 = \{(1,1,1), (0,0,1), (0,1,0), (1,0,0)\}$. The two permutations can be taken as $\omega^* = \begin{pmatrix} 0\ 0 & 0\ 0 & 0\ 0 \\ \rightarrow, & \rightarrow, & \rightarrow \\ 1\ 1 & 1\ 1 & 1\ 1 \end{pmatrix}$ and $\omega^{**} = \begin{pmatrix} 0\ 1 & 0\ 1 & 0\ 1 \\ \rightarrow, & \rightarrow, & \rightarrow \\ 1\ 0 & 1\ 0 & 1\ 0, \end{pmatrix}$. The relevant matrices are

$$\mathbf{X}_{\Gamma_1 1} = \begin{bmatrix} 1 & -1 & -1 & -1 \\ 1 & 1 & 1 & -1 \\ 1 & 1 & -1 & 1 \\ 1 & -1 & 1 & 1 \end{bmatrix}, \quad \mathbf{X}_{\Gamma_1 2} = \begin{bmatrix} 1 & 1 & 1 & -1 \\ 1 & -1 & -1 & -1 \\ -1 & 1 & -1 & -1 \\ -1 & -1 & 1 & -1 \end{bmatrix},$$

$$\mathbf{X}_{\Gamma_2 1} = \begin{bmatrix} 1 & 1 & 1 & 1 \\ 1 & -1 & -1 & 1 \\ 1 & -1 & 1 & -1 \\ 1 & 1 & -1 & -1 \end{bmatrix}, \quad \mathbf{X}_{\Gamma_2 2} = \begin{bmatrix} 1 & 1 & 1 & 1 \\ 1 & -1 & -1 & 1 \\ -1 & 1 & -1 & 1 \\ -1 & -1 & 1 & 1 \end{bmatrix},$$

and

$$\mathbf{P}_{\omega^{**}1} = \begin{bmatrix} 1 & 0 & 0 & 0 \\ 0 & -1 & 0 & 0 \\ 0 & 0 & -1 & 0 \\ 0 & 0 & 0 & -1 \end{bmatrix}, \quad \mathbf{P}_{\omega^{**}2} = \begin{bmatrix} 1 & 0 & 0 & 0 \\ 0 & 1 & 0 & 0 \\ 0 & 0 & 1 & 0 \\ 0 & 0 & 0 & -1 \end{bmatrix}.$$

Now, $\Sigma_{\omega \in \Omega} \mathbf{P}'_{\omega 1}(\mathbf{X}'_{\Gamma 1}\mathbf{X}_{\Gamma 1})^{-1}\mathbf{X}'_{\Gamma 1}\mathbf{X}_{\Gamma 2}\mathbf{P}_{\omega 2} = \mathbf{I}(\mathbf{X}'_{\Gamma 1}\mathbf{X}_{\Gamma 1})^{-1}\mathbf{X}'_{\Gamma 1}\mathbf{X}_{\Gamma 2}\mathbf{I} + \mathbf{P}'_{\omega^{**}1}(\mathbf{X}'_{\Gamma 1}\mathbf{X}_{\Gamma 1})^{-1}\mathbf{X}'_{\Gamma 1}\mathbf{X}_{\Gamma 2}\mathbf{P}_{\omega^{**}2} = \mathbf{A}_\Gamma + \mathbf{P}'_{\omega^{**}1}\mathbf{A}_\Gamma\mathbf{P}_{\omega^{**}2}$, where $\mathbf{A}_\Gamma$ is the alias matrix of Γ already calculated in Example 7.5, that is,

$$\mathbf{A}_\Gamma = \begin{bmatrix} 0 & 0 & 0 & -1 \\ 0 & 0 & -1 & 0 \\ 0 & -1 & 0 & 0 \\ -1 & 0 & 0 & 0 \end{bmatrix}.$$

Hence $\mathbf{A}_\Gamma + \mathbf{P}'_{\omega^{**}1}\mathbf{A}_\Gamma\mathbf{P}_{\omega^{**}2}$ is equal to

$$\begin{bmatrix} 0 & 0 & 0 & -1 \\ 0 & 0 & -1 & 0 \\ 0 & -1 & 0 & 0 \\ -1 & 0 & 0 & 0 \end{bmatrix} + \begin{bmatrix} 1 & 0 & 0 & 0 \\ 0 & -1 & 0 & 0 \\ 0 & 0 & -1 & 0 \\ 0 & 0 & 0 & -1 \end{bmatrix}$$

$$\times \begin{bmatrix} 0 & 0 & 0 & -1 \\ 0 & 0 & -1 & 0 \\ 0 & -1 & 0 & 0 \\ -1 & 0 & 0 & 0 \end{bmatrix} \begin{bmatrix} 1 & 0 & 0 & 0 \\ 0 & 1 & 0 & 0 \\ 0 & 0 & 1 & 0 \\ 0 & 0 & 0 & -1 \end{bmatrix}$$

$$= \begin{bmatrix} 0 & 0 & 0 & -1 \\ 0 & 0 & -1 & 0 \\ 0 & -1 & 0 & 0 \\ -1 & 0 & 0 & 0 \end{bmatrix} + \begin{bmatrix} 0 & 0 & 0 & 1 \\ 0 & 0 & 1 & 0 \\ 0 & 1 & 0 & 0 \\ 1 & 0 & 0 & 0 \end{bmatrix} = \begin{bmatrix} 0 & 0 & 0 & 0 \\ 0 & 0 & 0 & 0 \\ 0 & 0 & 0 & 0 \\ 0 & 0 & 0 & 0 \end{bmatrix}.$$

The unbiased estimator of $\boldsymbol{\beta}_1$ when $\boldsymbol{\beta}_2$ is not assumed to be zero is of the form

$$\hat{\boldsymbol{\beta}}_1 = \begin{bmatrix} \frac{1}{4} & 0 & 0 & 0 \\ 0 & \frac{1}{4} & 0 & 0 \\ 0 & 0 & \frac{1}{4} & 0 \\ 0 & 0 & 0 & \frac{1}{4} \end{bmatrix} \mathbf{X}'_{\omega(\Gamma)1}\mathbf{Y}_{\omega(\Gamma)},$$

where ω is either ω^* or ω^{**} and $\xi_1 = \xi_2 = \frac{1}{2}$.

The condition of admissibility of a parametric vector in obtaining the preceding results is a sufficient condition. Whether this is also a necessary condition has not been established yet. As seen in Example 9.5, the set of permutations does not necessarily yield $\prod k_i!$ distinct designs from an initial design. Several results are available in the literature regarding this matter, and the reader is referred to the literature list at the end of the chapter. Also, the results may be extended to estimability of linear parametric functions and possibly to other models. An interesting exercise for the reader is to develop the covariance matrix of the unbiased estimator of $\boldsymbol{\beta}_1$ and observe the practical implications.

9.4 AN APPLICATION OF $\Omega(\Gamma)$ WHEN $\boldsymbol{\beta}_2 = \mathbf{0}$

If the vector $\boldsymbol{\beta}_2$ is equal to $\mathbf{0}$, then for a fixed design Γ, the estimator $\hat{\boldsymbol{\beta}}_1 = (\mathbf{X}'_{\Gamma 1}\mathbf{X}_{\Gamma 1})^{-1}\mathbf{X}'_{\Gamma 1}\mathbf{Y}_{\Gamma}$ is an unbiased estimator as already concluded in Case 2 of Chapter 7. The set of permutations Ω provides us in this case with a set of designs that are optimal if the initial design is optimal, and the optimality criterion is taken as a functional on the spectrum of the information matrix. We may impose a further criterion, for example, economic or physical, to obtain an optimal design in this class relative to the chosen criterion.

Example 9.6. Consider the 2^4 factorial with a model similar to that already given for the 2^3 factorial in Example 9.3; that is, the design matrix for the minimal complete factorial is

$$\mathbf{X}_\rho = \mathbf{X}_2 \otimes \mathbf{X}_2 \otimes \mathbf{X}_2 \otimes \mathbf{X}_2,$$

where

$$\mathbf{X}_2 = \begin{bmatrix} 1 & -1 \\ 1 & 1 \end{bmatrix}.$$

A saturated D-optimal resolution III design is $\Gamma = \{(0,0,0,0), (0,1,1,1),(1,0,1,1),(1,1,0,1),(1,1,1,0)\}$. Using Ω, one obtains a class of sixteen saturated D-optimal resolution III designs. Taking the number of ones as a cost function one may show that the permuted design $\Gamma^* = \{(1,1,1,1),(1,0,0,0),(0,1,0,0),(0,0,1,0),(0,0,0,1)\}$ is the unique least cost D-optimal design in the class of permuted designs $\Omega(\Gamma)$.

9.5 REGULAR FRACTIONS OF THE PRIME POWERED FACTORIAL GENERATED BY Ω

Consider the s^t prime powered factorial, where s is a prime or power of a prime, and denote the levels of each factor by the symbols $0,1,2,\ldots,s-1$, which are made to correspond with the elements of the *Galois field*, $\mathrm{GF}(s)$. The complete set of treatment combinations (i.e., the minimal complete factorial design ρ), forms a group under componentwise addition over $\mathrm{GF}(s)$. In fact ρ can be viewed as a vector space F^t of dimension t over $F = \mathrm{GF}(s)$ or as a *finite Euclidean geometry* $\mathrm{EG}(t,s)$ of dimension t over

GF(s). The reader who is not familiar with finite mathematical structures, such as groups, rings, fields, and finite geometries, can find the definitions in the Appendix.

Let Γ be an FFD of the s^m factorial with parameters $m=n$ and all nonzero r_i equal to 1; that is, Γ consists of exactly m distinct treatment combinations. We adopt the following definition for a regular fraction.

Definition 9.4. Let Γ be an FFD as described previously. Then Γ is called *regular* if it is a subspace or coset of a subspace of the vector space F^t. A design Γ that is not regular is termed *irregular*.

From vector space theory it follows that the order m of a regular fraction Γ must be of the form s^α, $1 \le \alpha \le t-1$. Geometrically, a regular fraction Γ forms an α-flat in EG(t, s); that is, the s^α points in Γ satisfy a set of $(t-\alpha)$ independent and consistent equations in EG(t, s), which is of the form $\mathbf{AX}=\mathbf{Y}$, that is,

$$\begin{bmatrix} a_{11} & a_{12} & \cdots & a_{1t} \\ a_{21} & a_{22} & \cdots & a_{2t} \\ \cdots & \cdots & \cdots & \cdots \\ a_{(t-\alpha)1} & a_{(t-\alpha)2} & \cdots & a_{(t-\alpha)t} \end{bmatrix} \begin{bmatrix} x_1 \\ x_2 \\ \vdots \\ x_t \end{bmatrix} = \begin{bmatrix} y_1 \\ y_2 \\ \vdots \\ y_{t-\alpha} \end{bmatrix}. \tag{9.17}$$

If Γ is a regular FFD of order $m=s^\alpha$, then it can be deduced from group theory that the set $\Omega(\Gamma)$ of the FFD has order $s^{t-\alpha}$ and consists of a subgroup with all its cosets. In listing the FFD in $\Omega(\Gamma)$, it is customary to list the subgroup as the initial design. This subgroup is referred to in the literature as the *intrablock subgroup*.

Example 9.7. Consider the 3^3 factorial with $\rho = \{(0,0,0), (1,0,0), (0,1,0), \ldots, (2,2,2)\}$, that is, $\rho = \{(x_1, x_2, x_3),\ x_i \in \mathrm{GF}(3)\}$. Let $\Omega = \Omega_1 \otimes \Omega_2 \otimes \Omega_3$, where each Ω_i acts on the elements of GF(3). If $\Gamma = \{(0,0,0),(1,2,0),(2,1,0)\}$, then it may be verified that it is a subgroup. Hence $\Omega(\Gamma)$ consists of the $3^{3-1}=9$ fractions

$$\Gamma_1 = \{(0,0,0),(1,2,0),(2,1,0)\}, \qquad \Gamma_5 = \{(1,1,0),(2,0,0),(0,2,0)\},$$
$$\Gamma_2 = \{(0,0,2),(1,2,2),(2,1,2)\}, \qquad \Gamma_6 = \{(1,0,1),(2,2,1),(0,1,1)\},$$
$$\Gamma_3 = \{(0,1,0),(1,0,0),(2,2,0)\}, \qquad \Gamma_7 = \{(0,1,2),(1,0,2),(2,2,2)\},$$
$$\Gamma_4 = \{(0,0,1),(1,2,1),(2,1,1)\}, \qquad \Gamma_8 = \{(1,1,1),(2,0,1),(0,2,1)\},$$
$$\Gamma_9 = \{(1,1,2),(2,0,2),(0,2,2)\}.$$

Notice that each design in $\Omega(\Gamma)$ forms a 1-flat of EG(3,3). For example, design Γ_1 (which is the intrablock subgroup) satisfies the independent and consistent equations of (9.17), that is,

$$\begin{bmatrix} 1 & 1 & 0 \\ 0 & 0 & 1 \end{bmatrix} \begin{bmatrix} x_1 \\ x_2 \\ x_3 \end{bmatrix} = \begin{bmatrix} 0 \\ 0 \end{bmatrix}. \tag{9.18}$$

Consider the complete vector of effects $\boldsymbol{\beta}_p$ of the s^t prime powered factorial as a set whose elements are labeled in the following manner: $\boldsymbol{\beta}_p = \{\phi_1^{x_1}\phi_2^{x_2}\cdots\phi_t^{x_t}, x_i \in \mathrm{GF}(s)\}$. Delete the mean $\phi_1^0\phi_2^0\cdots\phi_t^0$ from $\boldsymbol{\beta}_p$ and call the resulting vector $\boldsymbol{\beta}_p^*$. Define two effects $\phi_1^{x_1}\phi_2^{x_2}\cdots\phi_t^{x_t}$ and $\phi_1^{z_1}\phi_2^{z_2}\cdots\phi_t^{z_t}$ in $\boldsymbol{\beta}_p^*$ to be *linearly dependent* if and only if $(z_1, z_2, \ldots, z_t) = c(x_1, x_2, \ldots, x_t)$, where c is a nonzero element of GF(s). This dependency relation is an equivalence relation over GF(s), and it can be shown that the set of equivalence classes $\boldsymbol{\beta}_p^*/\mathrm{GF}(s)$ is isomorphic to the *finite projective geometry* PG($t-1, s$) of dimension $t-1$ over GF(s). Notice that each equivalence class contains exactly $s-1$ effects and that the cardinality of $\boldsymbol{\beta}_p^*/\mathrm{GF}(s)$ is equal to $(s^t-1)/(s-1)$.

A set of r linearly independent (i.e., not linearly dependent) elements in $\boldsymbol{\beta}_p^*/\mathrm{GF}(s)$ [or points of PG($t-1, s$)] generates an $(r-1)$-flat in PG($t-$ $1, s$). In particular, corresponding to a set of $(t-\alpha)$ independent and consistent equations, as given in (9.17), we have the $(t-\alpha-1)$-flat of PG($t-1, s$),

$$\left\{\begin{array}{l} c_1(a_{11}, a_{12}, \cdots, a_{1t}) + c_2(a_{21}, a_{22}, \cdots, a_{2t}) + \cdots \\ \quad + c_{t-\alpha}(a_{(t-\alpha)1}, a_{(t-\alpha)2}, \cdots, a_{(t-\alpha)t}), \qquad c_i \in \mathrm{GF}(s) \\ \text{and not all simultaneously equal to 0} \end{array}\right\}. \tag{9.19}$$

Thus it is customary in the literature to depict a regular fraction by the shortened notation

$$\begin{aligned} \{ & (\phi_1^{a_{11}}\phi_2^{a_{12}}\cdots\phi_t^{a_{1t}})_{y_1}, (\phi_1^{a_{21}}\phi_2^{a_{22}}\cdots\phi_t^{a_{2t}})_{y_2}, \ldots, \\ & (\phi_1^{a_{(t-\alpha)1}}\phi_2^{a_{(t-\alpha)2}}\cdots\phi_t^{a_{(t-\alpha)t}})y_{t-\alpha}\}. \end{aligned} \tag{9.20}$$

In (9.20) only the generators are listed but one may also list the generated elements of $\boldsymbol{\beta}_p^*/\mathrm{GF}(s)$ with their appropriate y values.

It may be shown, under the geometric definition of effects, that if a *nonrandomized regular fraction* of the s^t factorial is selected, then $(s^{t-\alpha}-1)/(s-1)$ sets of $(s-1)$ effects each, determined by (9.19) or (9.20), will be confounded or aliased with the mean, which is the first element of $\boldsymbol{\beta}_1$. This means that the generalized defining relationship of the fraction will involve the mean and $(s^{t-\alpha}-1)/(s-1)$ sets of $(s-1)$ effects each. It is beyond the scope of this book to show that the generalized defining relationship of a given regular fraction is unique under any choice of $\boldsymbol{\beta}_1$ and $\boldsymbol{\beta}_2$, as long as $\boldsymbol{\beta}_1$ is estimable under the assumption that $\boldsymbol{\beta}_2$ is negligible. The term *defining contrast* is reserved in the literature for regular fractions and it refers to an element of the set $\boldsymbol{\beta}_\rho^*/\mathrm{GF}(s)$ that is present in the $(t-\alpha-1)$-flat given by (9.19). It follows that the set of defining contrasts for a regular fraction determines all the $(s^{t-\alpha}-1)$ effects, except the mean, that appear in the generalized defining relationship.

Example 9.8. Consider Example 9.7. Then the $t-\alpha-1=2-1=1$-flat of PG(2,3) for the design Γ_1, corresponding to equation (9.18), is found by utilizing (9.19). This is equal to

(9.21) $$\{(1,1,0),(0,0,1),(1,1,1),(1,1,2)\}.$$

From (9.20) we may describe Γ_1 compactly as

(9.22) $$\{(\phi_1^1\phi_2^1\phi_3^0)_0,(\phi_1^0\phi_2^0\phi_3^1)_0\},$$

or, equivalently, by

(9.23) $$\{(\phi_1^1\phi_2^1\phi_3^0)_0,(\phi_1^0\phi_2^0\phi_3^1)_0,(\phi_1^1\phi_2^1\phi_3^1)_0,(\phi_1^1\phi_2^1\phi_3^2)_0\}.$$

In (9.22) only the generators are written out, while in (9.23) all the elements of the 1-flat of PG(2,3) are utilized. The defining contrasts are the elements appearing in (9.23), and they determine the $(s^{t-\alpha}-1)/(s-1)=(3^2-1)/(3-1)=4$ sets of 2 effects each of the generalized defining relationship apart from the mean.

The notion of regular factorial designs can be tied up with factorial designs of arbitrary resolution, which is done frequently in the literature. Regular factorial designs also have intimate relationships with confounded incomplete block designs and lattice designs. For the geometric definition of effects, an explicit example of this, and other results see reference 25 in the list below. Other relevant references are also included.

9.6 SELECTED ADDITIONAL READING

Starred references are recommended for a first reading.

* 1. Addelman, S. (1961). Irregular fractions of the 2^m factorial experiments. *Technometrics*, **3**, 479–496.

2. Bailey, N. T. J. (1959). The use of linear algebra in deriving prime power factorial designs with confounding and fractional replication. *Sankhyā*, **21**, 345–354.

*3. Bose, R. C. (1947). Mathematical theory of the symmetrical factorial design. *Sankhyā*, **8**, 107–166.

4. Bose, R. C., and Srivastava, J. N. (1964). Analysis of irregular factorial fractions. *Sankhyā*, **26**, 117–144.

5. Brownlee, K. A., Kelly, B. K., and Loraine, P. K. (1948). Fractional replication arrangements for factorial experiments with factors at two levels. *Biometrika*, **25**, 268–276.

6. Burton, R. C., and Connor, W. S. (1957). On the identity relationship for fractional replicates in the 2^n series. *Ann. Math. Stat.*, **28**, 762–767.

7. Carmichael, R. D. (1956). *Introduction to the Theory of Groups of Finite Order*. (First published in 1937.) Dover, New York.

8. Dembowski, P. (1968). *Finite Geometries*. Springer-Verlag, Berlin.

9. Dempster, A. P. (1960). Random allocation designs, I. On general classes of estimation methods. *Ann. Math. Stat.*, **31**, 885–905.

10. Dempster, A. P. (1961). Random allocation designs, II. Approximation theory for single random allocations. *Ann. Math. Stat.*, **32**, 387–405.

*11. Ehrenfeld, S., and Zacks, S. (1961). Randomization and factorial experiments. *Ann. Math. Stat.*, **32**, 270–297.

12. Ehrenfeld, S., and Zacks, S. (1963). Optimal strategies in factorial experiments. *Ann. Math. Stat.*, **34**, 780–791.

13. Ehrenfeld, S., and Zacks, S. (1967). Testing hypotheses in randomized factorial experiments. *Ann. Math. Stat.*, **38**, 1494–1507.

14. Fraser, W., and Raktoe, B. L. (1976). A theorem on invariance of estimability of linear parametric functions in linear models. *Commun. Stat.* **A5**, 977–983.

15. Fujii, A., and Tsutagawa, N. (1965). Experimental analysis of process by the method of random combination of fractional factorial design. *Rep. Stat. Appl. Res.*, **12**, 165–177.

16. Katti, S. K. (1960). A direct method for constructing the intrablock subgroup. *Biometrics*, **16**, 691–694.

17. Kiefer, J. (1958). On the nonrandomized optimality and randomized nonoptimality of symmetrical designs. *Ann. Math. Stat.*, **29**, 675–699.

18. Kishen, K., and Srivastava, J. N. (1959). Mathematical theory of confounding in asymmetrical and symmetrical factorial designs. *J. Indian Soc. Agric. Stat.*, **11**, 73–110.

19. Nair, K. R. (1938). On a method of getting confounded arrangements in the general symmetrical type of experiments. *Sankhyā*, **4**, 121–138.

20. Ogawa, J. (1975). *Statistical Theory of the Analysis of Experimental Designs*. Marcel Dekker, New York.

21. Paik, U. B., and Federer, W. T. (1970). A randomized procedure of saturated main effect fractional replicates. *Ann. Math. Stat.*, **41**, 369–375.

*22. Paik, U. B., and Federer, W. T. (1972). On a randomized procedure for saturated fractional replicates in a 2^n-factorial. *Ann. Math. Stat.*, **43**, 1346–1351.

23. Pesotan, H., Raktoe, B. L., and Federer, W. T. (1975). On complexes of Abelian groups with applications to fractional factorial designs. *Ann. Inst. Stat. Math.*, **27**, 117–142.

24. Raktoe, B. L. (1967). Application of cyclic collineations to the construction of balanced l-restrictional prime powered lattice designs. *Ann. Math. Stat.*, **38**, 1127–1141.

*25. Raktoe, B. L., Pesotan, M., and Federer, W. T. (1980). On aliasing and generalized defining relationships of factorial arrangements. *Can. J. Stat.*, **8**, 65–77.

26. Raktoe, B. L, Rayner, A. A., and Chalton, D. O. (1978). On construction of confounded mixed factorial and lattice designs. *Aust. J. Stat.*, **20**, 209–218.

27. Satterthwaite, F. E. (1959). Random balance experimentation. *Technometrics*, **1**, 111–137.

*28. Srivastava, J. N., Raktoe, B. L., and Pesotan, H. (1976). On invariance and randomization in fractional replication. *Ann. Stat.*, **4**, 423–430.

29. Taguchi, G. (1962). A randomly combined design. *Rep. Stat. Appl. Res.*, **9**, 93–126.

30. Takeuchi, K. (1960). On a special class of regression problems and its applications: random combined fractional factorial designs. *Rep. Stat. Appl. Res.*, **7**, 166–198.

31. Takeuchi, K. (1961). On a special class of regression problems and its applications: random combined fractional factorial designs (continued). *Rep. Stat. Appl. Res.*, **8**, 1–6.

*32. Yamakawa, N. (1965). Random combined fractional factorial designs. I. Theory and application of the random combined fractional factorial designs (RACOFFD). *Bull. Math. Stat.*, **11**, 1–38.

33. Yamakawa, N. (1966). Random combined fractional factorial designs. II. Sampling theory from the finite population. *Bull. Math. Stat.*, **12**, 1–67.

*34. Zacks, S. (1963). On a complete case of linear unbiased estimators for randomized factorial experiments. *Ann. Math. Stat.*, **34**, 769–779.

35. Zacks, S. (1964). Generalized least squares estimators for randomized fractional replication designs. *Ann. Math. Stat.*, **35**, 696–704.

36. Zacks, S. (1966). Randomized fractional weighing designs. *Ann. Math. Stat.*, **37**, 1382–1395.

37. Zacks, S. (1968). Bayes sequential design of fractional factorial experiments for the estimation of a subgroup of pre-assigned parameters. *Ann. Math. Stat.*, **39**, 973–982.

CHAPTER 10

Factorial Designs of Resolution III

In Chapter 7 the partitioning of the parametric vector $\boldsymbol{\beta}_\rho$ into three parts led to four exhaustive cases. It was pointed out that designs of odd resolution belonged to Case 3 and designs of even resolution to Case 2. The objective of the present chapter is to provide a discussion of resolution III designs, which is a special case of odd resolution designs. In Chapter 11 designs of resolutions IV and V are developed. The references given at the end of Chapters 10 and 11 contain many examples of designs of resolution III, IV, and V.

10.1 THE ASSUMED MODEL

For the general $k_1 \times k_2 \times \cdots \times k_t$ factorial, we associate with the minimal complete factorial design the model

$$(10.1) \qquad E[\mathbf{Y}_\rho] = \mathbf{X}_\rho \boldsymbol{\beta}_\rho; \qquad \operatorname{Cov}(\mathbf{Y}_\rho) = \sigma^2 \mathbf{I}_N,$$

where $\mathbf{X}_\rho$ is either an orthonormal or columnwise orthogonal matrix, the parametric vector $\boldsymbol{\beta}_\rho$ representing the mean, main effects, and interaction effects. In the orthonormal case $\boldsymbol{\beta}_\rho = \mathbf{X}'_\rho E[\mathbf{Y}_\rho]$, where $E[\mathbf{Y}_\rho]$ is the $N \times 1$ vector of treatment means. In the columnwise orthogonal case $\boldsymbol{\beta}_\rho = \mathbf{D}^{-1}\mathbf{X}'_\rho E[\mathbf{Y}_\rho]$, where $\mathbf{D}$ is a diagonal matrix. It follows that in either case $\boldsymbol{\beta}_\rho$ is the vector consisting of the overall mean and a set of $N-1$ mutually orthogonal contrasts among treatment means. The main effects of the ith factor are represented by $k_i - 1$ mutually orthogonal contrasts, the interaction effects of the ith and jth factors are represented by $(k_i - 1)(k_j - 1)$

mutually orthogonal contrasts, and so on. Note that the orthogonal polynomial model is a special case of the preceding model.

10.2 ORTHOGONALITY AND BALANCEDNESS OF RESOLUTION III DESIGNS

Using the definition of a design of odd resolution, we see that in a resolution III design the vector $\boldsymbol{\beta}_1$ consists of the mean and $\Sigma_{i=1}^{t}(k_i - 1)$ main effects and $\boldsymbol{\beta}_3$ consists of second and higher order interaction effects. The model for a resolution III design Γ is implied by (10.1) and is equal to

$$E[\mathbf{Y}_\Gamma] = \mathbf{X}_{\Gamma 1}\boldsymbol{\beta}_1, \qquad \operatorname{Cov}(\mathbf{Y}_\Gamma) = \sigma^2 \mathbf{I}_n, \tag{10.2}$$

where $\mathbf{X}_{\Gamma 1}$ is an $n \times p$ $(= \Sigma k_i - t + 1)$ matrix read off from $\mathbf{X}_\rho$ and $\boldsymbol{\beta}_1$ is the $p \times 1$ vector of the mean and main effects. It is clear that a necessary condition for estimability of $\boldsymbol{\beta}_1$ is that the number of distinct treatment combinations in Γ is greater than or equal to p. If the rank of $\mathbf{X}_\Gamma$ is equal to p, then the BLUE of $\boldsymbol{\beta}_1$ is given by the well-known formula

$$\hat{\boldsymbol{\beta}}_1 = (\mathbf{X}'_{\Gamma 1}\mathbf{X}_{\Gamma 1})^{-1}\mathbf{X}'_{\Gamma 1}\mathbf{Y}_\Gamma, \tag{10.3}$$

with covariance matrix equal to $(\mathbf{X}'_{\Gamma 1}\mathbf{X}_{\Gamma 1})^{-1}\sigma^2$. A design Γ such that rank of $\mathbf{X}_\Gamma$ is less than p is a *singular-resolution III design*. A minimal resolution III design contains exactly $p+1$ distinct treatment combinations and hence is automatically saturated.

Let $\phi_1^0\phi_2^0\ldots\phi_i^{x_i}\ldots\phi_t^0$, $x_i \in \{1,2,\ldots,k_i - 1\}$, be the x_ith main effect corresponding to the ith factor and let $\phi_1^0\phi_2^0\ldots\phi_t^0$ be the mean. A *partially balanced resolution III design* then has the following structure for the information matrix:

$$\mathbf{X}'_{\Gamma 1}\mathbf{X}_{\Gamma 1} = \left[\begin{array}{c:c} n & c\mathbf{1}' \\ \hdashline c\mathbf{1} & a\mathbf{I} + b\mathbf{J} \end{array}\right], \tag{10.4}$$

where n is the number of treatment combinations in Γ, $\mathbf{1}$ is a $(p-1)\times 1$ vector of ones, $\mathbf{I}$ is a $(p-1)\times(p-1)$ identity matrix, $\mathbf{J}$ is $(p-1)\times(p-1)$ matrix of ones, and a, b, c are scalars.

In other situations, such as the quantitative case with the orthogonal polynomial model, a main effect $\phi_1^0\phi_2^0\ldots\phi_i^{x_i}\ldots\phi_t^0$ for the ith factor is said to be of degree x_i and one may impose partial balancedness on Γ by partitioning $\boldsymbol{\beta}_1$ as $\boldsymbol{\beta}'_1 = (\boldsymbol{\beta}_{10} \vdots \boldsymbol{\beta}_{11} \vdots \ldots \vdots \boldsymbol{\beta}_{1k})$ where $\boldsymbol{\beta}_{10} = \phi_1^0\phi_2^0\ldots\phi_t^0$ and $\boldsymbol{\beta}_{1j}$ is a vector containing all main effects of degree j, $1 \le j \le k = \max(k_1 - 1, k_2 - 1, \ldots, k_t - 1)$. The information matrix then has the form

$$\mathbf{X}'_{\Gamma 1}\mathbf{X}_{\Gamma 1} = \left[\begin{array}{c:c:c:c:c} n & c_1\mathbf{1}'_1 & c_2\mathbf{1}'_2 & & c_k\mathbf{1}'_k \\ \hdashline c_1\mathbf{1}_1 & a_1\mathbf{I}_1 + b_1\mathbf{J}_1 & a_{12}\mathbf{J}'_{12} & \cdots & a_{1k}\mathbf{J}'_{1k} \\ \hdashline c_2\mathbf{1}_2 & a_{12}\mathbf{J}_{12} & a_2\mathbf{I}_2 + b_2\mathbf{J}_2 & & a_{2k}\mathbf{J}'_{2k} \\ \hdashline & \vdots & & \ddots & \vdots \\ \hdashline c_k\mathbf{1}_k & a_{1k}\mathbf{J}_{1k} & a_{2k}\mathbf{J}_{2k} & & a_k\mathbf{I}_k + b_k\mathbf{J}_k \end{array}\right], \tag{10.5}$$

where the $\mathbf{1}_j$ are vectors of ones and the $\mathbf{I}_j$ and $\mathbf{J}_{ih}$ are identity matrices and matrices of ones of appropriate dimensions. Note that for the 2^t factorial a partially balanced design always has the form of (10.4).

A resolution III design is *orthogonal* if

$$\mathbf{X}'_{\Gamma 1}\mathbf{X}_{\Gamma 1} = \text{diagonal}\left(d_1, d_2, \ldots, d_p\right). \tag{10.6}$$

This implies that estimators of main effects and the mean will be uncorrelated.

Example 10.1. Consider the 2^3 factorial with the model $E[\mathbf{Y}_p] = [\mathbf{X}_2 \otimes \mathbf{X}_2 \otimes \mathbf{X}_2]\boldsymbol{\beta}_p$, where

$$\mathbf{X}_2 = \begin{bmatrix} 1 & -1 \\ 1 & 1 \end{bmatrix}.$$

Further consider the resolution III designs $\Gamma_1 = \{(0,0,0),(1,0,0),(0,1,0),(0,0,1)\}$ and $\Gamma_2 = \{(0,0,0),(1,1,0),(1,0,1),(0,1,1)\}$. The reader may verify that for $\boldsymbol{\beta}'_1 = (\phi_1^0\phi_2^0\phi_3^0, \phi_1^1\phi_2^0\phi_3^0, \phi_1^0\phi_2^1\phi_3^0, \phi_1^0\phi_2^0\phi_3^1)$, the information matrices have the forms, respectively,

$$\mathbf{X}'_{\Gamma_1 1}\mathbf{X}_{\Gamma_1 1} = \begin{bmatrix} 4 & -2 & -2 & -2 \\ -2 & 4 & 0 & 0 \\ -2 & 0 & 4 & 0 \\ -2 & 0 & 0 & 4 \end{bmatrix}, \qquad \mathbf{X}'_{\Gamma_2 1}\mathbf{X}_{\Gamma_2 1} = \begin{bmatrix} 4 & 0 & 0 & 0 \\ 0 & 4 & 0 & 0 \\ 0 & 0 & 4 & 0 \\ 0 & 0 & 0 & 4 \end{bmatrix}.$$

Hence Γ_1 is a minimal partially balanced resolution III design and Γ_2 is a minimal orthogonal completely balanced resolution III design. Moreover, from Chapter 5, we know that Γ_2 is D-optimal, A-optimal, and E-optimal in the class of minimal resolution III designs.

Example 10.2. Consider the $2^4 \times 3$ factorial with the model $E[\mathbf{Y}_\rho] = [\mathbf{X}_2 \otimes \mathbf{X}_2 \otimes \mathbf{X}_2 \otimes \mathbf{X}_2 \otimes \mathbf{X}_3]\boldsymbol{\beta}_\rho$, where

$$\mathbf{X}_2 = \begin{bmatrix} 1 & -1 \\ 1 & 1 \end{bmatrix}, \qquad \mathbf{X}_3 = \begin{bmatrix} 1 & -1 & 1 \\ 1 & 0 & -2 \\ 1 & 1 & 1 \end{bmatrix}.$$

Let the vector of the mean and the main effects be partitioned as $\boldsymbol{\beta}'_{11} = (\phi_1^1\phi_2^0\phi_3^0\phi_4^0\phi_5^0, \phi_1^0\phi_2^1\phi_3^0\phi_4^0\phi_5^0, \phi_1^0\phi_2^0\phi_3^1\phi_4^0\phi_5^0, \phi_1^0\phi_2^0\phi_3^0\phi_4^1\phi_5^0, \phi_1^0\phi_2^0\phi_3^0\phi_4^0\phi_5^1)$, and $\boldsymbol{\beta}'_{12} = (\phi_1^0\phi_2^0\phi_3^0\phi_4^0\phi_5^0, \phi_1^0\phi_2^0\phi_3^0\phi_4^0\phi_5^2)$. This is a partitioning into vectors of odd (sum of exponents is odd) and even (sum of exponents is even) effects. Let $\Gamma_1 = \{(1,1,1,0,0), (1,1,0,1,0), (1,1,0,0,1), (1,0,1,1,1), (0,1,1,1,1), (0,0,0,1,2), (0,0,1,0,2), (0,0,1,1,1), (0,1,0,0,1), (1,0,0,0,1)\}$ and $\Gamma_2 = \{(0,0,0,0,1), (0,1,0,1,2), (0,0,1,1,2), (0,1,1,0,0), (0,0,0,1,0), (1,1,1,1,1), (1,0,1,0,0), (1,1,0,0,0), (1,0,0,1,2), (1,1,1,0,2)\}$. Note that both Γ_1 and Γ_2 have 10 treatment combinations and in addition each satisfies the condition that when $(x_1, x_2, x_3, x_4, x_5)$ is contained in it then $(1-x_1, 1-x_2, 1-x_3, 1-x_4, 2-x_5)$ is also contained in it. The design matrices of Γ_1 and Γ_2 are, respectively,

$$\mathbf{X}_{\Gamma_1 1} = \left[\begin{array}{rrrrr:rr} 1 & 1 & 1 & -1 & -1 & 1 & 1 \\ 1 & 1 & -1 & 1 & -1 & 1 & 1 \\ 1 & 1 & -1 & -1 & 0 & 1 & -2 \\ 1 & -1 & 1 & 1 & 0 & 1 & -2 \\ -1 & 1 & 1 & 1 & 0 & 1 & -2 \\ -1 & -1 & -1 & 1 & 1 & 1 & 1 \\ -1 & -1 & 1 & -1 & 1 & 1 & 1 \\ -1 & -1 & 1 & 1 & 0 & 1 & -2 \\ -1 & 1 & -1 & -1 & 0 & 1 & -2 \\ 1 & -1 & -1 & -1 & 0 & 1 & -2 \end{array}\right],$$

$$\mathbf{X}_{\Gamma_2 1} = \left[\begin{array}{rrrrr:rr} -1 & -1 & -1 & -1 & 0 & -2 & 1 \\ -1 & 1 & -1 & 1 & 1 & 1 & 1 \\ -1 & -1 & 1 & 1 & 1 & 1 & 1 \\ -1 & 1 & 1 & -1 & -1 & 1 & 1 \\ -1 & -1 & -1 & 1 & -1 & 1 & 1 \\ 1 & 1 & 1 & 1 & 0 & 1 & -2 \\ 1 & -1 & 1 & -1 & -1 & 1 & 1 \\ 1 & 1 & -1 & -1 & -1 & 1 & 1 \\ 1 & -1 & -1 & 1 & 1 & 1 & 1 \\ 1 & 1 & 1 & -1 & 1 & 1 & 1 \end{array}\right].$$

The information matrices are

$$\mathbf{X}'_{\Gamma_1 1}\mathbf{X}_{\Gamma_1 1}=\left[\begin{array}{ccccc:cc} 10 & 2 & -2 & -2 & -4 & 0 & 0 \\ 2 & 10 & -2 & -2 & -4 & 0 & 0 \\ -2 & -2 & 10 & -2 & 0 & 0 & 0 \\ -2 & -2 & -2 & 10 & 0 & 0 & 0 \\ -4 & -4 & 0 & 0 & 4 & 0 & 0 \\ \hdashline 0 & 0 & 0 & 0 & 0 & 10 & -8 \\ 0 & 0 & 0 & 0 & 0 & -8 & 28 \end{array}\right],$$

$$\mathbf{X}'_{\Gamma_2 1}\mathbf{X}_{\Gamma_2 1}=\left[\begin{array}{ccccc:cc} 10 & 2 & 2 & -2 & 0 & 0 & 0 \\ 2 & 10 & 2 & -2 & 0 & 0 & 0 \\ 2 & 2 & 10 & -2 & 0 & 0 & 0 \\ -2 & -2 & -2 & 10 & 4 & 0 & 0 \\ 0 & 0 & 0 & 4 & 8 & 0 & 0 \\ \hdashline 0 & 0 & 0 & 0 & 0 & 10 & 4 \\ 0 & 0 & 0 & 0 & 0 & 4 & 12 \end{array}\right].$$

Both designs are neither partially balanced nor orthogonal relative to $\boldsymbol{\beta}_{11}$ and $\boldsymbol{\beta}_{12}$. It can be verified that $|\mathbf{X}'_{\Gamma_1 1}\mathbf{X}_{\Gamma_1 1}|=2^{15}(54)$ and $|\mathbf{X}'_{\Gamma_2 1}\mathbf{X}_{\Gamma_2 1}|=2^{15}(224)$. In the class of resolution III designs consisting of 10 treatment combinations with the condition that when $(x_1, x_2, x_3, x_4, x_5)$ is contained in a design, then $(1-x_1, 1-x_2, 1-x_3, 1-x_4, 2-x_5)$ is also contained in it, one may show that Γ_2 is a D-optimal design. Notice that both Γ_1 and Γ_2 have the property that estimators of $\boldsymbol{\beta}_{11}$ are uncorrelated with estimators of $\boldsymbol{\beta}_{12}$.

10.3 THE GENERAL COMBINATORIAL PROBLEM OF RESOLUTION III DESIGNS

We have seen that a resolution III design requires the estimation of the $p\times 1$ vector $\boldsymbol{\beta}_1$, which consists of $\Sigma k_i - t$ main effects and the mean. It follows that such a design must contain at least p distinct treatment combinations. In other words, if Γ is a resolution III design, then $m \geq p$. We know that a necessary and sufficient condition for estimating $\boldsymbol{\beta}_1$ is that rank of $\mathbf{X}_{\Gamma 1}$ is equal to p. If Γ contains exactly p treatment combinations, then the design is saturated and minimal relative to $\boldsymbol{\beta}_1$. If $n > p$ and rank of $\mathbf{X}_{\Gamma 1}$ is equal to p, then the variance σ^2 (which is usually unknown) is also estimable. Its estimator is then obtained by the usual formula

$$\hat{\sigma}^2 = \frac{\left[(\mathbf{Y}_\Gamma - \mathbf{X}_{\Gamma 1}\hat{\boldsymbol{\beta}}_1)'(\mathbf{Y}_\Gamma - \mathbf{X}_{\Gamma 1}\hat{\boldsymbol{\beta}}_1)\right]}{(n-p)}. \tag{10.7}$$

Since repetitions of treatment combinations in Γ can be ignored from the viewpoint of rank, all we need to consider for estimability of $\boldsymbol{\beta}_1$ is the set of distinct treatment combinations in Γ. Denote this set by $\bar{\Gamma}$, and call it a *reduced design* obtained from Γ.

Suppose now, that our problem is to construct a resolution III design $\bar{\Gamma}$ of given order m. Obviously there are many choices. Denote the class of possible designs by Δ_m, then clearly its cardinality is equal to

$$|\Delta_m| = \binom{N}{m} = \frac{N!}{(m!)(N-m)!}, \tag{10.8}$$

where $N = \prod k_i$. In general Δ_m will contain designs that are singular, so that they have to be excluded. We may denote the singular class by $\Delta_{m,0}$ and the nonsingular class by $\Delta_{m,p}$. The cardinality of $\Delta_{m,0}$ (or of $\Delta_{m,p}$) is known only for some specific cases. This problem is of extreme complexity even for the 2^t factorial, as will be shown shortly. There are methods available that always provide a $\bar{\Gamma}$ belonging to $\Delta_{m,p}$, but these designs may not be desirable when criteria are imposed. By varying m and applying replications, one may obtain the universal class of resolution III designs.

The most popular optimality criteria for selecting a design in $\Delta_{m,p}$ are usually based on the spectrum of the information matrix. Define two designs Γ_1 and Γ_2 to be of *equi-information* if and only if their underlying information matrices have the same spectrum, that is, if and only if $\mathbf{X}'_{\Gamma_1 1}\mathbf{X}_{\Gamma_1 1}$ and $\mathbf{X}'_{\Gamma_2 1}\mathbf{X}_{\Gamma_2 1}$ are *orthogonally similar*. This is an equivalence relation, which induces a partitioning of $\Delta_{m,p}$ into equivalence classes. Denote the set of equivalence classes by $\gamma_{m,p}$, then we are immediately confronted with another unresolved problem of determining the cardinality of $\gamma_{m,p}$ and finding representatives from each equivalence class. If a specific criterion, say the determinant of the information matrix, is chosen, this will lead to a partitioning of $\Delta_{m,p}$ into determinant equivalence classes. One may then pose the unresolved problem of finding: (i) the range of $|\mathbf{X}'_{\Gamma 1}\mathbf{X}_{\Gamma 1}|$ for $\Gamma \in \Delta_{m,p}$ and hence the cardinality of the set of determinant equivalence classes, (ii) the cardinalities of the equivalence classes themselves, and (iii) representatives from each equivalence class. A representative from the class with the highest determinantal value is then a D-optimal design with m treatment combinations for estimating $\boldsymbol{\beta}_1$. The problems associated with optimal designs can be further compounded by bringing in replications, that is, by working with the universal class of designs rather than $\Delta_{m,p}$.

We now illustrate the combinatorial problem as outlined previously for the 2^t factorial.

Example 10.3. Consider the 2^t factorial with the usual model $E[\mathbf{Y}_\rho] = [\mathbf{X}_2 \otimes \mathbf{X}_2 \otimes \ldots \otimes \mathbf{X}_2]\boldsymbol{\beta}_\rho$, where the Kronecker product is taken t times and

$$\mathbf{X}_2 = \begin{bmatrix} 1 & -1 \\ 1 & 1 \end{bmatrix}.$$

Suppose our interest lies in constructing the class of minimal resolution III designs. Each such design contains exactly $m = p = t+1$ treatment combinations and thus our class of designs is $\Delta_{t+1,t+1}$. The cardinality of $\Delta_{t+1,t+1}$ for general t is unknown as of the present and it has been enumerated for $t \leq 7$. These and other relevant numbers are as follows:

	t = 2	3	4	5	6	7
$\lvert\Delta_{t+1}\rvert = \binom{2^t}{t+1}$	4	70	4368	906192	621216192	1429702652400
$\lvert\Delta_{t+1,0}\rvert$	0	12	1360	350000	255036992	571462430224
$\lvert\Delta_{t+1,t+1}\rvert$	4	58	3008	556192	366179200	858240222176
$(\lvert\Delta_{t+1,0}\rvert)\Big/\binom{2^t}{t+1}$	0	0.1714	0.3114	0.3862	0.4105	0.3997
$(\lvert\Delta_{t+1,t+1}\rvert)\Big/\binom{2^t}{t+1}$	1	0.8286	0.6886	0.6138	0.5895	0.6003

Note that the proportion of singular designs increases up to $t=6$ but decreases the first time when $t=7$. It may be shown that the proportion of singular designs goes to zero as t goes to infinity. This means that for "large" t the class Δ_{t+1} consists of "almost all" nonsingular designs.

Using D-optimality, it is known that for any minimal design Γ of the 2^t factorial

(10.9) $$0 \leq |\det X_{\Gamma 1}| \leq \text{integer part of}\,(t+1)^{(t+1)/2}$$

or

(10.10) $$0 \leq \det X'_{\Gamma 1} X_{\Gamma 1} \leq (t+1)^{(t+1)},$$

where the upper bound is achieved if and only if $\mathbf{X}_{\Gamma 1}$ is a Hadamard matrix. A *Hadamard matrix* $\mathbf{H}$ of order h is a $(-1, 1)$-matrix such that $\mathbf{H}'\mathbf{H} = h\mathbf{I}_h$ (see also Chapter 13). A necessary condition for the existence of such a matrix (apart from $t+1=2$) is that $t+1 = 0 \pmod 4$. In the case $t+1 = 4d$, the upper bound in (10.10) is equal to $(4d)^{4d}$. The problem of constructing a minimal D-optimal design for the arbitrary 2^t factorial is not a trivial matter, and it is unresolved in all its generality.

10.4 CONSTRUCTION OF RESOLUTION III DESIGNS

The construction of resolution III designs for the general $k_1 \times k_2 \times \cdots \times k_t$ factorial depends on the specification of the design parameters m, n, and $r_1, r_2, \ldots, r_N$. We know already that $m \geq p = \Sigma k_i - t + 1$ and that rank of $\mathbf{X}_{\Gamma 1}$ for any resolution III design must be equal to p. Let us state at the outset that there are no general construction procedures to provide *optimal* (e.g., D-optimal) resolution III designs for the $k_1 \times k_2 \times \cdots \times k_t$ factorial and arbitrary parameters n, m, and $r_1, r_2, \ldots, r_N$. What is typically done in the literature is to give for a particular factorial (e.g., the 2^t factorial) optimal design constructions for selected values of n, m, $r_1, \ldots, r_N$ (e.g., $n = m = t + 1 = 4d$, so that all nonzero r_i are equal to 1).

To make some headway, let us tackle the construction of a minimal resolution III design, which has cardinality $n = m = p = \Sigma k_i - t + 1$. For, if we are able to construct one then we may generate a whole class of resolution III designs by augmenting and/or repeating treatment combinations without affecting the rank of the design matrix. We consider the following model (any other columnwise orthogonal matrix will lead to similar results) for the minimal complete factorial design:

$$E[\mathbf{Y}_\rho] = [\mathbf{X}_1 \otimes \mathbf{X}_2 \otimes \cdots \otimes \mathbf{X}_t]\boldsymbol{\beta}_\rho = \mathbf{X}_\rho \boldsymbol{\beta}_\rho, \tag{10.11}$$

when $\mathbf{X}_i$ is the following columnwise orthogonal matrix:

$$\mathbf{X}_i = \begin{bmatrix} 1 & 1 & 1 & \cdots & 1 \\ 1 & -1 & 1 & \cdots & 1 \\ 1 & 0 & -2 & \cdots & 1 \\ \vdots & \vdots & \vdots & \ddots & \\ 1 & 0 & 0 & \cdots & -(k_i - 1) \end{bmatrix}. \tag{10.12}$$

Such matrices arise when *Helmert polynomials* rather than orthogonal polynomials are considered, and they are easy to work with.

The design matrix $\mathbf{X}_{\Gamma 1}$ for any resolution III design can be read off from $\mathbf{X}_\rho$. However, because of the structure of $\mathbf{X}_i$, we may describe $\mathbf{X}_{\Gamma 1}$ explicitily. Let Γ be written out as an $n \times t$ matrix and let $\mathbf{A}'_{ih_i} = (a_{ih_i r})$, $i = 1, 2, \ldots, t$,

$h_i = 0,1,2,\ldots,k_i - 1$, $r = 1,2,\ldots,n$, be a vector such that

$$(10.13) \qquad a_{ih_ir} = \begin{cases} 1, & \text{if the } (r,i)\text{th element of } \Gamma \text{ is equal to } h \\ 0, & \text{otherwise.} \end{cases}$$

Note that the vectors $\mathbf{A}_{ih_i}$ are incidence vectors with the property

$$(10.14) \qquad \sum_{h_i=0}^{k_i-1} \mathbf{A}_{ih_i} = \mathbf{1}_{n\times 1},$$

$$\Gamma = \left[\sum_{h_1=0}^{k_1-1} h_1 \mathbf{A}_{1h_1} \vdots \sum_{h_2=0}^{k_2-1} h_2 \mathbf{A}_{2h_2} \vdots \cdots \vdots \sum_{h_t=0}^{k_t-1} h_t \mathbf{A}_{th_t} \right].$$

Writing the vector of main effects as $\boldsymbol{\beta}_1' = (\phi_1^0\phi_2^0 \ldots \phi_t^0, \phi_1^1\phi_2^0 \ldots \phi_t^0, \ldots, \phi_1^{k_1-1}\phi_2^0 \ldots \phi_t^0, \phi_1^0\phi_2^1 \ldots \phi_t^0, \ldots, \phi_1^0\phi_2^{k_2-1} \ldots \phi_t^0, \ldots, \phi_1^0\phi_2^0 \ldots \phi_t^1, \ldots, \phi_1^0\phi_2^0 \ldots \phi_t^{k_t-1})$, one may write

$$(10.15) \qquad \mathbf{X}_{\Gamma 1} = \left[\mathbf{1} \vdots \mathbf{H}_1 \vdots \mathbf{H}_2 \vdots \cdots \vdots \mathbf{H}_t \right],$$

where $\mathbf{1}$ is an $n \times 1$ column vector of ones and

$$(10.16) \qquad \mathbf{H}_i = \left[\mathbf{A}_{i0} - \mathbf{A}_{i1} \vdots \mathbf{A}_{i0} + \mathbf{A}_{i1} - 2\mathbf{A}_{i2} \vdots \cdots \vdots \sum_{h_i=0}^{k_i-1} \mathbf{A}_{ih_i} - (k_i - 1)\mathbf{A}_{ik_i-1} \right].$$

Example 10.4. Consider the $2^2 \times 3^2 \times 4$ factorial. Then the matrices $\mathbf{X}_i$ of (10.12) are, respectively,

$$\mathbf{X}_1 = \mathbf{X}_2 = \begin{bmatrix} 1 & 1 \\ 1 & -1 \end{bmatrix}, \qquad \mathbf{X}_3 = \mathbf{X}_4 = \begin{bmatrix} 1 & 1 & 1 \\ 1 & -1 & 1 \\ 1 & 0 & -2 \end{bmatrix},$$

and

$$\mathbf{X}_5 = \begin{bmatrix} 1 & 1 & 1 & 1 \\ 1 & -1 & 1 & 1 \\ 1 & 0 & -2 & 1 \\ 1 & 0 & 0 & -3 \end{bmatrix}.$$

Let Γ be the minimal resolution III design

$$\Gamma = \{(0,0,0,0,0),(1,0,0,0,0),(0,1,0,0,0),\\ (0,0,1,0,0),(0,0,2,0,0),(0,0,0,1,0),\\ (0,0,0,2,0),(0,0,0,0,1),(0,0,0,0,2),\\ (0,0,0,0,3)\}.$$

The incidence vectors of (10.13) are obtained as

$$\begin{aligned}
\mathbf{A}'_{10} &= (1\ 0\ 1\ 1\ 1\ 1\ 1\ 1\ 1\ 1), & \mathbf{A}'_{11} &= (0\ 1\ 0\ 0\ 0\ 0\ 0\ 0\ 0\ 0),\\
\mathbf{A}'_{20} &= (1\ 1\ 0\ 1\ 1\ 1\ 1\ 1\ 1\ 1), & \mathbf{A}'_{21} &= (0\ 0\ 1\ 0\ 0\ 0\ 0\ 0\ 0\ 0),\\
\mathbf{A}'_{30} &= (1\ 1\ 1\ 0\ 0\ 1\ 1\ 1\ 1\ 1), & \mathbf{A}'_{31} &= (0\ 0\ 0\ 1\ 0\ 0\ 0\ 0\ 0\ 0),\\
\mathbf{A}'_{32} &= (0\ 0\ 0\ 0\ 1\ 0\ 0\ 0\ 0\ 0), & \mathbf{A}'_{40} &= (1\ 1\ 1\ 1\ 1\ 0\ 0\ 1\ 1\ 1),\\
\mathbf{A}'_{41} &= (0\ 0\ 0\ 0\ 0\ 1\ 0\ 0\ 0\ 0), & \mathbf{A}'_{42} &= (0\ 0\ 0\ 0\ 0\ 0\ 1\ 0\ 0\ 0),\\
\mathbf{A}'_{50} &= (1\ 1\ 1\ 1\ 1\ 1\ 1\ 0\ 0\ 0), & \mathbf{A}'_{51} &= (0\ 0\ 0\ 0\ 0\ 0\ 0\ 1\ 0\ 0),\\
\mathbf{A}'_{52} &= (0\ 0\ 0\ 0\ 0\ 0\ 0\ 0\ 1\ 0), & \mathbf{A}'_{53} &= (0\ 0\ 0\ 0\ 0\ 0\ 0\ 0\ 0\ 1).
\end{aligned}$$

Observe that: $\mathbf{A}_{10}+\mathbf{A}_{11}=\mathbf{1}$, $\mathbf{A}_{20}+\mathbf{A}_{21}=\mathbf{1}$, $\mathbf{A}_{30}+\mathbf{A}_{31}+\mathbf{A}_{32}=\mathbf{1}$, $\mathbf{A}_{40}+\mathbf{A}_{41}+\mathbf{A}_{42}=\mathbf{1}$, and $\mathbf{A}_{50}+\mathbf{A}_{51}+\mathbf{A}_{52}+\mathbf{A}_{53}=\mathbf{1}$. Also $(\Sigma_{h_1=0}^{1} h_1\mathbf{A}_{1h_1})' = (0\ 1\ 0\ 0\ 0\ 0\ 0\ 0\ 0\ 0)$, $(\Sigma_{h_2=0}^{1} h_2\mathbf{A}_{2h_2})' = (0\ 0\ 1\ 0\ 0\ 0\ 0\ 0\ 0\ 0)$, $(\Sigma_{h_3=0}^{2} h_3\mathbf{A}_{3h_3})' = (0\ 0\ 0\ 1\ 2\ 0\ 0\ 0\ 0\ 0)$, $(\Sigma_{h_4=0}^{2} h_4\mathbf{A}_{4h_4})' = (0\ 0\ 0\ 0\ 0\ 1\ 2\ 0\ 0\ 0)$, $(\Sigma_{h_5=0}^{3} h_5\mathbf{A}_{5h_5})' = (0\ 0\ 0\ 0\ 0\ 0\ 0\ 1\ 2\ 3)$. It may now be verified that the columns of Γ consist of the preceding five vectors, which satisfy the second part of (10.14). The matrices in (10.16) are obtained by calculating

$$\begin{aligned}
\mathbf{A}'_{10}-\mathbf{A}'_{11} &= (1\ -1\ 1\ 1\ 1\ 1\ 1\ 1\ 1\ 1),\\
\mathbf{A}'_{20}-\mathbf{A}'_{21} &= (1\ 1\ -1\ 1\ 1\ 1\ 1\ 1\ 1\ 1),\\
\mathbf{A}'_{30}-\mathbf{A}'_{31} &= (1\ 1\ 1\ -1\ 0\ 1\ 1\ 1\ 1\ 1),\\
\mathbf{A}'_{30}+\mathbf{A}'_{31}-2\mathbf{A}'_{32} &= (1\ 1\ 1\ 1\ -2\ 1\ 1\ 1\ 1\ 1),\\
\mathbf{A}'_{40}-\mathbf{A}'_{41} &= (1\ 1\ 1\ 1\ 1\ -1\ 0\ 1\ 1\ 1),\\
\mathbf{A}'_{40}+\mathbf{A}'_{41}-2\mathbf{A}'_{42} &= (1\ 1\ 1\ 1\ 1\ 1\ -1\ 1\ 1\ 1),\\
\mathbf{A}'_{50}-\mathbf{A}'_{51} &= (1\ 1\ 1\ 1\ 1\ 1\ 1\ -1\ 0\ 0),\\
\mathbf{A}'_{50}+\mathbf{A}'_{51}-2\mathbf{A}'_{52} &= (1\ 1\ 1\ 1\ 1\ 1\ 1\ 1\ -2\ 0),\\
\mathbf{A}'_{50}+\mathbf{A}'_{51}+\mathbf{A}'_{52}-3\mathbf{A}'_{53} &= (1\ 1\ 1\ 1\ 1\ 1\ 1\ 1\ 1\ -3).
\end{aligned}$$

Hence

$$\mathbf{H}'_1 = [1 \ -1 \ 1 \ 1 \ 1 \ 1 \ 1 \ 1 \ 1 \ 1],$$

$$\mathbf{H}'_2 = [1 \ 1 \ -1 \ 1 \ 1 \ 1 \ 1 \ 1 \ 1 \ 1],$$

$$\mathbf{H}'_3 = \begin{bmatrix} 1 & 1 & 1 & -1 & 0 & 1 & 1 & 1 & 1 & 1 \\ 1 & 1 & 1 & 1 & -2 & 1 & 1 & 1 & 1 & 1 \end{bmatrix},$$

$$\mathbf{H}'_4 = \begin{bmatrix} 1 & 1 & 1 & 1 & 1 & -1 & 0 & 1 & 1 & 1 \\ 1 & 1 & 1 & 1 & 1 & 1 & -2 & 1 & 1 & 1 \end{bmatrix},$$

$$\mathbf{H}'_5 = \begin{bmatrix} 1 & 1 & 1 & 1 & 1 & 1 & 1 & -1 & 0 & 0 \\ 1 & 1 & 1 & 1 & 1 & 1 & 1 & 1 & -2 & 0 \\ 1 & 1 & 1 & 1 & 1 & 1 & 1 & 1 & 1 & -3 \end{bmatrix}.$$

Therefore, the design matrix of Γ according to (10.15) is equal to

$$\mathbf{X}_{\Gamma 1} = \begin{bmatrix} 1 & 1 & 1 & 1 & 1 & 1 & 1 & 1 & 1 & 1 \\ 1 & -1 & 1 & 1 & 1 & 1 & 1 & 1 & 1 & 1 \\ 1 & 1 & -1 & 1 & 1 & 1 & 1 & 1 & 1 & 1 \\ 1 & 1 & 1 & -1 & 1 & 1 & 1 & 1 & 1 & 1 \\ 1 & 1 & 1 & 0 & -2 & 1 & 1 & 1 & 1 & 1 \\ 1 & 1 & 1 & 1 & 1 & -1 & 1 & 1 & 1 & 1 \\ 1 & 1 & 1 & 1 & 1 & 0 & -2 & 1 & 1 & 1 \\ 1 & 1 & 1 & 1 & 1 & 1 & 1 & -1 & 1 & 1 \\ 1 & 1 & 1 & 1 & 1 & 1 & 1 & 0 & -2 & 1 \\ 1 & 1 & 1 & 1 & 1 & 1 & 1 & 0 & 0 & -3 \end{bmatrix}.$$

Using a suitable transformation matrix, we may transform the design matrix of any resolution III design to a (0,1)-matrix. For a minimal resolution III design we may (using Hadamard's theorem) then show that

$$(10.17) \qquad \prod_{i=1}^{t} (k_i!) \le |\det \mathbf{X}_{\Gamma 1}| \le \left(\prod_{i=1}^{t} k_i! \right) \text{integer part of } m^{m/2} \prod_{i=1}^{t} k_i^{-k_i/2}$$

or

$$(10.18) \qquad \prod_{i=1}^{t} (k_i!)^2 \le \det \mathbf{X}'_{\Gamma 1} \mathbf{X}_{\Gamma 1} \le \prod_{i=1}^{t} (k_i!)^2 m^m \prod_{i=1}^{t} k_i^{-k_i}.$$

The lower bound is attained by the so-called *one-at-a-time resolution III design*. This minimal design, say Γ_0, is obtained by holding the levels of all factors but one fixed and then running through all levels of the factor not fixed, that is,

$$\Gamma_0 = \{(0,0,0,\ldots,0),(1,0,0,\ldots,0),(2,0,0,\ldots,0),\ldots,(k_1-1,0,0,\ldots,0),\ldots,$$
$$(0,1,0,\ldots,0),(0,2,0,\ldots,0),\ldots,(0,k_2-1,0,\ldots,0),\ldots,$$
$$(0,0,0,\ldots,0,1),(0,0,0,\ldots,0,2),\ldots,(0,0,0,\ldots,0,k_t-1)\}.$$

It follows that when D-optimality is used for selecting an optimal design, then the one-at-a-time designs are least optimal.

As stated earlier, there exists no general construction procedure for D-optimal minimal resolution III designs. The lower bound in (10.17) or (10.18) is always attained by the one-at-a-time design. However, the upper bound is attained if and only if an orthogonal minimal resolution III design can be constructed. Note that the bounds in (10.17) and (10.18) are generalizations of those given in (10.9) and (10.10) for the 2^t factorial, and we know that the upper bound for the 2^t factorial is achieved when $\mathbf{X}_{\Gamma 1}$ is a Hadamard matrix. These matrices have been constructed for all $t+1 \leq 200$, where $t+1=0 \pmod 4$. For cases where $t+1 \neq 0 \pmod 4$ the construction of D-optimal designs has proceeded on a case-by-case or class-by-class basis.

We will not provide details of various constructions for specific factorials in this book. In the following examples we provide several designs to illustrate the development given previously. The reader is referred to the literature list at the end of this chapter and also to the techniques illustrated in Chapter 13, along with the literature cited there.

Example 10.5. The design given in Example 10.4 is a one-at-a-time design of the $2^2 \times 3^2 \times 4$ factorial. The reader may verify that $|\det \mathbf{X}_{\Gamma_0}| = 3456 = (2!)(2!)(3!)(3!)(4!)$, which agrees with (10.17). Hence in the class Δ_p consisting of $C(N, p) = C(144, 10)$ designs the one-at-a-time design is a least D-optimal unbiased resolution III design.

Example 10.6. In Example 10.3 we discussed the 2^t factorial from the viewpoint of singularity and nonsingularity, that is, in terms of the classes $\Delta_{t+1,0}$ and $\Delta_{t+1,t+1}$. For $2 \leq t \leq 7$, we have the following D-optimal designs:

t	D-optimal design
2	$\{(0,0),(1,0),(0,1)\}$
3	$\{(0,0,0),(1,1,0),(1,0,1),(0,1,1)\}$
4	$\{(0,0,0,0),(1,1,1,0),(1,1,0,1),(1,0,1,1),(0,1,1,1)\}$
5	$\{(0,0,0,0,0),(1,1,1,0,0),(1,1,0,1,0),(1,1,0,0,1),(1,0,1,1,1),$ $(0,1,1,1,1)\}$
6	$\{(0,0,0,0,0,0),(1,1,1,0,0,0),(1,1,0,1,1,0),(1,1,0,1,0,1),$ $(1,0,1,1,0,0),(1,0,1,0,1,1),(0,1,1,1,1,1)\}$
7	$\{(0,0,0,0,0,0,0),(1,1,1,1,0,0,0),(1,1,0,0,1,1,0),(1,0,1,0,1,0,1),$ $(1,0,0,1,0,1,1),(0,1,1,0,0,1,1),(0,1,0,1,1,0,1),(0,0,1,1,1,1,0)\}$.

The precise range of $|\det \mathbf{X}_{\Gamma 1}|$ is known for $t \leq 7$ and was studied for $t=8$, 9, and 10. For example, for $t=3$, the possible values of $|\det \mathbf{X}_{\Gamma 1}|$ are 0, 8, and 16. In this case the nonsingular class $\Delta_{4,4}$ consisting of 58 designs decomposes into two corresponding classes of cardinalities 56 and 2, respectively. From Chapter 9, we know that the vector of main effects and the mean form an admissible vector and that under the set of permutations Ω a permuted design is of equi-information to the given initial design. This implies that under Ω, $|\det \mathbf{X}_{\Gamma 1}|$ will be invariant as well. For the 2^t factorial, application of $\omega \in \Omega$ to a given design Γ is equivalent to componentwise addition modulo 2 of a fixed treatment combination to all the treatment combinations in Γ. Thus from the D-optimal design $\Gamma = \{(0,0,0),(1,1,0),(1,0,1),(0,1,1)\}$, we obtain the equi-information design $\omega(\Gamma) = \Gamma + (1,0,0) = \{(1,0,0),(0,1,0),(0,0,1),(1,1,1)\}$. The action of Ω produces only two designs from Γ. A representative of the 56 least D-optimal designs is the one-at-a-time plan $\Gamma = \{(0,0,0),(1,0,0),(0,1,0),(0,0,1)\}$. The action of Ω on Γ produces exactly eight equi-information designs. It may be shown that there exist seven designs among the 56 such that each of them generates exactly eight designs under Ω. Similarly, there exist six singular designs among the 12 singular ones such that each of them generates two designs under Ω. Hence the class of 70 designs can be decomposed into $1+7+6=14$ classes such that one representative from the first class (the D-optimal class) generates 2 designs under Ω, one representative from each of the next seven classes generates eight designs under Ω, and one representative from each of the remaining six classes generates two designs under Ω. A neat way of writing this result is $C(8,4) = 70 = 1(2) + 7(8) + 6(2)$ with the meaning that there are exactly two designs with $|\mathbf{X}_{\Gamma 1}| = 16$, which are generated by one generator, 56 designs with $|\mathbf{X}_{\Gamma 1}| = 8$, which are generated by seven generators, and 12 designs with $|\mathbf{X}_{\Gamma 1}| = 0$, which are generated by six generators. Such a decomposition is, in view of what has been said in Section 10.3, not known for the general 2^t factorial, let alone the general $k_1 \times k_2 \times \cdots \times k_t$ factorial.

Example 10.7. Consider the 3^4 factorial and the minimal resolution III design $\Gamma = \{(0,0,0,0),(0,1,1,1),(0,2,2,2),(1,0,1,2),(1,1,2,0),(1,2,0,1),(2,0,2,1),(2,1,0,2),(2,2,1,0)\}$. The model for this design is equal to

$$E\begin{bmatrix} Y_{(0,0,0,0)} \\ Y_{(0,1,1,1)} \\ Y_{(0,2,2,2)} \\ Y_{(1,0,1,2)} \\ Y_{(1,1,2,0)} \\ Y_{(1,2,0,1)} \\ Y_{(2,0,2,1)} \\ Y_{(2,1,0,2)} \\ Y_{(2,2,1,0)} \end{bmatrix} = \begin{bmatrix} 1 & 1 & 1 & 1 & 1 & 1 & 1 & 1 & 1 \\ 1 & 1 & 1 & -1 & 1 & -1 & 1 & -1 & 1 \\ 1 & 1 & 1 & 0 & -2 & 0 & -2 & 0 & -2 \\ 1 & -1 & 1 & 1 & 1 & -1 & 1 & 0 & -2 \\ 1 & -1 & 1 & -1 & 1 & 0 & -2 & 1 & 1 \\ 1 & -1 & 1 & 0 & -2 & 1 & 1 & -1 & 1 \\ 1 & 0 & -2 & 1 & 1 & 0 & -2 & -1 & 1 \\ 1 & 0 & -2 & -1 & 1 & 1 & 1 & 0 & -2 \\ 1 & 0 & -2 & 0 & -2 & -1 & 1 & 1 & 1 \end{bmatrix} \begin{bmatrix} \phi_1^0\phi_2^0\phi_3^0\phi_4^0 \\ \phi_1^1\phi_2^0\phi_3^0\phi_4^0 \\ \phi_1^2\phi_2^0\phi_3^0\phi_4^0 \\ \phi_1^0\phi_2^1\phi_3^0\phi_4^0 \\ \phi_1^0\phi_2^2\phi_3^0\phi_4^0 \\ \phi_1^0\phi_2^0\phi_3^1\phi_4^0 \\ \phi_1^0\phi_2^0\phi_3^2\phi_4^0 \\ \phi_1^0\phi_2^0\phi_3^0\phi_4^1 \\ \phi_1^0\phi_2^0\phi_3^0\phi_4^2 \end{bmatrix}.$$

It turns out that $|\det \mathbf{X}_{\Gamma 1}| = 34992 = (3!)^4(27)$, and it may be verified that the upper bound in (10.17) is achieved. Thus Γ is a D-optimal resolution III design. Further note that $\mathbf{X}'_{\Gamma 1}\mathbf{X}_{\Gamma 1} =$ diagonal (9 6 18 6 18 6 18 6 18) so that Γ is an orthogonal design. If $\boldsymbol{\beta}'_1$ is written as $(\phi_1^0\phi_2^0\phi_3^0\phi_4^0, \phi_1^1\phi_2^0\phi_3^0\phi_4^0, \phi_1^0\phi_2^1\phi_3^0\phi_4^0, \phi_1^0\phi_2^0\phi_3^1\phi_4^0, \phi_1^0\phi_2^0\phi_3^0\phi_4^1, \phi_1^2\phi_2^0\phi_3^0\phi_4^0, \phi_1^0\phi_2^2\phi_3^0\phi_4^0, \phi_1^0\phi_2^0\phi_3^2\phi_4^0, \phi_1^0\phi_2^0\phi_3^0\phi_4^2)$, then $\mathbf{X}'_{\Gamma 1}\mathbf{X}_{\Gamma 1} =$ diagonal (9 6 6 6 6 18 18 18 18) so that Γ is a partially balanced factorial design.

In the following list some of the references on resolution III designs also contain designs of other resolutions. In Chapter 11, designs of resolution IV and V are developed with a list of references, some of which also contain designs of resolution III.

10.5 SELECTED ADDITIONAL READING

Starred references are recommended for a first reading.

1. Addelman, S. (1962). Orthogonal main-effect plans for asymmetrical factorial experiments. *Technometrics*, **4**, 21–46.

*2. Addelman, S. (1963). Techniques for constructing fractional replicate plans. *J. Am. Stat. Assoc.*, **58**, 45–71.

*3. Anderson, D. A., and Federer, W. T. (1975). Possible absolute determinant values for square (0,1)-matrices useful in fractional replication. *Util. Math.*, **7**, 135–150.

4. Banerjee, A. K., Dey, A., and Saha, G. M. (1975). Some main effect plans for 3^n factorials. *Ann. Inst. Stat. Math.*, **27**, 159–165.

5. Banerjee, K. S. (1950). A note on fractional replication of factorial arrangements. *Sankhyā*, **10**, 87–94.

6. Banerjee, K. S., and Federer, W. T. (1967). On a special subset giving an irregular fractional replicate of a 2^n factorial experiment. *J. R. Stat. Soc.*, **B29**, 292–299.

*7. Box, G. E. P., and Hunter, J. S. (1961a). The 2^{k-p} fractional factorial designs. I. *Technometrics*, **3**, 311–351.

8. Box, G. E. P., and Hunter, J. S. (1961b). The 2^{k-p} fractional factorial designs. II. *Technometrics*, **3**, 449–458.

9. Connor, W. S., Clatworthy, W. H., and Zelen, M. (1957). Fractional factorial experiment designs for factors at two levels. *Natl. Bur. Stand. Appl. Math. Ser.*, **48**.

10. Connor, W. S., and Young, S. (1961). Fractional factorial designs for experiments with factors at two and three levels. *Natl. Bur. Stand. Appl. Math. Ser.*, **58**.

11. Connor, W. S., and Zelen, M. (1959). Fractional factorial experiment designs for factors at three levels. *Natl. Bur. Stand. Appl. Math. Ser.*, **54**.

12. Daniel, C. (1973). One-at-a-time plans. *J. Am. Stat. Assoc.*, **68**, 353–360.

13. Dey, A., and Saha, G. M. (1970). Main effect plans for k^n factorials with blocks. *Ann. Inst. Stat. Math.*, **22**, 381–388.

14. Draper, N. R., and Mitchell, T. J. (1967). The construction of saturated 2_R^{k-p} designs. *Ann. Math. Stat.*, **38**, 1110–1126.

15. Ehlich, H. (1964). Determinantenbeschatzungen für binäre matrizen. *Math. Z.*, **83**, 123–132.

16. Finney, D. J. (1945). The fractional replication of factorial experiments. *Ann. Eugen.*, **12**, 291–301.

17. Galil, Z. and Kiefer, J. (1980). D-optimum weighing designs. *Ann. Stat.*, **8**, 1293–1306.

18. Geramita, A. V., and Seberry, J. (1979). *Orthogonal Designs*. Marcel Dekker, New York.

19. Hedayat, A., and Wallis, W. D. (1978). Hadamard matrices and their applications. *Ann. Stat.*, **6**, 1184–1238.

20. Herzberg, A. M., and Cox, D. R. (1969). Recent work on the design of experiments: a bibliography and review. *J. R. Stat. Soc.*, **A132**, 29–67.

21. Huang, J. S., and Raktoe, B. L. (1977). A note on the proportion of singular (0, 1)-matrices in a class associated with fractional 2^n factorial designs. *J. Comb. Theory*, **A22**, 231–234.

*22. John, P. W. M. (1969). Some non-orthogonal fractions of 2^n designs. *J. R. Stat. Soc.*, **B31**, 270–275.

23. Kiefer, J. and Galil, Z. (1980). Optimum weighing designs. In *Recent Developments in Statistical Inference and Data Analysis*, K. Matusita, Ed. North Holland, Amsterdam, pp 183–189.

24. Komlos, J. (1967). On the determinants of (0, 1)-matrices. *Stud. Sci. Math. Hungar.*, **2**, 7–21.

*25. Margolin, B. H. (1968). Orthogonal main-effect 2^n3^m designs and two-factor interaction aliasing. *Technometrics*, **10**, 559–573.

26. Margolin, B. H. (1969). Orthogonal main-effect plans permitting estimation of all two-factor interactions for the 2^n3^m factorial series of design. *Technometrics*, **11**, 747–762.

*27. Margolin, B. H. (1972). Non-orthogonal main effect designs for asymmetrical factorial experiments. *J. R. Stat. Soc.*, **B34**, 431–440.

28. Metropolis, N., and Stein, P. R. (1967). On a class of (0, 1)-matrices with vanishing determinants. *J. Comb. Theory*, **3**, 191–198.

29. Paley, R. E. A. C. (1933). On orthogonal matrices. *J. Math. Phys.*, **12**, 311–320.

*30. Pesotan, H., and Raktoe, B. L. (1975). Remarks on extensions of the Anderson-Federer (0, 1)-matrix procedure for main effects to effects of a higher degree. *Commun. Stat.*, **4**, 797–811.

31. Pesotan, H., Raktoe, B. L., and Federer, W. T. (1975). On complexes of Abelian groups with applications to fractional factorial designs. *Ann. Inst. Stat. Math.*, **27**, 117–142.

*32. Pesotan, H., Raktoe, B. L., and Worthley, R. G. (1978). On main effect fold-over designs. *J. Stat. Plan. Inference*, **2**, 277–291.

*33. Plackett, R. L., and Burman, J. P. (1946). The design of optimum multifactorial experiments. *Biometrika*, **33**, 305–325.

34. Raktoe, B. L. (1974a). A geometrical formulation of an unsolved problem in fractional factorial designs. *Commun. Stat.*, **3**, 959–968.

35. Raktoe, B. L. (1974b). On classes of equi-information factorial arrangements. *Proceedings of the Eighth International Biometrics Conference*. Editura Academieii, Republicii Socialiste România.

36. Raktoe, B. L. (1979). How many degenerate simplices are generated by $n+1$ vertices of the unit n-cube? *Am. Math. Monthly*, **86**, 49.

37. Raktoe, B. L., and Federer, W. T. (1968). A unified approach for constructing a useful class of non-orthogonal main effect plans in k^n factorials. *J. R. Stat. Soc.*, **B30**, 371–380.

38. Raktoe, B. L., and Federer, W. T. (1969). Some non-orthogonal unsaturated main effect and resolution V plans derived from a one-restrictional lattice. *Ann. Inst. Stat. Math.*, **21**, 335–342.

*39. Rechtschaffner, R. L. (1967). Saturated fractions of 2^n and 3^n factorial designs. *Technometrics*, **9**, 569–575.

40. Ryser, H. J. (1963). *Combinatorial Mathematics* (Carus Monograph No. 14). Mathmatical Association of America, Oberlin, Ohio and Wiley, New York.

*41. Saha, G. M., and Mohanty, S. (1970). On non-orthogonal main effect plans for asymmetrical factorials. *Ann. Inst. Stat. Math.*, **22**, 159–169.

42. Srivastava, J. N., Raktoe, B. L., and Pesotan, H. (1976). On invariance and randomization in fractional replication. *Ann. Stat.*, **4**, 423–430.

43. Starks, T. H. (1964). A note on small orthogonal main-effect plans for factorial experiments. *Technometrics*, **6**, 220–222.

*44. Webb, S. R. (1971). Small incomplete factorial experiment designs for two-and three-level factors. *Technometrics*, **13**, 243–256.

45. Wells, M. B. (1971). *Elements of Combinatorial Computing*. Pergamon, New York.

CHAPTER 11

Factorial Designs of Resolutions IV and V

The problems associated with factorial designs of even resolution are compounded because in the partitioning (Cases 3 or 4) of Chapter 7, that is, $\boldsymbol{\beta}_p' = (\boldsymbol{\beta}_1' \vdots \boldsymbol{\beta}_2' \vdots \boldsymbol{\beta}_3')$, the $N_2 \times 1$ vector of parameters $\boldsymbol{\beta}_2$ behaves as a vector of *nuisance parameters*. In a resolution IV factorial design the role of nuisance parameters is played by the mean and $\Sigma_{u<v}(k_u - 1)(k_v - 1)$ two-factor interaction effects. On the other hand, the problems associated with odd resolution factorial designs, in particular resolution V factorial designs, are similar to those outlined for resolution III factorial designs, which have been fully discussed in Chapter 10. The only difference between the development for the resolution V case and the resolution III case is that the vector $\boldsymbol{\beta}_1$ now consists of the mean, the $\Sigma(k_i - 1)$ main effects, *and* the $\Sigma_{u<v}(k_u - 1)\ (k_v - 1)$ two-factor interaction effects. The purpose of this chapter is to discuss some aspects of resolution IV and V factorial designs as was done earlier for resolution III factorial designs.

11.1 THE ASSUMED MODEL FOR RESOLUTION IV FACTORIAL DESIGNS

In Chapter 7 we saw that a resolution IV design belongs to the class of even resolution designs. In a resolution IV design the objective is to estimate the $q = \Sigma_{i=1}^{t}(k_i - 1)$ main effects under the assumption that interaction effects of order three or higher are negligible. Note that the mean and the two-factor interaction effects are not assumed to be negligible and that they are not of interest with regard to estimation. Hence they play the role of *nuisance parameters*.

The model for the observation vector $\mathbf{Y}_\Gamma$ corresponding to a resolution IV design is given by

$$(11.1) \qquad E[\mathbf{Y}_\Gamma] = [\mathbf{X}_{\Gamma 1} \vdots \mathbf{X}_{\Gamma 2}] \begin{bmatrix} \boldsymbol{\beta}_1 \\ \cdots \\ \boldsymbol{\beta}_2 \end{bmatrix}, \qquad \operatorname{Cov}(\mathbf{Y}_\Gamma) = \sigma^2 \mathbf{I}_n,$$

where $\mathbf{X}_{\Gamma 1}$ is an $n \times q\ [=\Sigma_{i=1}^{t}(k_i - 1)]$ matrix obtained from $\mathbf{X}_\rho$ of (10.1), $\mathbf{X}_{\Gamma 2}$ is an $n \times r\ [=1+\Sigma_{u<v}(k_u - 1)(k_v - 1)]$ matrix also obtained from $\mathbf{X}_\rho$ of (10.1), $\boldsymbol{\beta}_1$ is the $q \times 1$ vector consisting of the main effects, and $\boldsymbol{\beta}_2$ is the $r \times 1$ vector of the mean and the two-factor interaction effects. In Chapter 7 it was shown that a necessary and sufficient condition for estimability of $\boldsymbol{\beta}_1$ is that the rank of $\mathbf{X}_{\Gamma 1}$ is equal to q and that no column of $\mathbf{X}_{\Gamma 2}$ is in the column space of $\mathbf{X}_{\Gamma 1}$ or vice versa. This necessary and sufficient condition for estimability is equivalent to the condition that the rank of the information matrix $[\mathbf{X}'_{\Gamma 1}\mathbf{X}_{\Gamma 1} - (\mathbf{X}'_{\Gamma 1}\mathbf{X}_{\Gamma 2})(\mathbf{X}'_{\Gamma 2}\mathbf{X}_{\Gamma 2})^- \mathbf{X}'_{\Gamma 2}\mathbf{X}_{\Gamma 1}]$ is equal to q. In the special case that $\mathbf{X}'_{\Gamma 1}\mathbf{X}_{\Gamma 2} = \mathbf{0}$ (i.e., the columns of $\mathbf{X}_{\Gamma 1}$ are orthogonal to the columns of $\mathbf{X}_{\Gamma 2}$) the estimability condition for $\boldsymbol{\beta}_1$ simplifies to the rank of $\mathbf{X}_{\Gamma 1}$ being equal to q. If Γ is such that $\boldsymbol{\beta}_1$ is estimable, then the BLUE of $\boldsymbol{\beta}_1$ is equal to

$$(11.2) \qquad \hat{\boldsymbol{\beta}}_1 = [\mathbf{X}'_{\Gamma 1}\mathbf{X}_{\Gamma 1} - (\mathbf{X}'_{\Gamma 1}\mathbf{X}_{\Gamma 2})(\mathbf{X}'_{\Gamma 2}\mathbf{X}_{\Gamma 2})^- \mathbf{X}'_{\Gamma 2}\mathbf{X}_{\Gamma 1}]^{-1} \times [\mathbf{X}'_{\Gamma 1} - (\mathbf{X}'_{\Gamma 1}\mathbf{X}_{\Gamma 2})(\mathbf{X}'_{\Gamma 2}\mathbf{X}_{\Gamma 2})^- \mathbf{X}'_{\Gamma 2}]\mathbf{Y}_\Gamma.$$

The covariance matrix of $\hat{\boldsymbol{\beta}}_1$ is equal to

$$[\mathbf{X}'_{\Gamma 1}\mathbf{X}_{\Gamma 1} - (\mathbf{X}'_{\Gamma 1}\mathbf{X}_{\Gamma 2})(\mathbf{X}'_{\Gamma 2}\mathbf{X}_{\Gamma 2})^- \mathbf{X}'_{\Gamma 2}\mathbf{X}_{\Gamma 1}]^{-1}\sigma^2.$$

A design Γ such that rank of $[\mathbf{X}'_{\Gamma 1}\mathbf{X}_{\Gamma 1} - (\mathbf{X}'_{\Gamma 1}\mathbf{X}_{\Gamma 2})(\mathbf{X}'_{\Gamma 2}\mathbf{X}_{\Gamma 2})^- \mathbf{X}'_{\Gamma 2}\mathbf{X}_{\Gamma 1}]$ is less than q is called a *singular-resolution IV design*. Balancedness, partial balancedness, and orthogonality of resolution IV designs are defined along the same lines as resolution III designs; that is, expressions (10.4), (10.5), and (10.6) are now related to the information matrix $[\mathbf{X}'_{\Gamma 1}\mathbf{X}_{\Gamma 1} - (\mathbf{X}'_{\Gamma 1}\mathbf{X}_{\Gamma 2})(\mathbf{X}'_{\Gamma 2}\mathbf{X}_{\Gamma 2})^- \mathbf{X}'_{\Gamma 2}\mathbf{X}_{\Gamma 1}]$ of a resolution IV design. Before proceeding further, let us illustrate the concepts developed so far by utilizing the 2^3 factorial.

Example 11.1. Consider the 2^3 factorial with $\mathbf{X}_\rho = \mathbf{X}_2 \otimes \mathbf{X}_2 \otimes \mathbf{X}_2$, where

$$\mathbf{X}_2 = \begin{bmatrix} 1 & -1 \\ 1 & 1 \end{bmatrix}$$

and the design $\Gamma = \{(1,0,0),(0,1,0),(0,0,1),(0,1,1),(1,0,1),(1,1,0)\}$. Then

$$E\begin{bmatrix} Y_{(1,0,0)} \\ Y_{(0,1,0)} \\ Y_{(0,0,1)} \\ Y_{(0,1,1)} \\ Y_{(1,0,1)} \\ Y_{(1,1,0)} \end{bmatrix} = \left[\begin{array}{rrr:rrrr} 1 & -1 & -1 & 1 & -1 & -1 & 1 \\ -1 & 1 & -1 & 1 & -1 & 1 & -1 \\ -1 & -1 & 1 & 1 & 1 & -1 & -1 \\ -1 & 1 & 1 & 1 & -1 & -1 & 1 \\ 1 & -1 & 1 & 1 & -1 & 1 & -1 \\ 1 & 1 & -1 & 1 & 1 & -1 & -1 \end{array}\right] \begin{bmatrix} \boldsymbol{\beta}_1 \\ \cdots \\ \boldsymbol{\beta}_2 \end{bmatrix}$$

$$= \mathbf{X}_{\Gamma 1}\boldsymbol{\beta}_1 + \mathbf{X}_{\Gamma 2}\boldsymbol{\beta}_2,$$

where $\boldsymbol{\beta}_1' = (\phi_1^1\phi_2^0\phi_3^0, \phi_1^0\phi_2^1\phi_3^0, \phi_1^0\phi_2^0\phi_3^1)$ and $\boldsymbol{\beta}_2' = (\phi_1^0\phi_2^0\phi_3^0, \phi_1^1\phi_2^1\phi_3^0, \phi_1^1\phi_2^0\phi_3^1, \phi_1^0\phi_2^1\phi_3^1)$. As stated in Chapter 7, the reader may verify that $\mathbf{X}_{\Gamma 1}'\mathbf{X}_{\Gamma 2} = 0$ and the rank of $\mathbf{X}_{\Gamma 1} = 3$. Hence the BLUE of $\boldsymbol{\beta}_1$ simplifies from equation (11.2) to

$$\hat{\boldsymbol{\beta}}_1 = (\mathbf{X}_{\Gamma 1}'\mathbf{X}_{\Gamma 1})^{-1}\mathbf{X}_{\Gamma 1}'\mathbf{Y}_\Gamma$$

$$= \begin{bmatrix} 6 & -2 & -2 \\ -2 & 6 & -2 \\ -2 & -2 & 6 \end{bmatrix}^{-1} \begin{bmatrix} 1 & -1 & -1 & -1 & 1 & 1 \\ -1 & 1 & -1 & 1 & -1 & 1 \\ -1 & -1 & 1 & 1 & 1 & -1 \end{bmatrix} \mathbf{Y}_\Gamma$$

$$= \begin{bmatrix} \frac{1}{4} & \frac{1}{8} & \frac{1}{8} \\ \frac{1}{8} & \frac{1}{4} & \frac{1}{8} \\ \frac{1}{8} & \frac{1}{8} & \frac{1}{4} \end{bmatrix} \begin{bmatrix} 1 & -1 & -1 & -1 & 1 & 1 \\ -1 & 1 & -1 & 1 & -1 & 1 \\ -1 & -1 & 1 & 1 & 1 & -1 \end{bmatrix} \mathbf{Y}_\Gamma$$

$$= \begin{bmatrix} 0 & -\frac{1}{4} & -\frac{1}{4} & 0 & \frac{1}{4} & \frac{1}{4} \\ -\frac{1}{4} & 0 & -\frac{1}{4} & \frac{1}{4} & 0 & \frac{1}{4} \\ -\frac{1}{4} & -\frac{1}{4} & 0 & \frac{1}{4} & \frac{1}{4} & 0 \end{bmatrix} \mathbf{Y}_\Gamma$$

$$= \begin{bmatrix} -\dfrac{y_{(0,1,0)}}{4} & -\dfrac{y_{(0,0,1)}}{4} & +\dfrac{y_{(1,0,1)}}{4} & +\dfrac{y_{(1,1,0)}}{4} \\ -\dfrac{y_{(1,0,0)}}{4} & -\dfrac{y_{(0,0,1)}}{4} & +\dfrac{y_{(0,1,1)}}{4} & +\dfrac{y_{(1,1,0)}}{4} \\ -\dfrac{y_{(1,0,0)}}{4} & -\dfrac{y_{(0,1,0)}}{4} & +\dfrac{y_{(0,1,1)}}{4} & +\dfrac{y_{(1,0,1)}}{4} \end{bmatrix}.$$

Notice that Γ is a resolution IV design relative to the vector $\boldsymbol{\beta}_1$ of main effects. It is balanced since the information matrix is equal to $8\mathbf{I}_3 - 2\mathbf{J}_3$. In the preceding development and in this example the mean is considered to be a part of $\boldsymbol{\beta}_2$. If the mean $\phi_1^0\phi_2^0\phi_3^0$ is included in $\boldsymbol{\beta}_1$, that is, if it is also to be estimated, then Γ becomes a singular-resolution IV design. This is verified by invoking the rank condition on the information matrix $[\mathbf{X}'_{\Gamma 1}\mathbf{X}_{\Gamma 1} - (\mathbf{X}'_{\Gamma 1}\mathbf{X}_{\Gamma 2})(\mathbf{X}'_{\Gamma 2}\mathbf{X}_{\Gamma 2})^{-}\mathbf{X}'_{\Gamma 2}\mathbf{X}_{\Gamma 1}]$ relative to the mean and main effects, which is equal to

$$\begin{bmatrix} 6 & 0 & 0 & 0 \\ 0 & 6 & -2 & -2 \\ 0 & -2 & 6 & -2 \\ 0 & -2 & -2 & 6 \end{bmatrix} - \begin{bmatrix} -2 & -2 & -2 \\ 0 & 0 & 0 \\ 0 & 0 & 0 \\ 0 & 0 & 0 \end{bmatrix} \begin{bmatrix} \frac{1}{4} & \frac{1}{8} & \frac{1}{8} \\ \frac{1}{8} & \frac{1}{4} & \frac{1}{8} \\ \frac{1}{8} & \frac{1}{8} & \frac{1}{4} \end{bmatrix}$$

$$\times \begin{bmatrix} -2 & 0 & 0 & 0 \\ -2 & 0 & 0 & 0 \\ -2 & 0 & 0 & 0 \end{bmatrix} = \begin{bmatrix} 0 & 0 & 0 & 0 \\ 0 & 6 & -2 & -2 \\ 0 & -2 & 6 & -2 \\ 0 & -2 & -2 & 6 \end{bmatrix}.$$

As can be seen, this matrix has rank equal to only 3.

11.2 THE GENERAL PROBLEM OF RESOLUTION IV FACTORIAL DESIGNS

In a resolution IV design the number of parameters to be estimated is equal to $q = \Sigma_{i=1}^{t}(k_i - 1)$. In the case of resolution III designs, it was easy to conclude that the minimum number of distinct treatment combinations is equal to $q+1$. No such simple statement is possible for resolution IV designs since there are nuisance parameters involved. Indeed, the problem of determining the *minimum number* of treatment combinations to be present in a resolution IV design for the general $k_1 \times k_2 \times \cdots \times k_t$ factorial is an unsolved problem. This number is known for some particular cases, for example, for the 2^t factorial, it is equal to $2t$. Using the necessary and sufficient condition for estimability of the previous section, one may show that when the general factorial is rewritten as the $2^{s_2} \times 3^{s_3} \times \cdots \times k^{s_k}$ factorial, where $s_i \geq 0$, a *lower bound* for the number of treatment combinations in a resolution IV design is equal to

$$k\left\{\left(\sum_{i=2}^{k}(i-1)s_i - (k-2)\right)\right\}. \tag{11.3}$$

For example,

Factorial	Lower Bound
2^t factorial: $k=2$, $s_2=t$	$2t$
$2^{t_1}\times 3^{t_2}$ factorial: $k=3$, $s_2=t_1$, $s_3=t_2$	$3(t_1+t_2-1)$
s^t factorial: $k=s$, $s_i=0$ for $i=1,2,\ldots,k-1$, and $s_k=t$	$s(s-1)t-s(s-2)$
etc.	etc.

The first problem that arises when one wants to construct a resolution IV design with given design parameters m, n, and $r_1, r_2, \ldots, r_N$ is the nonsingularity problem. In rank considerations, the repetitions do not play a role. Hence one needs only the calculation of the rank of the information matrix $[\mathbf{X}'_{\bar{\Gamma}1}\mathbf{X}_{\bar{\Gamma}1}-(\mathbf{X}'_{\bar{\Gamma}1}\mathbf{X}_{\bar{\Gamma}2})(\mathbf{X}'_{\bar{\Gamma}2}\mathbf{X}_{\bar{\Gamma}2})^{-}\mathbf{X}'_{\bar{\Gamma}2}\mathbf{X}_{\bar{\Gamma}1}]$ of the reduced design $\bar{\Gamma}$ consisting of the m distinct treatment combinations. For given m, there will be a class Δ^*_m of possible designs with cardinality equal to $|\Delta^*_m| = \binom{N}{m} = (N!)/(m!)(N-m)!$. As was the case with resolution III designs, it is not known how many designs in Δ^*_m will be resolution IV designs. Exhibition of resolution IV designs has proceeded on a case-by-case or series-by-series basis.

Since most of the popular optimality criteria are based on the spectrum of the information matrix, the same concept of equi-information presented in Chapter 10 will lead to a disjoint partitioning of the class of resolution IV designs contained in Δ^*_m. Instead of using equi-information one may partition the class of resolution IV designs using *optimality criterion equivalence*, for example, D-optimality. An optimal design is then found by exhibiting a resolution IV design from the appropriate equivalence class, for example, equivalence class corresponding to the highest value of the determinant of the information matrix.

One should note that the discussion of the general problem of resolution IV designs is basically the same as that for resolution III designs, the only difference being that the information matrix is more complicated due to the presence of nuisance parameters in the model. Finally, note that a design of resolution IV is automatically a design of resolution III.

We shall now illustrate the general problem of resolution IV designs through two examples.

Example 11.2. Let us continue with the 2^3 factorial of Example 11.1. Suppose it is desired to find a resolution IV design with minimal number of treatment combinations and no replication. Hence the number of treatment combinations required is equal to $m=2t=2(3)=6$. It follows that the class of possible designs will consist of $|\Delta_6^*| = \binom{8}{6} = 28$ designs. It is an interesting exercise to find out how many of these will be resolution IV designs. Using D-optimality as the selection criterion, it is an equally interesting exercise to show whether the design $\Gamma = \{(1,0,0), (0,1,0), (0,0,1), (0,1,1), (1,0,1), (1,1,0)\}$ is a D-optimal resolution IV design or not. For the design Γ, we already know that the information matrix is equal to

$$\mathbf{X}'_{\Gamma 1}\mathbf{X}_{\Gamma 1} = \begin{bmatrix} 6 & -2 & -2 \\ -2 & 6 & -2 \\ -2 & -2 & 6 \end{bmatrix} = 8\mathbf{I}_3 - 2\mathbf{J}_3 .$$

Hence it follows that Γ is a balanced and nonorthogonal resolution IV design.

Example 11.3. Consider the 3^4 factorial with the model $E[\mathbf{Y}_\rho] = [\mathbf{X}_3 \otimes \mathbf{X}_3 \otimes \mathbf{X}_3 \otimes \mathbf{X}_3]\,\boldsymbol{\beta}_\rho$ for the minimal complete factorial ρ, where

$$\mathbf{X}_3 = \begin{bmatrix} 1 & -1 & 1 \\ 1 & 0 & -2 \\ 1 & 1 & 1 \end{bmatrix}.$$

Suppose it is desired to exhibit a resolution IV design consisting of $m=n=27$ treatment combinations. The class Δ_{27}^* will in this case consist of $\binom{81}{27} = (81!)/(27!)(54!)$ possible designs, which is indeed a very large number. The *regular* design

$$\Gamma = \left\{ \begin{array}{l} (0,0,0,0),(1,2,0,0),(1,0,2,0),(1,0,0,2),(0,1,2,0),(0,1,0,2), \\ (0,0,1,2),(2,1,0,0),(2,0,1,0),(2,0,0,1),(0,2,1,0),(0,2,0,1), \\ (0,0,2,1),(1,1,1,0),(1,1,0,1),(1,0,1,1),(0,1,1,1),(2,2,2,0), \\ (2,2,0,2),(2,0,2,2),(0,2,2,2),(1,1,2,2),(1,2,1,2),(1,2,2,1), \\ (2,2,1,1),(2,1,2,1),(2,1,1,2) \end{array} \right\}$$

is an orthogonal resolution IV design, which is D-, A-, and E-optimal. The reader is invited to find its information matrix.

It should be noted that for this same 3^4 factorial the minimum run requirement is not known. From what has been said earlier, a lower bound on the cardinality of a resolution IV design in this case is 21. Hence a major problem of exhibiting a resolution IV design of the 3^4 factorial, which is minimal *and* optimal (in the sense of either D-, A-, or E-optimality) is still an unresolved issue.

11.3 CONSTRUCTION OF RESOLUTION IV FACTORIAL DESIGNS

The construction of a resolution IV design depends on the specified design parameters $k_1, k_2, \ldots, k_t, m, n, r_1, r_2, \ldots, r_N$. There is no general construction method for the $k_1 \times k_2 \times \cdots \times k_t$ factorial. Successful attempts have been made in constructing resolution IV designs for certain factorials, for example, the 2^t and the 3^t factorials.

We have seen that in a resolution IV design $q = \Sigma_{i=1}^{t}(k_i - 1)$ parameters must be estimated in the presence of $1 + \Sigma_{u<v}(k_u - 1)(k_v - 1)$ nuisance parameters. Suppose that the minimum number of treatments for a resolution IV design is equal to ν. As remarked in Section 11.2, this number is only known for certain cases. It follows that for a resolution IV design $m \geq \nu$. Once such an m is specified, the problem of constructing a resolution IV design Γ boils down to coming up with a set of m treatment combinations from among the $\binom{N}{m}$ possible sets such that the rank of the resulting information matrix $[\mathbf{X}'_{\Gamma 1}\mathbf{X}_{\Gamma 1} - (\mathbf{X}'_{\Gamma 1}\mathbf{X}_{\Gamma 2})(\mathbf{X}'_{\Gamma 2}\mathbf{X}_{\Gamma 2})^{-}\mathbf{X}'_{\Gamma 2}\mathbf{X}_{\Gamma 1}]$ is equal to q. Such a search could be made for "small enough" N, and indeed a search for an optimal (in any of the senses) design could be done in such cases. Obviously this type of search is not practical nor feasible in general. What is usually done in practice is to propose a construction method for a certain series of factorials known to yield resolution IV designs of certain values of m. The aim in these constructions is usually to let m range between ν and N and be "near" an optimal (in any of the senses) design of that size. Comparisons with an optimal design are in general not possible because of the lack of constructions of optimal designs of order m for most factorials.

We shall now illustrate with examples the construction problem without going into specific methods. This is done in Chapter 13 and in some of the references mentioned in the reading list at the end of this chapter.

Example 11.4. Consider the 2^t factorial with the usual $\mathbf{X}_\rho = \mathbf{X}_2 \otimes \mathbf{X}_2 \otimes \cdots \otimes \mathbf{X}_2$, where

$$\mathbf{X}_2 = \begin{bmatrix} 1 & -1 \\ 1 & 1 \end{bmatrix}.$$

If $(x_1, x_2, \ldots, x_t)$ is a treatment combination in $\rho = \{(x_1, x_2, \ldots, x_t),$

$x_i \in \{0,1\}\}$, then the treatment combination $(x_1^*, x_2^*, \ldots, x_t^*)$ in ρ is said to be the *foldover* of $(x_1, x_2, \ldots, x_t)$ if $(x_1, x_2, \ldots, x_t) + (x_1^*, x_2^*, \ldots, x_t^*) = (11.....1)$. A design Γ consisting of treatment combinations such that every treatment combination in Γ is a foldover of some treatment combination in Γ is known as a *foldover design*. Note that the cardinality of a foldover design Γ for the 2^t factorial is equal to an even number. It can be shown that $\nu = 2t$ and that every minimal resolution IV design of the $2t$ factorial is a foldover design. For minimal resolution IV design Γ, we can then easily establish that $\mathbf{X}'_{\Gamma 1}\mathbf{X}_{\Gamma 2} = \mathbf{0}$, so that for estimability of $\boldsymbol{\beta}_1$ a necessary and sufficient condition then becomes that the rank of $\mathbf{X}_{\Gamma 1}$ or $\mathbf{X}'_{\Gamma 1}\mathbf{X}_{\Gamma 1}$ is equal to $q = t$. Because of the foldover nature of every minimal resolution IV design Γ, we may decompose it as $\Gamma = \Gamma^* \cup \Gamma^{**}$, where Γ^{**} is the foldover of Γ^*, that is, the ith treatment combination in Γ^{**} is a foldover of the ith treatment combination in Γ^*, and $|\Gamma^*| = |\Gamma^{**}| = t$. Using this decomposition the design matrix $\mathbf{X}_\Gamma$ then assumes a simple form, that is,

$$\mathbf{X}_\Gamma = \left[\begin{array}{c|c} \mathbf{X}_{\Gamma^*1} & \mathbf{X}_{\Gamma^*2} \\ \hline -\mathbf{X}_{\Gamma^*1} & \mathbf{X}_{\Gamma^*2} \end{array}\right], \tag{11.4}$$

where $\mathbf{X}_{\Gamma^*1}$ is the $t \times t$ design matrix of Γ^* relative to the t main effects and $\mathbf{X}_{\Gamma^*2}$ is the $t \times \binom{t}{2}$ design matrix of Γ^* relative to the $\binom{t}{2}$ two factor interaction effects. It follows that a foldover factorial design will be a minimal resolution IV design if and only if rank of $\mathbf{X}_{\Gamma^*1}$ is equal to t.

In Example 11.1 for the 2^3 factorial, we considered the minimal resolution IV design $\Gamma = \{(1,0,0), (0,1,0), (0,0,1), (0,1,1), (1,0,1), (1,1,0)\}$. Taking $\Gamma^* = \{(1,0,0), (0,1,0), (0,0,1)\}$ and $\Gamma^{**} = \{(0,1,1), (1,0,1), (1,1,0)\}$, we then have that $\Gamma = \Gamma^* \cup \Gamma^{**}$ so that $\mathbf{X}_\Gamma$ is equal to

$$\mathbf{X}_\Gamma = \left[\begin{array}{rrr|rrrr} 1 & -1 & -1 & 1 & -1 & -1 & 1 \\ -1 & 1 & -1 & 1 & -1 & 1 & -1 \\ -1 & -1 & 1 & 1 & 1 & -1 & -1 \\ \hline -1 & 1 & 1 & 1 & -1 & -1 & 1 \\ 1 & -1 & 1 & 1 & -1 & 1 & -1 \\ 1 & 1 & -1 & 1 & 1 & -1 & -1 \end{array}\right] = \left[\begin{array}{c|c} \mathbf{X}_{\Gamma^*1} & \mathbf{X}_{\Gamma^*2} \\ \hline -\mathbf{X}_{\Gamma^*1} & \mathbf{X}_{\Gamma^*2} \end{array}\right].$$

Since rank of $\mathbf{X}_{\Gamma^*1}$ is equal to 3 it follows that Γ is a minimal resolution IV design.

Because of the simple structure of the design matrix in (11.4), the BLUE of $\boldsymbol{\beta}_1$ for a minimal resolution IV design is given by

$$\hat{\boldsymbol{\beta}}_1 = \tfrac{1}{2}\mathbf{X}_{\Gamma^*1}^{-1}(\mathbf{Y}_{\Gamma^*} - \mathbf{Y}_{\Gamma^{**}}), \tag{11.5}$$

where $\mathbf{Y}_{\Gamma^*}$ and $\mathbf{Y}_{\Gamma^{**}}$ are $t \times 1$ observation vectors for the designs Γ^* and Γ^{**}, respectively.

For the design $\Gamma = \{(1,0,0),(0,1,0),(0,0,1),(0,1,1),(1,0,1),(1,1,0)\}$ of the 2^3 factorial the BLUE of $\boldsymbol{\beta}_1$ is equal to

$$\hat{\boldsymbol{\beta}}_1 = \begin{bmatrix} \widehat{\phi_1^1\phi_2^0\phi_3^0} \\ \widehat{\phi_1^0\phi_2^1\phi_3^0} \\ \widehat{\phi_1^0\phi_2^0\phi_3^1} \end{bmatrix} = \frac{1}{2}\begin{bmatrix} 1 & -1 & -1 \\ -1 & 1 & -1 \\ -1 & -1 & 1 \end{bmatrix}^{-1} \begin{bmatrix} Y_{(1,0,0)} - Y_{(0,1,1)} \\ Y_{(0,1,0)} - Y_{(1,0,1)} \\ Y_{(0,0,1)} - Y_{(1,1,0)} \end{bmatrix}.$$

Using the decomposition $\Gamma = \Gamma^* \cup \Gamma^{**}$ the reader is invited to find the covariance matrix of $\hat{\boldsymbol{\beta}}_1$ in general and for this example.

In the literature list at the end of this chapter there are many examples of resolution IV designs. Chapter 13 also provides examples.

11.4 RESOLUTION V FACTORIAL DESIGNS

The problems associated with resolution V designs are identical to those of resolution III with the only difference being that $\boldsymbol{\beta}_1$ now consists of the mean, the main effects, and the two-factor interaction effects. The number of parameters in $\boldsymbol{\beta}_1$ for the general $k_1 \times k_2 \times \cdots \times k_t$ factorial, therefore, is equal to $s = 1 + \Sigma(k_i - 1) + \Sigma_{u<v}(k_u - 1)(k_v - 1)$. The model for a resolution V design follows directly from Case 2 of Chapter 7, that is, $\boldsymbol{\beta}_p' = (\boldsymbol{\beta}_1 \vdots \boldsymbol{\beta}_3)$, where $\boldsymbol{\beta}_1$ is the vector consisting of the mean, the main effects, and two-factor interaction effects and $\boldsymbol{\beta}_3$ is the vector consisting of three-factor and higher order interaction effects. For a resolution V design Γ, we have the model

$$E[\mathbf{Y}_\Gamma] = \mathbf{X}_{\Gamma 1}\boldsymbol{\beta}_1, \qquad \mathrm{Cov}(\mathbf{Y}_\Gamma) = \sigma^2 \mathbf{I}_n,$$

where $\mathbf{Y}_\Gamma$ is the $n \times 1$ observation vector associated with the n treatment combinations in Γ, $\mathbf{X}_{\Gamma 1}$ is the $n \times s[= 1 + \Sigma(k_i - 1) + \Sigma_{u<v}(k_u - 1)(k_v - 1)]$ design matrix.

A necessary and sufficient condition for estimability is that rank of $\mathbf{X}_{\Gamma 1}$ is equal to s. If this is the case, then the BLUE of $\boldsymbol{\beta}_1$ is equal to

$$\hat{\boldsymbol{\beta}}_1 = (\mathbf{X}_{\Gamma 1}'\mathbf{X}_{\Gamma 1})^{-1}\mathbf{X}_{\Gamma 1}'\mathbf{Y}_\Gamma, \tag{11.7}$$

and the covariance matrix of $\hat{\boldsymbol{\beta}}_1$ is equal to $\text{Cov}(\hat{\boldsymbol{\beta}}_1) = (\mathbf{X}'_{\Gamma 1}\mathbf{X}_{\Gamma 1})^{-1}\sigma^2$. Since $s = 1 + \Sigma(k_i - 1) + \Sigma_{u<v}(k_u - 1)(k_v - 1)$ parameters have to be estimated in a resolution V design, it follows that a minimal resolution V must contain exactly s distinct treatment combinations. Hence $n \geq m \geq s$ for a resolution V design. If the rank of $\mathbf{X}_{\Gamma 1}$ in a resolution V design is less than s, then it will be called a *singular-resolution V design*.

The combinatorial problems associated with resolution V designs can be formulated along the lines indicated in Chapter 10 and the reader is encouraged to do so. Optimality criteria (e.g., D-optimality) and other properties (e.g., balancedness) can be imposed on a class of competing resolution V designs for design selection purposes.

The construction of a resolution V design depends on the specified design parameters $k_1, k_2, \ldots, k_t, m, n, r_1, r_2, \ldots, r_N$. There is no general construction method for the $k_1 \times k_2 \times \cdots \times k_t$ factorial for arbitrary m and n. For $n = m = s = 1 + \Sigma(k_i - 1) + \Sigma_{u<v}(k_u - 1)(k_u - 1)$, that is, the *saturated* or *minimal* case, one may always exhibit the *one-at-a-time resolution V design*,

$$\Gamma_0 = \left\{ \begin{matrix} (x_1, x_2, \ldots, x_t) \in \rho \text{ such that} \\ 0, 1, \text{ or } 2 \text{ coordinates are nonzero} \end{matrix} \right\},$$

which can be analyzed using, for example, the Kronecker product model $\mathbf{X}_1 \otimes \mathbf{X}_2 \otimes \cdots \otimes \mathbf{X}_k$, each $\mathbf{X}_i$ arising from orthogonal or Helmert polynomials. It can be shown that the design Γ_0 under these two models is a resolution V design.

We now provide two examples to illustrate the construction of resolution V designs.

Example 11.5. Consider the 2^3 factorial with $\mathbf{X}_\rho = \mathbf{X}_2 \otimes \mathbf{X}_2 \otimes \mathbf{X}_2$, where $\mathbf{X}_2 = \begin{bmatrix} 1 & -1 \\ 1 & 1 \end{bmatrix}$. The number of parameters in $\boldsymbol{\beta}_1$ is equal to $s = 1 + 3 + 3 = 7$. Hence $n \geq m \geq 7$ for a resolution V design in this case. Suppose $n = m = s = 7$, then the class of competing minimal designs consists exactly of $\binom{8}{7} = 8$ designs. Via a simple matrix algebraic argument it can be shown that: (i) each design is nonsingular and (ii) each design has the same spectrum for the information matrix. Hence there are eight choices for a A-, D-, or E-optimal design and one of these is the one-at-a-time resolution V design

$$\Gamma_0 = \{(0,0,0), (1,0,0), (0,1,0), (0,0,1), (1,1,0), (1,0,1), (0,1,1)\}.$$

By allowing repetitions, an infinite class of unbiased designs is generated from the Γ_0 or any other of the remaining seven resolution V designs. The design matrix $\mathbf{X}_{\Gamma_0 1}$ is obtained in the usual way and for the one-at-a-time resolution V design above the model is equal to

$$E\begin{bmatrix} Y_{(0,0,0)} \\ Y_{(1,0,0)} \\ Y_{(0,1,0)} \\ Y_{(0,0,1)} \\ Y_{(1,1,0)} \\ Y_{(1,0,1)} \\ Y_{(0,1,1)} \end{bmatrix} = \begin{bmatrix} 1 & -1 & -1 & -1 & 1 & 1 & 1 \\ 1 & 1 & -1 & -1 & -1 & -1 & 1 \\ 1 & -1 & 1 & -1 & -1 & 1 & -1 \\ 1 & -1 & -1 & 1 & 1 & -1 & -1 \\ 1 & 1 & 1 & -1 & 1 & -1 & -1 \\ 1 & 1 & -1 & 1 & -1 & 1 & -1 \\ 1 & -1 & 1 & 1 & -1 & -1 & 1 \end{bmatrix} \begin{bmatrix} \phi_1^0\phi_2^0\phi_3^0 \\ \phi_1^1\phi_2^0\phi_3^0 \\ \phi_1^0\phi_2^1\phi_3^0 \\ \phi_1^0\phi_2^0\phi_3^1 \\ \phi_1^1\phi_2^1\phi_3^0 \\ \phi_1^1\phi_2^0\phi_3^1 \\ \phi_1^0\phi_2^1\phi_3^1 \end{bmatrix}$$

$$\mathbf{Y}_{\Gamma_0} \quad = \quad \mathbf{X}_{\Gamma_0 1} \quad \boldsymbol{\beta}_1$$

The reader may verify that the rank of $\mathbf{X}_{\Gamma_0}$ is equal to 7 so that Γ_0 is a resolution V design. The BLUE of $\boldsymbol{\beta}_1$ is obtained in the usual way by calculating $\hat{\boldsymbol{\beta}}_1 = (\mathbf{X}'_{\Gamma_0 1}\mathbf{X}_{\Gamma_0 1})^{-1}\mathbf{X}'_{\Gamma_0 1}\mathbf{Y}_{\Gamma_0}$. By the way, the set of eight spectrum equivalent designs is generated by applying the group of level permutations Ω to the foregoing design Γ_0.

Example 11.6. Now consider the general 2^t factorial with $\mathbf{X}_\rho = \mathbf{X}_2 \otimes \mathbf{X}_2 \otimes \cdots \otimes \mathbf{X}_2$, and

$$\mathbf{X}_2 = \begin{bmatrix} 1 & -1 \\ 1 & 1 \end{bmatrix}.$$

Suppose it is desired to obtain a minimal resolution V design, which is partially balanced according to Definition 8.2. It can be shown that the design

$$\Gamma = \left\{ (x_1, x_2, \ldots, x_t) \in \rho \text{ such that } \sum_{i=1}^{t} x_i = 0, 1, \text{or } t-2 \right\},$$

consisting of $1 + t + C(t,2)$ treatment combinations, is a minimal partially balanced resolution V design. For example, for the 2^5 factorial, the design consists of the following $s = 1 + 5 + 10 = 16$ treatment combinations:

$$\begin{aligned} \Gamma = \{ &(0,0,0,0,0), (1,0,0,0,0), (0,1,0,0,0), (0,0,1,0,0), (0,0,0,1,0), \\ &(0,0,0,0,1), (1,1,1,0,0), (1,1,0,1,0), (1,1,0,0,1), (1,0,1,1,0), \\ &(1,0,1,0,1), (1,0,0,1,1), (0,1,1,1,0), (0,1,1,0,1), (0,1,0,1,1), \\ &(0,0,1,1,1) \}. \end{aligned}$$

The reader is invited to obtain the information matrix of Γ and to verify that it is partially balanced according to Definition 8.2 of Chapter 8.

In the literature list at the end of this chapter there are many examples of resolution V designs. Several of the references in Chapters 10 and 13 also contain examples of resolution V designs.

11.5 SELECTED ADDITIONAL READING

Starred references are recommended for a first reading.

*1. Addelman, S. (1963). Techniques for constructing fractional replicate plans. *J. Am. Stat. Assoc.*, **58**, 45–71.

2. Anderson, D. A., and Srivastava, J. N. (1972). Resolution IV designs of the $2^n \times 3$ series. *J. R. Stat. Soc.*, **B34**, 377–384.

3. Anderson, D. A., and Thomas, A. M. (1979). Resolution IV fractional factorial designs for the general asymmetric factorial. *Commun. Stat.*, **A8**, 931–943.

*4. Anderson, D. A., and Thomas, A. M. (1979). Near minimal resolution IV designs for the s^n factorial experiment. *Technometrics*, **21**, 331–336.

5. Banerjee, K. S., and Federer, W. T. (1967). On a special subset giving an irregular fractional replicate of a 2^n factorial experiment. *J. R. Stat. Soc.*, **B29**, 292–299.

6. Birkes, D., and Seely, J. (1976). Three-way classification designs of resolutions III, IV and V. *Tech. Report No. 50*, Department of Statistics, Oregan State University, Corvallis.

*7. Box, G. E. P., and Hunter, J. S. (1961). The 2^{k-p} fractional factorial designs, I and II. *Technometrics*, **3**, 311–351, 449–458.

8. Box, G. E. P., and Wilson, K. B. (1951). On the experimental attainment of optimum conditions. *J. R. Stat. Soc.*, **B13**, 1–45.

*9. Chopra, D. V. (1975a). Balanced optimal 2^8 fractional factorial designs of resolution V, $52 \leq N \leq 59$. In *A Survey of Statistical Designs and Linear Models*, J. N. Srivastava, Ed. North Holland, Amsterdam, pp. 91–99.

10. Chopra, D. V. (1975b). Optimal balanced 2^8 fractional factorial designs of resolution V, with 60 to 65 runs. *Proc. Int. Stat. Inst.* (*Warsaw*), 164–168.

11. Chopra, D. V., and Srivastava, J. N. (1973a). Optimal balanced 2^7 fractional factorial designs of resolution V with $N \leq 42$. *Ann. Inst. Stat. Math.*, **25**, 587–604.

12. Chopra, D. V., and Srivastava, J. N. (1973b). Optimal balanced 2^7 fractional factorial designs of resolution V, $49 \leq N \leq 55$. *Commun. Stat.*, **2**, 59–84.

13. Chopra, D. V., and Srivastava, J. N. (1974). Optimal balanced 2^8 fractional factorial designs of resolution V, $37 \leq N \leq 51$. *Sankhyā*, **A36**, 41–52.

*14. Daniel, C. (1973). One-at-a-time plans. *J. Am. Stat. Assoc.*, **68**, 353–360.

*15. Draper, N. R., and Mitchell, T. J. (1967). The construction of saturated 2_R^{k-p} designs. *Ann. Math. Stat.*, **38**, 1110–1126.

*16. Gulati, B. R. (1972). Construction of two-level saturated symmetrical factorial designs of resolution VI. *Ann. Math. Stat.*, **43**, 1652–1663.

17. Hoke, A. T. (1975). The characteristic polynomial of the information matrix for second order models. *Ann. Stat.*, **3**, 780–786.

18. John, P. W. M. (1962). Three-quarter replicates of 2^n designs. *Biometrics.*, **18**, 172–184.

19. Kuwada, M., (1979). Optimal balanced fractional 3^m designs of resolution V and balanced third-order designs. *Hiroshima Math. J.*, **9**, 347–450.

20. Margolin, B. H. (1969a). Resolution IV fractional factorial designs. *J. R. Stat. Soc.*, **B31**, 514–523.

*21. Margolin, B. H. (1969b). Results on factorial designs of resolution IV for the 2^n and 2^n3^m series. *Technometrics*, **11**, 431–444.

22. Mitchell, T. J. (1974). Computer construction of "*D*-optimal" first-order designs. *Technometrics*, **16**, 211–220.

23. Pesotan, H., and Raktoe, B. L. (1975). Remarks on extensions of the Anderson-Federer (0, 1)-matrix procedure for main effects to effects of a higher degree. *Commun. Stat.*, **4**, 797–811.

24. Raktoe, B. L., and Federer, W. T. (1969). Some non-orthogonal unsaturated main effect and resolution V plans derived from a one-restrictional lattice. *Ann. Inst. Stat. Math.*, **21**, 335–342.

*25. Shirakura, T. (1976). Optimal balanced fractional 2^m factorial designs of resolution VII, $6 \leq m \leq 8$. *Ann. Stat.*, **4**, 515–531.

26. Srivastava, J. N., and Chopra, D. V. (1971a). On the comparison of certain classes of balanced 2^8 fractional factorial designs of resolution V, with respect to the trace criterion. *J. Indian Soc. Agric. Stat.*, **23**, (2), 124–131.

27. Srivastava, J. N., and Chopra, D. V. (1971b). On the characteristic roots of the information matrix of 2^m balanced factorial designs of resolution V, with applications. *Ann. Math. Stat.*, **42**, 722–734.

28. Srivastava, J. N., and Chopra, D. V. (1971c). Balanced optimal 2^m fractional factorial designs of resolution V, $m \leq 6$. *Technometrics*, **13**, 257–269.

29. Srivastava, J. N., and Chopra, D. V. (1974). Balanced trace optimal 2^8 fractional factorial designs of resolution V with 56 to 68 runs. *Util. Math.* **5**, 263–279.

30. Srivastava, J. N., and Ghosh, S. (1977). Balanced 2^m factorial designs of resolution V which allow search and estimation of one extra unknown effect, $4 \leq m \leq 8$. *Commun. Stat.*, **A6**, 141–166.

*31. Webb, S. R. (1968). Non-orthogonal designs of even resolution. *Technometrics*, **10**, 291–299.

CHAPTER 12

Search Factorial Designs*

This chapter is concerned with the relatively new topic of "search linear models". When assumptions about the negligibility of some of the parameters in $\boldsymbol{\beta}_3$ in the general partitioning $\boldsymbol{\beta}_\rho' = (\boldsymbol{\beta}_1' \vdots \boldsymbol{\beta}_2' \vdots \boldsymbol{\beta}_3')$ of Chapter 7 are in doubt, then the problem is to search these out and make inferences about these and other parameters in the model. Depending on the assumptions on the three components in the partitioning, several families of competing models arise and are discussed. A necessary condition for a design to allow the search of nonnegligible effects in $\boldsymbol{\beta}_3$ is provided. In case $\sigma^2 = 0$, this condition is also sufficient. Examples are provided to illustrate the developments and open problems are pointed out.

12.1 SEARCH LINEAR MODELS

In Chapter 7 we introduced a general partitioning of the total parametric vector $\boldsymbol{\beta}_\rho$, that is,

$$\boldsymbol{\beta}_\rho' = \left(\boldsymbol{\beta}_1' \vdots \boldsymbol{\beta}_2' \vdots \boldsymbol{\beta}_3'\right) \tag{12.1}$$

where $\boldsymbol{\beta}_1$ is an $N_1 \times 1$ vector of parameters to be estimated, $\boldsymbol{\beta}_2$ is an $N_2 \times 1$ vector of unknown parameters not of interest, and $\boldsymbol{\beta}_3$ is an $N_3 \times 1$ vector of parameters assumed to be known (which without loss of generality can be taken to be zero), such that $1 \leq N_1 \leq N$, $0 \leq N_2 \leq N-1$ and $0 \leq N_3 \leq N - N_1 - N_2 \leq N-1$. Explicitly, the following cases occur:

(i) $N_1 = N$, $N_2 = N_3 = 0$.
(ii) $N_2 = 0$, $N_3 \neq 0$.
(iii) $N_2 \neq 0$, $N_3 \neq 0$.
(iv) $N_2 \neq 0$, $N_3 = 0$.

*This chapter is based on the University of Guelph technical report "Search Factorial Designs" by B. L. Raktoe and H. Pesotan.

Henceforth we are going to depart from the above four cases and consider the general partitioning (12.1) with the condition $N_1 \geq 0$, $N_2 \geq 0$, $N_3 \geq 0$, $N_1 + N_2 + N_3 = N$. A tacit assumption underlying the statistical analysis of the general partitioning (12.1) as applied in Chapter 7 was that a *fixed* linear model was the *correct* one. Thus in each of the four preceding cases *only one* correct model was assumed. Under this fixed and correct model approach, the BLUE of $\boldsymbol{\beta}_1$ under the assumption $N_1 \geq 1$ and the basic setting of Chapter 7 is equal to

$$\hat{\boldsymbol{\beta}}_1 = \left[\mathbf{X}'_{\Gamma 1}\mathbf{X}_{\Gamma 1} - \mathbf{X}'_{\Gamma 1}\mathbf{X}_{\Gamma 2}(\mathbf{X}'_{\Gamma 2}\mathbf{X}_{\Gamma 2})^{-}\mathbf{X}'_{\Gamma 2}\mathbf{X}_{\Gamma 1}\right]^{-1} \times \left[\mathbf{X}'_{\Gamma 1} - \mathbf{X}'_{\Gamma 1}\mathbf{X}_{\Gamma 2}(\mathbf{X}'_{\Gamma 2}\mathbf{X}_{\Gamma 2})^{-}\mathbf{X}'_{\Gamma 2}\right]\mathbf{Y}_{\Gamma}. \tag{12.2}$$

Note that in the partitioning (12.1) and subsequent formula (12.2) the assumption is that the experimenter knows what parameters constitute the vectors $\boldsymbol{\beta}_1$, $\boldsymbol{\beta}_2$, and $\boldsymbol{\beta}_3$. There is a problem inherent in the fixed model approach. This is that, if $N_3 \neq 0$, the assumption that all the elements of $\boldsymbol{\beta}_3$ are zero might be false. Indeed, there is the possibility that at most l_3 of the elements in $\boldsymbol{\beta}_3$ are nonzero for some suitable l_3 such that $0 \leq l_3 \leq N_3$. As a consequence of this problem, we will have a family of linear models for a fixed design Γ. In order to define this family, let l_3 be a given number such that $0 \leq l_3 \leq N_3$ and let $P_{l_3}(\boldsymbol{\beta}_3)$ be the set of all subsets of $\boldsymbol{\beta}_3$ of cardinality *at most* l_3. For a given fixed design Γ, the general partitioning (12.1), and a given l_3, $0 \leq l_3 \leq N_3$, we have the following *family* of competing models:

$$E(\mathbf{Y}_{\Gamma}) \in \mathfrak{M}_{\Gamma} = \{\mathbf{X}_{\Gamma 1}\boldsymbol{\beta}_1 + \mathbf{X}_{\Gamma 2}\boldsymbol{\beta}_2 + \mathbf{X}_{\Gamma 3}\boldsymbol{\gamma}_3 : \boldsymbol{\gamma}_3 \in P_{l_3}(\boldsymbol{\beta}_3)\}, \quad \text{with } \operatorname{Cov}(\mathbf{Y}_{\Gamma}) = \sigma^2 \mathbf{I}_n. \tag{12.3}$$

The statistical inference problem is to determine (or "*search*") the correct model in the family $\mathfrak{M}_{\Gamma}$ given in (12.3) and subsequently to draw inferences on $\boldsymbol{\beta}_1$ and the *correct* $\boldsymbol{\gamma}_3$.

Several observations can now be made regarding the family of models $\mathfrak{M}_{\Gamma}$ given in (12.3):

(a) $|\mathfrak{M}_{\Gamma}| = \Sigma_{i=0}^{l_3} \binom{N_3}{i}$.

(b) If $l_3 = 0$, then the family $\mathfrak{M}_{\Gamma}$ reduces to a single model and we are back in the classical cases (ii) or (iii) according as $N_2 = 0$ or $N_2 \neq 0$.

(c) If $l_3 \neq 0$, $N_2 = 0$, then the family $\mathfrak{M}_{\Gamma}$ reduces to a class of *linear models without nuisance parameters*.

(d) If $N_2 \neq 0$ and $l_3 \neq 0$, then the family $\mathfrak{M}_{\Gamma}$ reduces to a class of *linear models with nuisance parameters*.

(e) If $N_1=0$, $N_2=0$, and $l_3 \neq 0$, then the family $\mathcal{M}_\Gamma$ reduces to a class of "*pure search*" linear models. In this case statistical inference, after determining the correct model, is made with respect to γ_3 only.
(f) If $N_1=0$, $N_2 \neq 0$, and $l_3 \neq 0$, then the family $\mathcal{M}_\Gamma$ reduces to a class of "*pure search*" *linear models with nuisance parameters*.
(g) The search problem as outlined in (c) can be tied up with the odd resolution setting and, in (d), with the even resolution setting.
(h) If we have no further information regarding the models in $\mathcal{M}_\Gamma$, we will assume that each model in $\mathcal{M}_\Gamma$ is equally likely to be the correct one.

We now formally define the concept of a search linear model for a fixed design Γ under the basic setting of Chapter 7.

Definition 12.1. Let Γ be a design in the $k_1 \times k_2 \times \ldots \times k_t$ factorial and let the vector $\boldsymbol{\beta}_\rho$ be partitioned as in (12.1) with $\text{Cov}(\mathbf{Y}_\Gamma)=\sigma^2\mathbf{I}_n$. Then the family $\mathcal{M}_\Gamma$ given in (12.3) will be called the family of *search linear models* with the understanding that statistical inference has to be made on $\boldsymbol{\beta}_1$ and γ_3 after the correct model has been determined.

There are according to this definition two stages involved in the search problem:

(i) The development of a *search procedure* to determine the correct model.
(ii) After determination of the correct model the estimation of $\boldsymbol{\beta}_1$ and the *correct* γ_3 or linear functions of these.

We will now illustrate the preceding ideas with an example.

Example 12.1. (i) Consider the 2^5 factorial with $\mathbf{X}_\rho=\mathbf{X}_2 \otimes \mathbf{X}_2 \otimes \mathbf{X}_2 \otimes \mathbf{X}_2 \otimes \mathbf{X}_2$, where

$$\mathbf{X}_2=\begin{bmatrix} 1 & -1 \\ 1 & 1 \end{bmatrix}.$$

Let Γ be a design of the 2^5 factorial and consider the situation where the partitioning in (12.1) is such that $N_1=N_2=0$ and $N_3=N=32$. In this setting we have assumed that all the 32 parameters are negligible, that is to say, that $E(\mathbf{Y}_\Gamma)=\mathbf{0}$. If this assumption is false and say at most $l_3=2$ parameters in $\boldsymbol{\beta}_3$ are nonnegligible, then we have the following family of *pure search* linear models for Γ:

$$E(\mathbf{Y}_\Gamma) \in \mathcal{M}_\Gamma=\{\mathbf{X}_{\Gamma 3}\gamma_3 : \gamma_3 \in P_2(\boldsymbol{\beta}_\rho)\}.$$

Then note that

$$|\mathfrak{M}_\Gamma| = \sum_{i=0}^{2} \binom{32}{i} = 529.$$

Hence the search problem is to discover which of these 529 models is the correct one and subsequently to estimate the correct γ_3 or linear functions thereof.

(ii) Consider the case where Γ is a resolution V design of the 2^5 factorial. Then $N_1 = 16$, $N_2 = 0$, and $N_3 = 16$. For a resolution V design in the earlier setting, the 16 parameters in $\boldsymbol{\beta}_3$ consisting of the third and higher order interaction effects are assumed to be zero. If this assumption is false and say at most $l_3 = 1$ of these is nonnegligible, then we have the following family of search linear models for Γ:

$$E(\mathbf{Y}_\Gamma) \in \mathfrak{M}_\Gamma = \{\mathbf{X}_{\Gamma 1}\boldsymbol{\beta}_1 + \mathbf{X}_{\Gamma 3}\boldsymbol{\gamma}_3 : \boldsymbol{\gamma}_3 \in P_1(\boldsymbol{\beta}_3)\}.$$

Then $|\mathfrak{M}_\Gamma| = \Sigma_{i=0}^{1} \binom{16}{i} = 17$, so that the search linear model problem is to discover which of the 17 models is the correct one and subsequently to estimate $\boldsymbol{\beta}_1$ and the correct γ_3 or linear functions thereof. This example belongs to (c), that is, *search linear models without nuisance parameters.*

(iii) Next let Γ be a resolution IV design of the 2^5 factorial. Then $N_1 = 5$, $N_2 = 11$, $N_3 = 16$. For a resolution IV design in the earlier setting the $N_2 = 11$ parameters consisting of the mean and two-factor interactions would be considered as nuisance parameters and the $N_3 = 16$ parameters consisting of the three and higher order interaction effects would be considered to be zero. If this latter assumption is false and say at most $l_3 = 2$ of the interactions in $\boldsymbol{\beta}_3$ are nonnegligible, then we have the following family of search linear models for Γ:

$$E(\mathbf{Y}_\Gamma) \in \mathfrak{M}_\Gamma = \{\mathbf{X}_{\Gamma 1}\boldsymbol{\beta}_1 + \mathbf{X}_{\Gamma 2}\boldsymbol{\beta}_2 + \mathbf{X}_{\Gamma 3}\boldsymbol{\gamma}_3 : \boldsymbol{\gamma}_3 \in P_2(\boldsymbol{\beta}_3)\}.$$

Then $|\mathfrak{M}_\Gamma| = \Sigma_{i=0}^{2} \binom{16}{i} = 137$, so that the search linear model problem is to discover which of these 137 models is the correct one and then to estimate $\boldsymbol{\beta}_1$ and the correct γ_3 or linear functions thereof. This example belongs to Case 4, that is, *search linear models with nuisance parameters.*

(iv) The reader is invited to formulate for the 2^5 factorial, in a resolution IV design, the *pure search linear model problem with nuisance parameters*, as outlined in (f).

12.2 REDUCING BIAS WITH SEARCH LINEAR MODELS

Suppose that for the general partitioning (12.1), in the family of competing models (12.2), the correct γ_3 is γ_3^* and the corresponding design matrix is $\mathbf{X}_{\Gamma 3}^*$ so that the correct model is

$$E(\mathbf{Y}_\Gamma) = \mathbf{X}_{\Gamma 1}\boldsymbol{\beta}_1 + \mathbf{X}_{\Gamma 2}\boldsymbol{\beta}_2 + \mathbf{X}_{\Gamma 3}^*\gamma_3^*. \tag{12.4}$$

Note that in our notation $\gamma_3^* \in P_{l_3}(\boldsymbol{\beta}_3)$. In the usual fixed model setting $\boldsymbol{\beta}_3$ is assumed to be zero (and hence $l_3 = 0$), and under this assumption the BLUE of $\boldsymbol{\beta}_1$ is given by (12.2). If this assumption is false, that is, if (12.4) is the correct model, then

$$E(\hat{\boldsymbol{\beta}}_1) = \boldsymbol{\beta}_1 + \mathbf{A}_\Gamma^*\gamma_3^*, \tag{12.5}$$

where $\mathbf{A}_\Gamma^*$ is the *alias matrix* for the given design Γ relative to $\boldsymbol{\beta}_1$, $\boldsymbol{\beta}_2$, γ_3^* and the corresponding design matrix $\mathbf{X}_{\Gamma 3}^*$; that is,

$$\begin{aligned} \mathbf{A}_\Gamma^* = {} & \left[\mathbf{X}'_{\Gamma 1}\mathbf{X}_{\Gamma 1} - \mathbf{X}'_{\Gamma 1}\mathbf{X}_{\Gamma 2}(\mathbf{X}'_{\Gamma 2}\mathbf{X}_{\Gamma 2})^{-}\mathbf{X}'_{\Gamma 2}\mathbf{X}_{\Gamma 1}\right]^{-1} \\ & \times \left[\mathbf{X}'_{\Gamma 1} - \mathbf{X}'_{\Gamma 1}\mathbf{X}_{\Gamma 2}(\mathbf{X}'_{\Gamma 2}\mathbf{X}_{\Gamma 2})^{-}\mathbf{X}'_{\Gamma 2}\right]\mathbf{X}_{\Gamma 3}^*. \end{aligned} \tag{12.6}$$

The following comments can now be made regarding the alias matrix $\mathbf{A}_\Gamma^*$:

(i) If $N_1 \neq 0$, $N_2 = 0$, $l_3 \neq 0$, that is, if we are in the case of the search linear model without nuisance parameters, then $\mathbf{A}_\Gamma^*$ becomes

$$\mathbf{A}_\Gamma^* = (\mathbf{X}'_{\Gamma 1}\mathbf{X}_{\Gamma 1})^{-1}\mathbf{X}'_{\Gamma 1}\mathbf{X}_{\Gamma 3}^*. \tag{12.7}$$

(ii) If $N_1 = 0$, $N_2 = 0$, $l_3 \neq 0$, which is the case of pure search linear models, then since $\boldsymbol{\beta}_1$ is not being estimated, there is no alias matrix in this case. Hence the concept of bias does not appear in this case.

(iii) If $N_1 \neq 0$, $N_2 \neq 0$, $l_3 \neq 0$, the case of search linear models with nuisance parameters, then $\mathbf{A}_\Gamma^*$ is given by (12.6).

(iv) If $N_1 = 0$, $N_2 \neq 0$, $l_3 \neq 0$, the case of pure search linear models with nuisance parameters, then the reader may verify that the concept of alias matrix does not arise and hence the concept of bias does not appear.

It thus follows that there are only two cases, (i) and (iii), when bias is present and is a problem to be handled. In (i), the exact amount of bias depends on the design Γ, $\boldsymbol{\beta}_1$ (and hence $\mathbf{X}_{\Gamma 1}$), and γ_3^* (and hence $\mathbf{X}_{\Gamma 3}^*$). In (iii), the exact amount of bias depends additionally on $\boldsymbol{\beta}_2$ (and hence $\mathbf{X}_{\Gamma 2}$).

We will limit ourselves to the most popular case, search linear models without nuisance parameters [i.e., (i)]. As noted earlier the bias matrix is then given by (12.7), that is, $\mathbf{A}_\Gamma^* = (\mathbf{X}_{\Gamma 1}'\mathbf{X}_{\Gamma 1})^{-1}\mathbf{X}_{\Gamma 1}'\mathbf{X}_{\Gamma 3}^*$. The bias will be identically zero if and only if the columns of $\mathbf{X}_{\Gamma 1}$ are orthogonal to the columns of $\mathbf{X}_{\Gamma 3}^*$, that is, $\mathbf{X}_{\Gamma 1}'\mathbf{X}_{\Gamma 3}^* = \mathbf{0}$. This is equivalent to requiring that the factorial design Γ be one or more copies of the minimal complete factorial ρ. If the experimenter used one or more copies of ρ, this would involve a huge waste of resources since there are N_1 plus at most l_3 parameters to be estimated. The objective of search linear models is to determine γ_3^* and $\mathbf{X}_{\Gamma 3}^*$ and thus correct for the bias. If $\sigma^2 = 0$ (i.e., if the observations are nonstochastic), the search problem is deterministic and the evaluation of γ_3^* and $\mathbf{X}_{\Gamma 3}^*$ is exact if Γ satisfies some specified conditions (see Section 12.3). When $\sigma^2 > 0$, which is the most practical case, the problem is stochastic, so that γ_3^* and $\mathbf{X}_{\Gamma 3}^*$ would be determined with a probability that is less than one and is dependent on σ^2. The reader is invited to explore the implications in the remaining case, (iii).

12.3 STATISTICAL INFERENCE UNDER SEARCH LINEAR MODELS

The development of the subject of statistical inference under search linear models is in its infancy. However, techniques have been developed to meet certain basic needs of the theory. Before we proceed further we should address ourselves to the following important question: What conditions should be imposed on the factorial design Γ so that the correct model can be determined or searched out? The following theorem points out one important necessary condition for (i) in Section 12.2.

Theorem 12.1. In order that the correct model in the family of search linear models

$$\mathfrak{M}_\Gamma = \{\mathbf{X}_{\Gamma 1}\boldsymbol{\beta}_1 + \mathbf{X}_{\Gamma 3}\gamma_3 : \gamma_3 \in P_{l_3}(\boldsymbol{\beta}_3)\} \tag{12.8}$$

be determined, a *necessary* condition on Γ is that

$$\operatorname{rank}\left[\mathbf{X}_{\Gamma 1} \vdots \mathbf{B}_{\Gamma 3}\right] = N_1 + 2l_3, \tag{12.9}$$

for every submatrix $\mathbf{B}_{\Gamma 3}$ of order $n \times 2l_3$ from $\mathbf{X}_{\Gamma 3}^{(0)}$, the design matrix of Γ relative to $\boldsymbol{\beta}_3$. (A proof of this theorem may be found in reference [6].)

We would like to emphasize the following points:

(i) Condition (12.9) is only a *necessary* condition.
(ii) Since (12.9) has to be satisfied by Γ we require that $n \geq N_1 + 2l_3$.
(iii) Condition (12.9) becomes *sufficient* in the degenerate case in which $\sigma^2 = 0$. In this case exactly one correct model in the family (12.8) is determined, that is, there is a *unique solution* for γ_3 and the corresponding $\mathbf{X}_{\Gamma 3}$.
(iv) In order to establish a set of sufficient conditions on Γ (in the case $\sigma^2 > 0$) to determine the correct model, we must first specify our statistical procedure and the associated inference framework. To date, no satisfactory theoretical work has been done in this context (see Remark at the end of this chapter).

We now illustrate Theorem 12.1 with the following example.

Example 12.2. Consider the 2^5 factorial with $\mathbf{X}_\rho = \mathbf{X}_2 \otimes \mathbf{X}_2 \otimes \mathbf{X}_2 \otimes \mathbf{X}_2 \otimes \mathbf{X}_2$, where

$$\mathbf{X}_2 = \begin{bmatrix} 1 & -1 \\ 1 & 1 \end{bmatrix}.$$

Let $\boldsymbol{\beta}_1$ be the 16×1 vector consisting of the mean, main effects and two-factor interaction effects and let $\boldsymbol{\beta}_3$ be the 16×1 vector of three and higher order interaction effects. Consider the following design in 21 treatment combinations:

$$\text{Treatment Combinations}$$
$$\Gamma: \begin{matrix} 0&1&0&0&0&0&1&1&1&1&0&0&0&0&0&0&1&1&1&1&0 \\ 0&0&1&0&0&0&1&0&0&0&1&1&1&0&0&0&1&1&1&0&1 \\ 0&0&0&1&0&0&0&1&0&0&1&0&0&1&1&0&1&1&0&1&1 \\ 0&0&0&0&1&0&0&0&1&0&0&1&0&1&0&1&1&0&1&1&1 \\ 0&0&0&0&0&1&0&0&0&1&0&0&1&0&1&1&0&1&1&1&1 \end{matrix}$$

It can be verified that Γ is a balanced array of strength 5 (see Section 13.7) and a resolution V design. In addition Γ satisfies the necessary condition (12.9) for $N_1 = 16$, $N_3 = 16$, $l_3 = 1$; that is, the rank $[\mathbf{X}_{\Gamma 1} \vdots \mathbf{B}_{\Gamma 3}] = 18$ for every submatrix $\mathbf{B}_{\Gamma 3}$ of order 21×2 from $\mathbf{X}_{\Gamma 3}^{(0)}$, the design matrix of order 21×16 of Γ relative to $\boldsymbol{\beta}_3$. When the observation vector $\mathbf{Y}_\Gamma'$ is written out as

$$\mathbf{Y}_\Gamma' = \left(Y_{(0,0,0,0,0)}, Y_{(1,0,0,0,0)}, \ldots, Y_{(1,0,1,1,1)}, Y_{(0,1,1,1,1)}\right),$$

then the parts in $E(\mathbf{Y}_\Gamma) = [\mathbf{X}_{\Gamma 1}\boldsymbol{\beta}_1 + \mathbf{X}_{\Gamma 3}^{(0)}\boldsymbol{\beta}_3]$ are explicitly given as

$$
\underset{\mathbf{X}_{\Gamma 1}}{\begin{bmatrix}
1 & -1 & -1 & -1 & -1 & -1 & 1 & 1 & 1 & 1 & 1 & 1 & 1 & 1 & 1 & 1 \\
1 & 1 & -1 & -1 & -1 & -1 & -1 & -1 & -1 & -1 & 1 & 1 & 1 & 1 & 1 & 1 \\
1 & -1 & 1 & -1 & -1 & -1 & -1 & 1 & 1 & 1 & -1 & -1 & -1 & 1 & 1 & 1 \\
1 & -1 & -1 & 1 & -1 & -1 & 1 & -1 & 1 & 1 & -1 & 1 & 1 & -1 & -1 & 1 \\
1 & -1 & -1 & -1 & 1 & -1 & 1 & 1 & -1 & 1 & 1 & -1 & 1 & -1 & 1 & -1 \\
1 & -1 & -1 & -1 & -1 & 1 & 1 & 1 & 1 & -1 & 1 & 1 & -1 & 1 & -1 & -1 \\
1 & 1 & 1 & -1 & -1 & -1 & 1 & -1 & -1 & -1 & -1 & -1 & -1 & 1 & 1 & 1 \\
1 & 1 & -1 & 1 & -1 & -1 & -1 & 1 & -1 & -1 & -1 & 1 & 1 & -1 & -1 & 1 \\
1 & 1 & -1 & -1 & 1 & -1 & -1 & -1 & 1 & -1 & 1 & -1 & 1 & -1 & 1 & -1 \\
1 & 1 & -1 & -1 & -1 & 1 & -1 & -1 & -1 & 1 & 1 & 1 & -1 & 1 & -1 & -1 \\
1 & -1 & 1 & 1 & -1 & -1 & -1 & -1 & 1 & 1 & 1 & -1 & -1 & -1 & -1 & 1 \\
1 & -1 & 1 & -1 & 1 & -1 & -1 & 1 & -1 & 1 & -1 & 1 & -1 & -1 & 1 & -1 \\
1 & -1 & 1 & -1 & -1 & 1 & -1 & 1 & -1 & 1 & -1 & 1 & -1 & 1 & -1 & -1 \\
1 & -1 & -1 & 1 & 1 & -1 & 1 & -1 & -1 & 1 & -1 & -1 & 1 & 1 & -1 & -1 \\
1 & -1 & -1 & 1 & -1 & 1 & 1 & -1 & 1 & -1 & -1 & 1 & -1 & -1 & 1 & -1 \\
1 & -1 & -1 & -1 & 1 & 1 & 1 & 1 & -1 & -1 & 1 & -1 & -1 & -1 & -1 & 1 \\
1 & 1 & 1 & 1 & 1 & -1 & 1 & 1 & 1 & -1 & 1 & 1 & -1 & 1 & -1 & -1 \\
1 & 1 & 1 & 1 & -1 & 1 & 1 & 1 & -1 & 1 & 1 & -1 & 1 & -1 & 1 & -1 \\
1 & 1 & 1 & -1 & 1 & 1 & 1 & -1 & 1 & 1 & -1 & 1 & 1 & -1 & -1 & 1 \\
1 & 1 & -1 & 1 & 1 & 1 & -1 & 1 & 1 & 1 & -1 & -1 & -1 & 1 & 1 & 1 \\
1 & -1 & 1 & 1 & 1 & 1 & -1 & -1 & -1 & -1 & 1 & 1 & 1 & 1 & 1 & 1
\end{bmatrix}}
\underset{\boldsymbol{\beta}_1}{\begin{bmatrix}
\phi_1^0\phi_2^0\phi_3^0\phi_4^0\phi_5^0 \\
\phi_1^1\phi_2^0\phi_3^0\phi_4^0\phi_5^0 \\
\phi_1^0\phi_2^1\phi_3^0\phi_4^0\phi_5^0 \\
\phi_1^0\phi_2^0\phi_3^1\phi_4^0\phi_5^0 \\
\phi_1^0\phi_2^0\phi_3^0\phi_4^1\phi_5^0 \\
\phi_1^0\phi_2^0\phi_3^0\phi_4^0\phi_5^1 \\
\phi_1^1\phi_2^1\phi_3^0\phi_4^0\phi_5^0 \\
\phi_1^1\phi_2^0\phi_3^1\phi_4^0\phi_5^0 \\
\phi_1^1\phi_2^0\phi_3^0\phi_4^1\phi_5^0 \\
\phi_1^1\phi_2^0\phi_3^0\phi_4^0\phi_5^1 \\
\phi_1^0\phi_2^1\phi_3^1\phi_4^0\phi_5^0 \\
\phi_1^0\phi_2^1\phi_3^0\phi_4^1\phi_5^0 \\
\phi_1^0\phi_2^1\phi_3^0\phi_4^0\phi_5^1 \\
\phi_1^0\phi_2^0\phi_3^1\phi_4^1\phi_5^0 \\
\phi_1^0\phi_2^0\phi_3^1\phi_4^0\phi_5^1 \\
\phi_1^0\phi_2^0\phi_3^0\phi_4^1\phi_5^1
\end{bmatrix}}
$$

and

$$
\underset{\mathbf{X}_{\Gamma 3}^{(0)}}{\begin{bmatrix}
-1 & -1 & -1 & -1 & -1 & -1 & -1 & -1 & -1 & -1 & 1 & 1 & 1 & 1 & 1 & 1 \\
1 & 1 & 1 & 1 & 1 & 1 & -1 & -1 & -1 & -1 & -1 & -1 & -1 & -1 & 1 & 1 \\
1 & 1 & 1 & -1 & -1 & -1 & 1 & 1 & 1 & -1 & -1 & -1 & -1 & 1 & -1 & 1 \\
1 & -1 & -1 & 1 & 1 & -1 & 1 & 1 & -1 & 1 & -1 & -1 & 1 & -1 & -1 & 1 \\
-1 & 1 & -1 & 1 & -1 & 1 & 1 & -1 & 1 & 1 & -1 & 1 & -1 & -1 & -1 & 1 \\
-1 & -1 & 1 & -1 & 1 & 1 & -1 & 1 & 1 & 1 & 1 & -1 & -1 & -1 & -1 & 1 \\
-1 & -1 & -1 & 1 & 1 & 1 & 1 & 1 & 1 & -1 & 1 & 1 & 1 & -1 & -1 & -1 \\
-1 & 1 & 1 & -1 & -1 & 1 & 1 & 1 & -1 & 1 & 1 & 1 & -1 & 1 & -1 & -1 \\
1 & -1 & 1 & -1 & 1 & -1 & 1 & -1 & 1 & 1 & 1 & -1 & 1 & 1 & -1 & -1 \\
1 & 1 & -1 & 1 & -1 & -1 & -1 & 1 & 1 & 1 & -1 & 1 & 1 & 1 & -1 & -1 \\
-1 & 1 & 1 & 1 & 1 & -1 & -1 & -1 & 1 & 1 & 1 & 1 & -1 & -1 & 1 & -1 \\
1 & -1 & 1 & 1 & -1 & 1 & -1 & 1 & -1 & 1 & 1 & -1 & 1 & -1 & 1 & -1 \\
1 & 1 & -1 & -1 & 1 & 1 & 1 & -1 & -1 & 1 & -1 & 1 & 1 & -1 & 1 & -1 \\
1 & 1 & -1 & -1 & 1 & 1 & -1 & 1 & 1 & -1 & 1 & -1 & -1 & 1 & 1 & -1 \\
1 & -1 & 1 & 1 & -1 & 1 & 1 & -1 & 1 & -1 & -1 & 1 & -1 & 1 & 1 & -1 \\
-1 & 1 & 1 & 1 & 1 & -1 & 1 & 1 & -1 & -1 & -1 & -1 & 1 & 1 & 1 & -1 \\
1 & 1 & 1 & 1 & 1 & 1 & 1 & 1 & 1 & 1 & 1 & -1 & -1 & -1 & -1 & 1 \\
1 & 1 & 1 & 1 & 1 & 1 & 1 & 1 & 1 & 1 & -1 & 1 & -1 & -1 & -1 & 1 \\
1 & 1 & 1 & 1 & 1 & 1 & 1 & 1 & 1 & 1 & -1 & -1 & 1 & -1 & -1 & 1 \\
1 & 1 & 1 & 1 & 1 & 1 & 1 & 1 & 1 & 1 & -1 & -1 & -1 & 1 & -1 & 1 \\
1 & 1 & 1 & 1 & 1 & 1 & 1 & 1 & 1 & 1 & -1 & -1 & -1 & -1 & 1 & 1
\end{bmatrix}}
\underset{\boldsymbol{\beta}_3}{\begin{bmatrix}
\phi_1^1\phi_2^1\phi_3^1\phi_4^0\phi_5^0 \\
\phi_1^1\phi_2^1\phi_3^0\phi_4^1\phi_5^0 \\
\phi_1^1\phi_2^1\phi_3^0\phi_4^0\phi_5^1 \\
\phi_1^1\phi_2^0\phi_3^1\phi_4^1\phi_5^0 \\
\phi_1^1\phi_2^0\phi_3^1\phi_4^0\phi_5^1 \\
\phi_1^1\phi_2^0\phi_3^0\phi_4^1\phi_5^1 \\
\phi_1^0\phi_2^1\phi_3^1\phi_4^1\phi_5^0 \\
\phi_1^0\phi_2^1\phi_3^1\phi_4^0\phi_5^1 \\
\phi_1^0\phi_2^1\phi_3^0\phi_4^1\phi_5^1 \\
\phi_1^0\phi_2^0\phi_3^1\phi_4^1\phi_5^1 \\
\phi_1^1\phi_2^1\phi_3^1\phi_4^1\phi_5^0 \\
\phi_1^1\phi_2^1\phi_3^1\phi_4^0\phi_5^1 \\
\phi_1^1\phi_2^1\phi_3^0\phi_4^1\phi_5^1 \\
\phi_1^1\phi_2^0\phi_3^1\phi_4^1\phi_5^1 \\
\phi_1^0\phi_2^1\phi_3^1\phi_4^1\phi_5^1 \\
\phi_1^1\phi_2^1\phi_3^1\phi_4^1\phi_5^1
\end{bmatrix}}
$$

Remark. Referring to the family (12.3) of competing search linear models an ad hoc procedure, borrowed from the stepwise regression method, has been proposed in the literature for searching out a γ_3, whose probability of being the correct one is unknown.

The first stage in this procedure searches out a γ_3, say $\gamma_3^{(+)}$, such that the corresponding sum of squares of least squares residuals, that is,

$$\left[\mathbf{Y}_\Gamma - \mathbf{X}_{\Gamma 1}\hat{\boldsymbol{\beta}}_1 - \mathbf{X}_{\Gamma 2}\hat{\boldsymbol{\beta}}_2 - \mathbf{X}_{\Gamma 3}^{(+)}\gamma_3^{(+)}\right]'\left[\mathbf{Y}_\Gamma - \mathbf{X}_{\Gamma_1}\hat{\boldsymbol{\beta}}_1 - \mathbf{X}_{\Gamma 2}\hat{\boldsymbol{\beta}}_2 - \mathbf{X}_{\Gamma 3}^{(+)}\gamma_3^{(+)}\right]$$

is minimized over all possible choices of γ_3 in $P_{l_3}(\boldsymbol{\beta}_3)$. Here $\hat{\boldsymbol{\beta}}_1$ and $\hat{\boldsymbol{\beta}}_2$ indicate least squares solutions for $\boldsymbol{\beta}_1$ and $\boldsymbol{\beta}_2$, respectively. The second stage assumes that the correct model is given by

$$E[\mathbf{Y}_\Gamma] = \mathbf{X}_{\Gamma 1}^{(+)}\boldsymbol{\beta}_1^{(+)} + \mathbf{X}_{\Gamma 2}\boldsymbol{\beta}_2, \qquad \operatorname{Cov}(\mathbf{Y}_\Gamma) = \sigma^2\mathbf{I}_n,$$

where

$$\mathbf{X}_{\Gamma 1}^{(+)} = \begin{bmatrix} \mathbf{X}_{\Gamma 1} \\ \cdots \\ \mathbf{X}_{\Gamma 3}^{(+)} \end{bmatrix} \text{ and } \boldsymbol{\beta}_1^{(+)} = \begin{bmatrix} \boldsymbol{\beta}_1 \\ \cdots \\ \gamma_3^{(+)} \end{bmatrix}.$$

Least squares estimators are then obtained of $\boldsymbol{\beta}_1^{(+)}$ using the least squares formulas developed for the various cases in Chapter 7. The statistical properties of this search procedure and of the subsequent estimator of $\boldsymbol{\beta}_1^{(+)}$ have not been worked out.

In the following references most of the results relate to the case $\sigma^2 = 0$. In the case $\sigma^2 > 0$ various ad hoc procedures have been proposed whose theoretical properties have not been worked out. Thus we conclude that in the stochastic case the problem of search linear models is wide open. Apart from the search, estimation, and inferential aspects, there are the problems of optimal (in any of the senses) design selection.

12.4 SELECTED ADDITIONAL READING

Starred references are recommended for a first reading.

1. Anderson, D. A., and Thomas, A. M. (1980). Weakly resolvable IV. 3 search designs for the p^n factorial experiment. *J. Stat. Plan. Inf.* **4**, 299–312.
2. Ghosh, S. (1975). Search designs. Unpublished Ph.D. thesis, Colorado State University, Fort Collins, Colo.
3. Jain, N. C. (1980). *q*-coverings of finite affine spaces. Unpublished Ph.D. thesis, Colorado State University, Fort Collins, Colo.
4. Katona, G. O. H., and Srivastava, J. N. (1980). Studies on the property P_t ($t \le 7$) of $(1, -1)$ matrices with a group structure. Tech. Report, Department of Statistics, Colorado State University, Fort Collins, Colo.

5. Mallenby, D. G. (1977). Inferences for the search linear models. Unpublished Ph.D. thesis, Colorado State University, Fort Collins, Colo.

*6. Srivastava, J. N. (1975). Designs for searching non-negligible effects. In *A Survey of Statistical Design and Linear Models*, J. N. Srivastava, Ed. North-Holland, Amsterdam, pp. 507–519.

7. Srivastava, J. N. (1976). Some further theory of search linear models. In *Contributions to Applied Statistics*, W. J. Ziegler, Ed. Birkhäuser Vêrlag, Basel and Stuttgart, pp. 249–256.

8. Srivastava, J. N. (1977). Optimal search designs, or designs optimal under bias free optimality criteria. In *Statistical Decision Theory and Related Topics*, S. S. Gupta and D. S. Moore, Eds. Academic, New York, pp. 2, 375–409.

9. Srivastava, J. N. (1977). On the linear independence of sets of 2^q columns of certain $(1,-1)$ matrices with a group structure and its connection with finite geometries. In *Proceeding of the International Symposium on Combinatorial Mathematics*, A. D. Holton and J. Seberry, Eds. Springer-Verlag, New York, pp. 79–88.

*10. Srivastava, J. N. (1980). Some basic results on search linear models with nuisance parameters. In *Contributions in Statistics*, A. Bartoszynski, Ed. Banach Center of Mathematics, Warsaw, Poland, pp. 309–313.

11. Srivastava, J. N., and Ghosh, S. (1976). A series of balanced 2^m factorial designs of resolution V which allow search and estimation of one extra unknown effect. *Sankhyā*, **B38**, 280–289.

*12. Srivastava, J. N., and Ghosh, S. (1977a). Balanced 2^m factorial designs of resolution V which allow search and estimation of one extra unknown effect, $4 \le m \le 8$. *Commun. Stat.*, **A6**, 141–166.

13. Srivastava, J. N., and Ghosh, S. (1977b). On the existence of search designs with continuous factors. *Ann. Inst. Stat. Math.*, **29**, 301–306.

14. Srivastava, J. N., and Ghosh, S. (1980). On the enumeration and representation of non-isomorphic bipartite graphs arising in factorial search designs. *Ann. Discrete Math.*, **6**, 315–322.

15. Srivastava, J. N. and Gupta, B. C. (1979). Main effect plan for 2^m factorials which allow search and estimation of one unknown effect. *J. Stat. Plan. Inf.*, **3**, 259–265.

*16. Srivastava, J. N., and Mallenby, D. G. (1981). Some studies of a new method of search in search linear models. *J. Multivar. Anal.* (to appear).

CHAPTER 13

Some Known Methods for Constructing Factorial Designs

A detailed description of all the known methods for the construction of factorial designs would call for a book of at least a couple of hundred pages. However, this is not the purpose of the book. What we shall do is list known methods together with illustrations of some of them, and provide the reader with a generous list of selected literature citations at the end of the chapter. We believe this approach is not only compatible with the spirit of this book, but also does not bias the reader in favor of the techniques preferred by the authors. For illustration purposes, a discussion of some well-known methods is provided.

13.1 A LIST OF CONSTRUCTION METHODS

The following methods of constructing factorial designs appear in the literature:

(i) Orthogonal arrays.
(ii) Balanced arrays.
(iii) Latin squares and orthogonal Latin squares.
(iv) Hadamard matrices.
(v) Finite geometries.
(vi) Confounding.
(vii) Group theory.

(viii) Algebraic decomposition.
(ix) Combinatorial topology.
(x) Foldover.
(xi) Collapsing of levels.
(xii) Composition (direct product and direct sum).
(xiii) Codes.
(xiv) Block designs.
(xv) *F*-squares.
(xvi) Weighing designs.
(xvii) Lattice designs.
(xviii) Finite graphs.
(xix) One-at-a-time.
(xx) Trial and error.
(xxi) Others.

We will discuss and illustrate techniques based on orthogonal arrays, Latin squares and orthogonal Latin squares, and Hadamard matrices. A brief introduction is also provided for balanced arrays and direct product composition method. Throughout we shall assume the orthogonal polynomial model for the data.

13.2 ORTHOGONAL ARRAY METHODS

A great deal of symmetry can be identified within a minimal complete symmetric factorial design based on t factors each at s levels. Presenting the s^t treatment combinations in a $t \times s^t$ array with rows identifying the factors and the columns the treatments, we observe that each $d \le t$ rows of the array is precisely $\lambda^* = s^{t-d}$ copies of the minimal complete symmetric factorial design based on the chosen d factors. For example, the minimal complete factorial design based on $t=4$ factors each at $s=2$ levels (exhibited below in a 4×16 array) has the property that any d rows, $2 \le d \le 4$, of the array contain $\lambda^* = 2^{4-d}$ copies of the minimal complete factorial design based on d factors each at two levels.

$$\begin{matrix} 1 & 0 & 0 & 0 & 0 & 1 & 1 & 1 & 0 & 1 & 1 & 1 & 1 & 0 & 0 & 0 \\ 0 & 1 & 0 & 0 & 1 & 0 & 1 & 1 & 0 & 1 & 0 & 0 & 1 & 0 & 1 & 1 \\ 0 & 0 & 1 & 0 & 1 & 1 & 0 & 1 & 0 & 0 & 1 & 0 & 1 & 1 & 1 & 0 \\ 0 & 0 & 0 & 1 & 1 & 1 & 1 & 0 & 0 & 0 & 0 & 1 & 1 & 1 & 0 & 1 \end{matrix} . \tag{13.1}$$

The fact that for any d factors the minimal complete symmetric factorial contains λ^* copies of the minimal complete factorial based on d factors implies many desirable statistical properties in analyzing data from such a design. One such property related to the estimability of factorial effects is explored later. Since these useful properties depend mainly on the fact that there is a *fixed number* (not necessarily λ^*) of copies of the minimal complete factorial design for any d factors inside the design leads us to ask the following important question: Is there any symmetric factorial design, apart from the minimal complete symmetric factorial design or copies of it, based on t factors each at s levels, with the property that there exists at least one $d<t$ such that any d rows of the array, which represents the design, are precisely λ copies of the minimal complete symmetric factorial design based on the chosen d factors? Fortunately, the answer to this question is yes, and the corresponding array for such a design is now known as an *orthogonal array*. These arrays have been found to have many applications apart from the field of experimental design. For this reason it is now customary to define an orthogonal array independent of the terminology of factorial designs.

Definition 13.1. A $t\times n$ matrix $\mathbf{A}$ (array) with entries from a set S of s symbols is called an *orthogonal array* of size n, t constraints, s levels, strength d, and index λ if any $d\times n$ submatrix of $\mathbf{A}$ contains all s^d possible $d\times 1$ column vectors based on s symbols of S with the same frequency λ. Such an array is denoted by $\mathrm{OA}(n,t,s,d;\lambda)$. We refer to n,t,s,d,λ as the *parameters* of the array. (In case λ is a power of s, the array is also known as a *hypercube of strength d*).

It is clear from the definition of an orthogonal array that: (i) $n=\lambda s^t$ in any $\mathrm{OA}(n,t,s,d;\lambda)$; (ii) an orthogonal array of strength d is also an orthogonal array of strength $l<d$; (iii) an orthogonal array remains an orthogonal array if we permute rows and/or columns of the array. The following global inequalities for the parameters of an $\mathrm{OA}(n,t,s,d;\lambda)$ array are well-known if $d\geq 2$

(13.2)

$$n\geq 1+\binom{t}{1}(s-1)+\cdots+\binom{t}{u}(s-1)^u, \qquad \text{if } d=2u$$

$$n\geq 1+\binom{t}{1}(s-1)+\cdots+\binom{t}{u}(s-1)^u+\binom{t-1}{u}(s-1)^{u+1},$$

$$\text{if } d=2u+1.$$

As can be seen these inequalities do not involve λ and thus could be improved in particular cases by considering the index of the array. The literature of orthogonal arrays contains many useful bounds. Our purpose here is not to review the literature. The reader should consult our selected bibliography at the end of this chapter for this purpose. In passing, we mention the following known useful results. In an OA$(n, t, s, d; \lambda)$ array:

$$
\begin{array}{lll}
 & t \le s+d-1, & \text{if } s \text{ is even and } \lambda = 1 \\
(13.3) & t \le s+d-2, & \text{if } s \text{ is odd}, d \ge 3, \text{ and } \lambda = 1 \\
 & t \le d+1, & \text{if } s = 2 \text{ and } \lambda \text{ is odd}.
\end{array}
$$

Before we proceed further we reiterate that the concept of orthogonal arrays evolved from factorial designs and an OA$(n, t, s, d; \lambda)$ is simply a factorial design consisting of n (not necessarily distinct) treatment combinations based on t factors each at s levels with the property that with respect to any d factors the design contains λ copies of the minimal complete symmetric factorial design for those d factors.

In the following development we present several examples for two and three levels.

Example 13.1. The following is an OA(8,4,2,3;1) array.

$$
\begin{matrix}
1 & 0 & 0 & 0 & 0 & 1 & 1 & 1 \\
0 & 1 & 0 & 0 & 1 & 0 & 1 & 1 \\
0 & 0 & 1 & 0 & 1 & 1 & 0 & 1 \\
0 & 0 & 0 & 1 & 1 & 1 & 1 & 0
\end{matrix}.
$$

Note that this is one-half of the 16 treatment combinations in the 2^4 factorial. The remaining eight treatment combinations also form an OA(8,4,2,3;1) array.

Example 13.2. An OA(16,5,2,4;1) array follows:

$$
\begin{matrix}
0 & 0 & 0 & 0 & 0 & 0 & 0 & 1 & 1 & 1 & 1 & 0 & 1 & 1 & 1 & 1 \\
0 & 0 & 0 & 0 & 1 & 1 & 1 & 0 & 0 & 0 & 1 & 1 & 0 & 1 & 1 & 1 \\
0 & 0 & 1 & 1 & 0 & 0 & 1 & 0 & 0 & 1 & 0 & 1 & 1 & 0 & 1 & 1 \\
0 & 1 & 0 & 1 & 0 & 1 & 0 & 0 & 1 & 0 & 0 & 1 & 1 & 1 & 0 & 1 \\
0 & 1 & 1 & 0 & 1 & 0 & 0 & 1 & 0 & 0 & 0 & 1 & 1 & 1 & 1 & 0
\end{matrix}.
$$

Again observe that this is one-half of the 32 treatment combinations in the 2^5 factorial. The remaining 16 treatment combinations also form an OA(16,5,2,4;1) array.

Examples 13.1 and 13.2 indicate a general phenomenon, namely, if $n < s^t$ distinct treatment combinations of the s^t factorial design form an orthogonal array, then the remaining $s^t - n$ treatment combinations also form an orthogonal array. The derivation of the parameters of the latter array is left as an exercise to the reader.

For given values of n and s it is possible, in general, to construct orthogonal arrays for a large value of t if we allow the value of d to decrease. The following example in conjunction with Example 13.1 explains this point.

Example 13.3. The following is an OA(8,7,2,2;2) array:

$$\begin{array}{cccccccc}
0 & 0 & 0 & 0 & 1 & 1 & 1 & 1 \\
0 & 0 & 1 & 1 & 1 & 1 & 0 & 0 \\
0 & 1 & 0 & 1 & 1 & 0 & 1 & 0 \\
0 & 0 & 1 & 1 & 0 & 0 & 1 & 1 \\
0 & 1 & 0 & 1 & 0 & 1 & 0 & 1 \\
0 & 1 & 1 & 0 & 0 & 1 & 1 & 0 \\
0 & 1 & 1 & 0 & 1 & 0 & 0 & 1
\end{array}.$$

In subsequent sections, we shall give a few examples of orthogonal arrays based on three symbols.

Example 13.4. The following array is an OA(9,4,3,2;1) array.

$$\begin{array}{ccccccccc}
0 & 0 & 0 & 1 & 1 & 1 & 2 & 2 & 2 \\
0 & 1 & 2 & 0 & 1 & 2 & 0 & 1 & 2 \\
0 & 1 & 2 & 1 & 2 & 0 & 2 & 0 & 1 \\
0 & 1 & 2 & 2 & 0 & 1 & 1 & 2 & 0
\end{array}.$$

Example 13.5. An OA(18,7,3,2;2) array follows:

$$\begin{array}{cccccccccccccccccc}
0 & 0 & 0 & 1 & 1 & 1 & 2 & 2 & 2 & 0 & 0 & 0 & 1 & 1 & 1 & 2 & 2 & 2 \\
0 & 1 & 2 & 0 & 1 & 2 & 0 & 1 & 2 & 0 & 1 & 2 & 0 & 1 & 2 & 0 & 1 & 2 \\
0 & 1 & 2 & 1 & 2 & 0 & 2 & 0 & 1 & 2 & 0 & 1 & 0 & 1 & 2 & 1 & 2 & 0 \\
0 & 2 & 1 & 1 & 0 & 2 & 2 & 1 & 0 & 1 & 0 & 2 & 2 & 1 & 0 & 0 & 2 & 1 \\
0 & 1 & 2 & 1 & 2 & 0 & 1 & 2 & 0 & 0 & 1 & 2 & 2 & 0 & 1 & 2 & 0 & 1 \\
0 & 1 & 2 & 2 & 0 & 1 & 0 & 1 & 2 & 1 & 2 & 0 & 2 & 0 & 1 & 1 & 2 & 0 \\
0 & 1 & 2 & 0 & 1 & 2 & 2 & 0 & 1 & 1 & 2 & 0 & 1 & 2 & 0 & 2 & 0 & 1
\end{array}.$$

Example 13.6. The following is an OA(27,4,3,3;1) array:

$$\begin{array}{ccccccccccccccccccccccccccc}
0 & 1 & 2 & 0 & 1 & 2 & 0 & 1 & 2 & 0 & 1 & 2 & 0 & 1 & 2 & 0 & 1 & 2 & 0 & 1 & 2 & 0 & 1 & 2 & 0 & 1 & 2 \\
0 & 1 & 2 & 1 & 2 & 0 & 2 & 0 & 1 & 1 & 2 & 0 & 2 & 0 & 1 & 0 & 1 & 2 & 2 & 0 & 1 & 0 & 1 & 2 & 1 & 2 & 0 \\
0 & 1 & 2 & 2 & 0 & 1 & 1 & 2 & 0 & 1 & 2 & 0 & 0 & 1 & 2 & 2 & 0 & 1 & 2 & 0 & 1 & 1 & 2 & 0 & 0 & 1 & 2 \\
0 & 0 & 0 & 0 & 0 & 0 & 0 & 0 & 0 & 1 & 1 & 1 & 1 & 1 & 1 & 1 & 1 & 1 & 2 & 2 & 2 & 2 & 2 & 2 & 2 & 2 & 2
\end{array}.$$

Holding $n=27$, $s=3$, $d=3$, then $t=4$ is the maximum possible value in an orthogonal array. The verification of this fact is left as a challenging exercise to the reader. However, if we let $d=2$, then we can increase t to 13 as the following example shows.

Example 13.7. Here is an orthogonal array with $n=27$, $t=13$, $s=3$, $d=2$, and $\lambda=3$.

$$\begin{array}{ccccccccccccccccccccccccccc}
0&1&2&0&1&2&0&1&2&0&1&2&0&1&2&0&1&2&0&1&2&0&1&2&0&1&2\\
0&1&2&1&2&0&0&1&2&1&2&0&2&0&1&2&0&1&0&1&2&2&0&1&1&2&0\\
0&1&2&2&1&2&1&2&0&2&0&1&2&0&1&0&1&2&2&0&1&1&2&0&1&2&0\\
0&1&2&1&2&0&2&0&1&2&0&1&0&1&2&2&0&1&1&2&0&1&2&0&0&1&2\\
0&1&2&2&0&1&2&0&1&0&1&2&2&0&1&1&2&0&1&2&0&0&1&2&1&2&0\\
0&1&2&2&0&1&0&1&2&2&0&1&1&2&0&1&2&0&0&1&2&1&2&0&2&0&1\\
0&1&2&0&1&2&2&0&1&1&2&0&1&2&0&0&1&2&1&2&0&2&0&1&2&0&1.\\
0&1&2&2&0&1&1&2&0&1&2&0&0&1&2&1&2&0&2&0&1&2&0&1&0&1&2\\
0&1&2&1&2&0&1&2&0&0&1&2&1&2&0&2&0&1&2&0&1&0&1&2&2&0&1\\
0&0&0&0&0&0&0&0&0&1&1&1&1&1&1&1&1&1&2&2&2&2&2&2&2&2&2\\
0&0&0&1&1&1&2&2&2&0&0&0&1&1&1&2&2&2&0&0&0&1&1&1&2&2&2\\
0&0&0&1&1&1&2&2&2&1&1&1&2&2&2&0&0&0&2&2&2&0&0&0&1&1&1\\
0&0&0&2&2&2&1&1&1&1&1&1&0&0&0&2&2&2&2&2&2&1&1&1&0&0&0
\end{array}$$

It is also a very challenging exercise to verify that $t=13$ in the preceding array is maximum; that is, there is no OA$(27,t,3,2;3)$ array with $t>13$.

13.3 FACTORIAL DESIGNS THAT ARE ORTHOGONAL ARRAYS

As we mentioned earlier in Section 13.2, if we consider the columns of an OA$(n,t,s,d;\lambda)$ array as treatment combinations then such an array is a factorial design containing n (not necessarily distinct) treatment combinations from the s^t factorial. Moreover, such a factorial design has the property that with respect to any d factors, $d<t$, the design contains λ copies of the minimal complete factorial for the chosen d factors. Because of this property, as we may expect, a factorial design that is an orthogonal array has many desirable statistical properties. One such property is detailed in the following theorem.

Theorem 13.1. If Γ is a factorial design that is an OA$(n,t,s,d;\lambda)$ array, then under the orthogonal polynomial model (4.10) Γ is a resolution $d+1$

orthogonal factorial design with the additional property that the general mean or intercept is also estimable whether or not $d+1$ is even or odd.

Proof. We consider two cases:

Case 1. The strength of the array is even; that is, d is even. Since $d+1$ is odd we have to show that under the orthogonal polynomial model (4.10) the general mean and all factorial effects (main effects and interaction effects) involving $d/2$ or fewer factors (also called effects of degree or order $d/2$ or less, see Chapter 4) are estimable and the covariance of the BLUEs of any two such estimators is zero if we can assume that all other factorial effects are zero. Or, in model notation,

$$(13.4) \qquad \mathbf{Y}_\Gamma = \mathbf{X}_{\Gamma 1}\boldsymbol{\beta}_1 + \boldsymbol{\epsilon}_\Gamma, \qquad \text{Cov}(\boldsymbol{\epsilon}_\Gamma) = \sigma^2 \mathbf{I}_n$$

has the property that $\mathbf{X}'_{\Gamma 1}\mathbf{X}_{\Gamma 1}$ is a diagonal matrix, where $\boldsymbol{\beta}_1$ consists of the general mean and those factorial effects involving $d/2$ or fewer factors, that is, in the notation of (4.12) all those parameters in $\boldsymbol{\beta}_\rho$ having $d/2$ or fewer nonzero superscripts. To see this, consider any two columns of $\mathbf{X}_{\Gamma 1}$, say $\mathbf{x}_i$ and $\mathbf{x}_j$. These two columns are related to two parameters in $\boldsymbol{\beta}_1$ such that the number of nonzero superscripts together for these two effects is $h_1 \leq d/2 + d/2 = d$. Choose any d factors out of t factors that contain these h_1 factors. To observe that $\mathbf{x}'_i\mathbf{x}_j = 0$, rearrange the entries of $\mathbf{Y}_\Gamma$ (and correspondingly the rows of $\mathbf{X}_{\Gamma 1}$) in a way such that Γ is partitioned in λ parts and each part is a s^d minimal complete factorial for these d factors. If we now recall the definition of $\mathbf{X}_\rho$ (see 4.11), which has been used to produce $\mathbf{X}_{\Gamma 1}$, we see that $\mathbf{x}'_i\mathbf{x}_j$ is the sum of λ times zero; that is, each column vector $\mathbf{x}_i$ and $\mathbf{x}_j$ is partitioned into λ subvectors each with s^d components and the corresponding portions of these vectors in $\mathbf{x}_i$ and $\mathbf{x}_j$ have inner product zero. Hence $\mathbf{x}'_i\mathbf{x}_j = 0$.

Case 2. The strength of the array is odd; that is, d is odd. Since in this case $d+1$ is even we have to show that under the orthogonal polynomial model (4.10), the general mean and all factorial effects (main effects and interactions) involving $(d-1)/2$ or fewer factors are estimable and the covariance of the BLUEs of any two such estimators is zero if we can assume that all factorial effects *apart* from those factorial effects involving *precisely* $(d+1)/2$ factors are zero. Or, in model notation,

$$(13.5) \qquad \mathbf{Y}_\Gamma = \mathbf{X}_{\Gamma 1}\boldsymbol{\beta}_1 + \mathbf{X}_{\Gamma 2}\boldsymbol{\beta}_2 + \boldsymbol{\epsilon}_\Gamma, \qquad \text{Cov}(\boldsymbol{\epsilon}_\Gamma) = \sigma^2 \mathbf{I}_n$$

has the property that $\mathbf{X}'_{\Gamma 1}\mathbf{X}_{\Gamma 1}$ is a diagonal matrix and $\mathbf{X}'_{\Gamma 1}\mathbf{X}_{\Gamma 2} = \mathbf{0}$, where

$\boldsymbol{\beta}_1$ consists of the general mean and those factorial effects involving $(d-1)/2$ or fewer factors and $\boldsymbol{\beta}_2$ consists of all factorial effects involving precisely $(d+1)/2$ factors. If $\mathbf{x}_i$ and $\mathbf{x}_j$ are two columns of $\mathbf{X}_{\Gamma 1}$, then their corresponding effects in $\boldsymbol{\beta}_1$ together involve $h_2 \leq (d-1)/2+(d-1)/2=d-1$ factors. Now argue as in Case 1 to show that $\mathbf{X}'_{\Gamma 1}\mathbf{X}_{\Gamma 1}$ is diagonal. If $\mathbf{x}_i$ is in $\mathbf{X}_{\Gamma 1}$ and $\mathbf{x}_j$ is in $\mathbf{X}_{\Gamma 2}$, then the corresponding effects in $\boldsymbol{\beta}_1$ and $\boldsymbol{\beta}_2$ together involve $h_3 \leq (d-1)/2+(d+1)/2=d$ factors. Again argue as in Case 1. Note however that if $\mathbf{x}_i$ and $\mathbf{x}_j$ are both in $\mathbf{X}_{\Gamma 2}$, then their corresponding effects in $\boldsymbol{\beta}_2$ together involve $h_4 \leq (d+1)/2+(d+1)/2=d+1$ factors and in case $h_4=d+1$ we cannot argue as in Case 1 since the design does not necessarily have the property that with respect to these $d+1$ factors it is λ/s copies of the minimal s^{d+1} factorial. Indeed, there is no guarantee that $\mathbf{x}'_i\mathbf{x}_j=0$ in this case.

Remark. We should emphasize that the converse statement in Theorem 13.1 is false; that is, there are resolution $d+1$ factorial designs that are not orthogonal arrays. For example, consider the following factorial design based on the 2^3 factorial.

$$\Gamma = \begin{matrix} 1 & 0 & 0 \\ 0 & 1 & 0 \\ 1 & 1 & 0 \\ 0 & 1 & 1 \\ 1 & 0 & 1 \end{matrix}.$$

It is easy to show that (left as an exercise) under the orthogonal polynomial model Γ is a resolution IV design but obviously it is not an orthogonal array of strength 3. Indeed, with this design the general mean $\phi_1^0\phi_2^0\phi_3^0$ is not even estimable, as was demonstrated in Chapter 11.

We shall now give two examples to elucidate the content of Theorem 13.1.

Example 13.8. Let Γ be the following factorial design consisting of nine treatment combinations of the 3^4 factorial.

$$\Gamma = \begin{matrix} 0 & 0 & 0 & 0 \\ 0 & 1 & 1 & 1 \\ 0 & 2 & 2 & 2 \\ 1 & 0 & 1 & 2 \\ 1 & 1 & 2 & 0 \\ 1 & 2 & 0 & 1 \\ 2 & 0 & 2 & 1 \\ 2 & 1 & 0 & 2 \\ 2 & 2 & 1 & 0 \end{matrix}.$$

This factorial design is the same OA(9,4,3,2;1) array given in Example 13.4 (see also Example 10.7). The strength of this array is $d=2$, so our claim is that Γ is a resolution III orthogonal factorial design under the orthogonal polynomial model (4.10). Thus we have to verify that the general mean and all main effects are estimable if we can assume that all other effects are zero. Using (4.10), (4.11), and (4.12) with the assumption that all interaction effects are zero, we obtain

$$
E\begin{bmatrix} Y_{(0,0,0,0)} \\ Y_{(0,1,1,1)} \\ Y_{(0,2,2,2)} \\ Y_{(1,0,1,2)} \\ Y_{(1,1,2,0)} \\ Y_{(1,2,0,1)} \\ Y_{(2,0,2,1)} \\ Y_{(2,1,0,2)} \\ Y_{(2,2,1,0)} \end{bmatrix} = \begin{bmatrix} 1 & 1 & 1 & 1 & 1 & 1 & 1 & 1 & 1 \\ 1 & 1 & 1 & -1 & 1 & -1 & 1 & -1 & 1 \\ 1 & 1 & 1 & 0 & -2 & 0 & -2 & 0 & -2 \\ 1 & -1 & 1 & 1 & 1 & -1 & 1 & 0 & -2 \\ 1 & -1 & 1 & -1 & 1 & 0 & -2 & 1 & 1 \\ 1 & -1 & 1 & 0 & -2 & 1 & 1 & -1 & 1 \\ 1 & 0 & -2 & 1 & 1 & 0 & -2 & -1 & 1 \\ 1 & 0 & -2 & -1 & 1 & 1 & 1 & 0 & -2 \\ 1 & 0 & -2 & 0 & -2 & -1 & 1 & 1 & 1 \end{bmatrix} \begin{bmatrix} \phi_1^0\phi_2^0\phi_3^0\phi_4^0 \\ \phi_1^1\phi_2^0\phi_3^0\phi_4^0 \\ \phi_1^2\phi_2^0\phi_3^0\phi_4^0 \\ \phi_1^0\phi_2^1\phi_3^0\phi_4^0 \\ \phi_1^0\phi_2^2\phi_3^0\phi_4^0 \\ \phi_1^0\phi_2^0\phi_3^1\phi_4^0 \\ \phi_1^0\phi_2^0\phi_3^2\phi_4^0 \\ \phi_1^0\phi_2^0\phi_3^0\phi_4^1 \\ \phi_1^0\phi_2^0\phi_3^0\phi_4^2 \end{bmatrix}
$$

$$
\mathbf{Y}_\Gamma \qquad = \qquad \mathbf{X}_{\Gamma 1} \qquad \boldsymbol{\beta}_1
$$

Since

$$
\mathbf{X}'_{\Gamma 1}\mathbf{X}_{\Gamma 1} = \begin{bmatrix} 9 & & & & & & & & \\ & 6 & & & & & \mathbf{0} & & \\ & & 18 & & & & & & \\ & & & 6 & & & & & \\ & & & & 18 & & & & \\ & & & & & 6 & & & \\ & & & & & & 18 & & \\ & \mathbf{0} & & & & & & 6 & \\ & & & & & & & & 18 \end{bmatrix}
$$

we conclude that Γ is indeed a resolution III orthogonal factorial design.

Example 13.9. Consider the following factorial design consisting of eight treatment combinations of the 2^4 factorial.

$$
\Gamma = \begin{matrix} 1 & 0 & 0 & 0 \\ 0 & 1 & 0 & 0 \\ 0 & 0 & 1 & 0 \\ 0 & 0 & 0 & 1 \\ 0 & 1 & 1 & 1 \\ 1 & 0 & 1 & 1 \\ 1 & 1 & 0 & 1 \\ 1 & 1 & 1 & 0 \end{matrix}.
$$

This factorial design is the same OA(8,4,2,3; 1) array that we presented in Example 13.1. The strength of this array is $d=3$, so our claim is that this factorial design is a resolution IV orthogonal factorial design under the orthogonal polynomial model (4.10) with the additional property that the general mean is also estimable. Thus we have to verify that the general mean and all main effects are estimable in the presence of all two-factor interaction effects if we can assume that all three and higher factor interaction effects are zero. Using (4.10), (4.11), and (4.12) with the assumption that all three- and four-factor interaction effects are zero, we have

$$
\underset{\mathbf{Y}_\Gamma}{E\begin{bmatrix} Y_{(1,0,0,0)} \\ Y_{(0,1,0,0)} \\ Y_{(0,0,1,0)} \\ Y_{(0,0,0,1)} \\ Y_{(0,1,1,1)} \\ Y_{(1,0,1,1)} \\ Y_{(1,1,0,1)} \\ Y_{(1,1,1,0)} \end{bmatrix}}
=
\left[\begin{array}{ccccc|cccccc}
1 & 1 & -1 & -1 & -1 & -1 & -1 & -1 & 1 & 1 & 1 \\
1 & -1 & 1 & -1 & -1 & -1 & 1 & 1 & -1 & -1 & 1 \\
1 & -1 & -1 & 1 & -1 & 1 & -1 & 1 & -1 & 1 & -1 \\
1 & -1 & -1 & -1 & 1 & 1 & 1 & -1 & 1 & -1 & -1 \\
1 & -1 & 1 & 1 & 1 & -1 & -1 & -1 & 1 & 1 & 1 \\
1 & 1 & -1 & 1 & 1 & -1 & 1 & 1 & -1 & -1 & 1 \\
1 & 1 & 1 & -1 & 1 & 1 & -1 & 1 & -1 & 1 & -1 \\
1 & 1 & 1 & 1 & -1 & 1 & 1 & -1 & 1 & -1 & -1
\end{array}\right]
\begin{bmatrix}
\phi_1^0\phi_2^0\phi_3^0\phi_4^0 \\
\phi_1^1\phi_2^0\phi_3^0\phi_4^0 \\
\phi_1^0\phi_2^1\phi_3^0\phi_4^0 \\
\phi_1^0\phi_2^0\phi_3^1\phi_4^0 \\
\phi_1^0\phi_2^0\phi_3^0\phi_4^1 \\
\phi_1^1\phi_2^1\phi_3^0\phi_4^0 \\
\phi_1^1\phi_2^0\phi_3^1\phi_4^0 \\
\phi_1^1\phi_2^0\phi_3^0\phi_4^1 \\
\phi_1^0\phi_2^1\phi_3^1\phi_4^0 \\
\phi_1^0\phi_2^1\phi_3^0\phi_4^1 \\
\phi_1^0\phi_2^0\phi_3^1\phi_4^1
\end{bmatrix}
\begin{matrix} \\ \\ \boldsymbol{\beta}_1 \\ \\ \\ \\ \\ \\ \boldsymbol{\beta}_2 \\ \\ \\ \end{matrix}
$$

$$
\mathbf{X}_{\Gamma 1} \qquad\qquad\qquad\qquad \mathbf{X}_{\Gamma 2}
$$

For this design and model

$$
\mathbf{X}'_\Gamma\mathbf{X}_\Gamma = \left[\begin{array}{c|c} \mathbf{X}'_{\Gamma 1}\mathbf{X}_{\Gamma 1} & \mathbf{X}'_{\Gamma 1}\mathbf{X}_{\Gamma 2} \\ \hline \mathbf{X}'_{\Gamma 2}\mathbf{X}_{\Gamma 1} & \mathbf{X}'_{\Gamma 2}\mathbf{X}_{\Gamma 2} \end{array}\right]
=
\left[\begin{array}{ccccc|cccccc}
8 & 0 & 0 & 0 & 0 & 0 & 0 & 0 & 0 & 0 & 0 \\
0 & 8 & 0 & 0 & 0 & 0 & 0 & 0 & 0 & 0 & 0 \\
0 & 0 & 8 & 0 & 0 & 0 & 0 & 0 & 0 & 0 & 0 \\
0 & 0 & 0 & 8 & 0 & 0 & 0 & 0 & 0 & 0 & 0 \\
0 & 0 & 0 & 0 & 8 & 0 & 0 & 0 & 0 & 0 & 0 \\
\hline
0 & 0 & 0 & 0 & 0 & 8 & 0 & 0 & 0 & 0 & -8 \\
0 & 0 & 0 & 0 & 0 & 0 & 8 & 0 & 0 & -8 & 0 \\
0 & 0 & 0 & 0 & 0 & 0 & 0 & 8 & -8 & 0 & 0 \\
0 & 0 & 0 & 0 & 0 & 0 & 0 & -8 & 8 & 0 & 0 \\
0 & 0 & 0 & 0 & 0 & 0 & -8 & 0 & 0 & 8 & 0 \\
0 & 0 & 0 & 0 & 0 & -8 & 0 & 0 & 0 & 0 & 8
\end{array}\right]
$$

showing that the general mean and all main effects are estimable orthogonally since $\mathbf{X}'_{\Gamma 1}\mathbf{X}_{\Gamma 2}=\mathbf{0}$ and $\mathbf{X}'_{\Gamma 1}\mathbf{X}_{\Gamma 1}=8\mathbf{I}_5$. For this design the two-factor interaction effects are not individually estimable (are confounded), since

$\mathbf{X}'_{\Gamma 2}\mathbf{X}_{\Gamma 2}$ is not a full rank matrix. Since

$$(\mathbf{X}'_{\Gamma}\mathbf{X}_{\Gamma})^{-1} = \begin{bmatrix} \frac{1}{8}\mathbf{I}_5 & \mathbf{0} \\ \mathbf{0} & \frac{1}{2}\mathbf{I}_6 \end{bmatrix}$$

we can easily obtain the BLUEs of the components of $\boldsymbol{\beta}_1$ and estimable linear parametric functions of $\boldsymbol{\beta}_2$. The reader is invited to derive all such estimators.

Before we leave this section we leave the reader the following important exercise: Investigate the optimality properties of a factorial design that is an OA$(n, t, s, d; \lambda)$ array from the s^t factorial in the class of all factorial designs with n treatment combinations from the s^t factorial under the orthogonal polynomial model.

13.4 CONSTRUCTION OF ORTHOGONAL ARRAYS

Since factorial designs that are orthogonal arrays have many desirable statistical properties, some elementary techniques for the construction of such arrays are useful. The literature of orthogonal arrays contains many interesting geometrical and algebraic techniques for the construction of these arrays. We refer the reader to the selected list of bibliography at the end of this chapter for this purpose. Here we shall present some techniques that serve as an introduction to this subject.

(i) *Complementation.* If Γ is an orthogonal array with $n < s^t$ treatment combinations such that no treatment is duplicated, then the set of remaining $s^t - n$ treatment combinations is also an orthogonal array.

(ii) *Juxtaposition.* If Γ_i is an OA$(n_i, t, s, d; \lambda_i)$ array based on the s^t factorial, $i = 1, 2, \ldots, l$. Then the factorial design

$$\Gamma = \bigcup_{i=1}^{l} \Gamma_i$$

is an OA$(n, t, s, d; \lambda)$ array where $n = n_1 + \cdots + n_l$ and $\lambda = \lambda_1 + \ldots + \lambda_l$. The verification is obvious by observing that if $\mathbf{A}_i$ is the matrix representation of Γ_i then

$$\mathbf{A} = \mathbf{A}_1 \vdots \mathbf{A}_2 \vdots \ldots \vdots \mathbf{A}_l$$

is the array representation of Γ, which is clearly an orthogonal array with the given parameters.

(iii) *Multiplication.* If Γ_i is an OA$(n_i, t, s_i, d; \lambda_i)$ array based on $S_i = \{0, 1, \dots, s_i - 1\}$, $i = 1, 2$. Then we can construct a factorial design Γ that is an OA$(n, t, s, d; \lambda)$ array based on the s^t factorial where $n = n_1 n_2$ and $s = s_1 s_2$. To do this, let $\mathbf{A} = (a_{ij})$ and $\mathbf{B} = (b_{ij})$ be the matrix representations of Γ_1 and Γ_2, respectively. Then form the matrix

$$\mathbf{C} = \begin{bmatrix} (a_{11}, b_{11}) & \dots & (a_{11}, b_{1n_2}) & \dots & (a_{1n_1}, b_{11}) & \dots & (a_{1n_1}, b_{1n_2}) \\ (a_{21}, b_{21}) & \dots & (a_{21}, b_{2n_2}) & \dots & (a_{2n_1}, b_{21}) & \dots & (a_{2n_1}, b_{2n_2}) \\ \vdots & \vdots & \vdots & \vdots & \vdots & \vdots & \vdots \\ (a_{t1}, b_{t1}) & \dots & (a_{t1}, b_{tn_2}) & \dots & (a_{tn_1}, b_{t1}) & \dots & (a_{tn_1}, b_{tn_2}) \end{bmatrix}$$

which can be easily verified to be the desired orthogonal array on $S = S_1 \times S_2$. By relabeling the elements of $\mathbf{C}$ to $0, 1, \dots, s_1 s_2 - 1$, we can utilize it for a factorial design.

In an obvious fashion this procedure can be extended to three or more orthogonal arrays by constructing it sequentially with two arrays at a time.

Sections 13.5 and 13.6 deal with factorial designs based on Latin squares, sets of orthogonal Latin squares and Hadamard matrices, respectively. As we shall see all such factorial designs are orthogonal arrays. Therefore, we can also construct orthogonal arrays via Latin squares, sets of orthogonal Latin squares and Hadamard matrices. Because of the importance of these topics we shall present them in different sections.

13.5 FACTORIAL DESIGNS BASED ON LATIN SQUARES AND SETS OF ORTHOGONAL LATIN SQUARES

Let S be a set consisting of s elements. Then a *Latin square* of order s (or size s) is a square matrix of order s with entries from S such that every element of S appears exactly once in each row and once in each column. Two Latin squares L_1 and L_2 on S are said to be *orthogonal* (designated by $L_1 \perp L_2$) if, on superposition of L_1 and L_2, the resulting s^2 pairs of entries are all distinct. By a set of *t pairwise orthogonal* Latin squares of order s on S we mean a set of t Latin squares $\{L_1, L_2, \dots, L_t\}$ such that $L_i \perp L_j$, $i \neq j$. We shall designate such a set by POL(s, t). The following are well-known facts: (i) $t=1$ if $s=2$ or 6; (ii) maximum value of $t=s-1$; (iii) a POL$(s,s-1)$ exists if s is a prime or power of a prime number; (iv) for every $s\neq 2$, 6 there is a POL(s,t) with $t\geq 2$; and (v) no POL$(s,s-1)$ is known if s is not a prime or power of a prime.

We shall now indicate the relationship between Latin squares, a set of pairwise orthogonal Latin squares and orthogonal arrays.

Theorem 13.2. (a) A Latin square of order s is equivalent to a factorial design, based on the s^3 factorial, that is an OA$(s^2,3,s,2;1)$. (b) A POL(s,t) is equivalent to a factorial design, based on the s^{t+2} factorial, that is an OA$(s^2,t+2,s,2;1)$.

Proof. (a) If L_1 is a Latin square of order s on S, then index the rows and columns of L_1 by the elements of S in any arbitrary fashion. Now form the $3\times s^2$ array A_1 as follows. Label the first row of A_1 by R, the second row by C, and the third row by L_1. Arrange the s^2 entries of L_1 in any fashion in the last row of A labeled by L_1. For each such entry coming from L_1, put its row index and column index right above it in row C and row R of A_1. It is easy to verify that A is an OA$(s^2,3,s,2;1)$. Now by the result of Section 13.2 A_1 is the desired factorial design if we consider the columns of A_1 as treatment combinations. The converse is obvious if we follow the reverse of this procedure.

Example 13.10. Let $S=\{0,1,2,3\}$. Let L_1 be the following Latin square of order 4 on S:

$$L_1=\begin{matrix}0&1&2&3\\1&0&3&2\\2&3&0&1\\3&2&1&0\end{matrix}.$$

We now index the rows and columns of L_1 by the elements of S

		Columns			
	Rows	0	1	2	3
	0	0	1	2	3
$L_1=$	1	1	0	3	2
	2	2	3	0	1
	3	3	2	1	0

Then

$A_1=$																	
	R	0	0	0	0	1	1	1	1	2	2	2	2	3	3	3	3
	C	0	1	2	3	0	1	2	3	0	1	2	3	0	1	2	3
	L_1	0	1	2	3	1	0	3	2	2	3	0	1	3	2	1	0

which is clearly an OA(16,3,4,2; 1) array.

(b) If $\{L_1, L_2, \ldots, L_t\}$ is a POL(s,t) on S, then we construct the required array as follows. Index the rows and columns of L_i by the elements of S in any arbitrary fashion, but use the same indexing for all $i=1,2,\ldots,t$. Form a $(t+2)\times s^2$ array A_t as follows: Label the first row of A_t by R, the second row by C, the third row by L_1, the fourth row by L_2, and so on. As in case (a) fill out the first three rows of A_t. While filling out the third row of A_t with the elements of L_1, recall the pattern used to pick up the entries of L_1 for inserting in A_t. Perhaps the easiest procedure is to start in row one from left to right, then move across row two from left to right, and so on. Fill out the fourth row of A_t by the entries of L_2 in the same order used for L_1; continue this procedure with the remaining $t-2$ Latin squares. Again it is easy to verify that A_t is an OA$(s^2, t+2, s, 2; 1)$ array. This process is clearly reversible. We leave the details of the proof as an exercise.

Example 13.11. Let $S=\{0,1,2,3\}$. Then $\{L_1, L_2, L_3\}$ is a POL(4,3) on S where

$$L_1=\begin{matrix}0&1&2&3\\1&0&3&2\\2&3&0&1\\3&2&1&0\end{matrix},\qquad L_2=\begin{matrix}0&1&2&3\\2&3&0&1\\3&2&1&0\\1&0&3&2\end{matrix},\qquad L_3=\begin{matrix}0&1&2&3\\3&2&1&0\\1&0&3&2\\2&3&0&1\end{matrix}.$$

Now index the rows and columns of L_1, L_2, and L_3 in the same fashion by the elements of S.

	Column					Column					Column			
Row	0	1	2	3	Row	0	1	2	3	Row	0	1	2	3
$L_1=$ 0	0	1	2	3	$L_2=$ 0	0	1	2	3	$L_3=$ 0	0	1	2	3
1	1	0	3	2	1	2	3	0	1	1	3	2	1	0.
2	2	3	0	1	2	3	2	1	0	2	1	0	3	2
3	3	2	1	0	3	1	0	3	2	3	2	3	0	1

Then

$A_3=$

R	0	0	0	0	1	1	1	1	2	2	2	2	3	3	3	3
C	0	1	2	3	0	1	2	3	0	1	2	3	0	1	2	3
L_1	0	1	2	3	1	0	3	2	2	3	0	1	3	2	1	0
L_2	0	1	2	3	2	3	0	1	3	2	1	0	1	0	3	2
L_3	0	1	2	3	3	2	1	0	1	0	3	2	2	3	0	1

which is clearly an OA(16,5,4,2; 1) array.

Note that A_1 is a subarray of A_3. However, we would like to point out that not every OA(16,3,4,2; 1) array can be extended to an OA(16, k,4,2; 1) array with $k>3$. Here is an example.

Example 13.12. The following Latin square of order 4 based on $S=\{0,1,2,3\}$ produces an OA(16,3,4,2; 1), which cannot be extended to an OA(16, k,4,2; 1) with $k>3$.

$$L=\begin{matrix} 0 & 1 & 2 & 3 \\ 3 & 0 & 1 & 2 \\ 2 & 3 & 0 & 1 \\ 1 & 2 & 3 & 0 \end{matrix}$$

and the corresponding OA(16,3,4,2; 1) array is

$A=$

R	0	0	0	0	1	1	1	1	2	2	2	2	3	3	3	3
C	0	1	2	3	0	1	2	3	0	1	2	3	0	1	2	3
L	0	1	2	3	3	0	1	2	2	3	0	1	1	2	3	0

In a similar fashion, one can obtain certain factorial designs from a generalization of Latin squares and orthogonal Latin squares, that is, F-squares and orthogonal F-squares.

13.6 FACTORIAL DESIGNS BASED ON HADAMARD MATRICES

Recall from Chapter 10 that a matrix of order n, $\mathbf{H}_n$, is said to be a *Hadamard matrix* if its entries are $+1$ or -1 provided that its rows are pairwise orthogonal; in other words,

$$\mathbf{H}_n\mathbf{H}_n' = n\mathbf{I}_n. \tag{13.6}$$

Equation (13.6) implies that $\mathbf{H}_n$ is nonsingular and has the inverse $n^{-1}\mathbf{H}_n'$;

consequently,

$$\mathbf{H}_n'\mathbf{H}_n = n\mathbf{I}_n.$$

This tells us that the columns of a Hadamard matrix are also pairwise orthogonal. As an example, the following matrix is a Hadamard matrix of order 4.

$$\mathbf{H}_4 = \begin{bmatrix} +1 & +1 & +1 & +1 \\ +1 & +1 & -1 & -1 \\ +1 & -1 & +1 & -1 \\ +1 & -1 & -1 & +1 \end{bmatrix}.$$

There exist Hadamard matrices of orders 1 and 2, but it can be shown that every other Hadamard matrix has order $4t$ for some positive integer t. Hadamard matrices of infinitely many orders have been constructed, and it has been conjectured that one exists for every t, but no general proof is available, and the number of unsettled orders is infinite. As far as we know, the smallest order that is undecided is 268. (A historical review and various applications of Hadamard matrices may be found in reference [51].)

Next we shall indicate how we can utilize such matrices to produce a family of optimal factorial designs by transforming these matrices to factorial designs which are orthogonal arrays. Before we proceed we would like to observe that a Hadamard matrix remains a Hadamard matrix if any row or column is multiplied by -1. This observation allows us to convert any Hadamard matrix to a Hadamard matrix whose first row consists of all ones. Such a Hadamard matrix is called a *seminormalized Hadamard matrix*.

Theorem 13.3. The existence of a Hadamard matrix of order $4t$ is equivalent to the existence of:

(a) A factorial design, based on the 2^{4t-1} factorial, that is an OA$(4t, 4t-1, 2, 2; t)$ array.

(b) A factorial design, based on the 2^{4t} factorial, that is an OA$(8t, 4t, 2, 3; t)$ array.

Proof. (i) Suppose $\mathbf{H}_{4t}$ is a seminormalized Hadamard matrix. Delete the first row of $\mathbf{H}_{4t}$. The remaining matrix is the desired factorial design, after relabeling -1 and 1 to 0 and 1 to indicate the two levels of each factor. Conversely, given a factorial design that is an OA$(4t, 4t-1, 2, 2; t)$ array, relabel its elements as $+1$ and -1. Then this procedure can be reversed. (ii) The following array provides a factorial design that is an OA$(8t, 4t, 2, 3; t)$ array:

$$\mathbf{A} = \left[-\mathbf{H}_{4t} \vdots \mathbf{H}_{4t}\right].$$

The details of the proof in both cases are left as an exercise.

Example 13.14. For simplicity we use $+$ and $-$ to indicate $+1$ and -1. Based on

$$
\mathbf{H}_{12}=\begin{bmatrix}
+ & + & + & + & + & + & + & + & + & + & + & + \\
+ & + & - & + & + & + & - & - & - & + & - & - \\
+ & - & + & - & + & + & + & - & - & - & + & - \\
+ & - & - & + & - & + & + & + & - & - & - & + \\
+ & + & - & - & + & - & + & + & + & - & - & - \\
+ & - & + & - & - & + & - & + & + & + & - & - \\
+ & - & - & + & - & - & + & - & + & + & + & - \\
+ & - & - & - & + & - & - & + & - & + & + & + \\
+ & + & - & - & - & + & - & - & + & - & + & + \\
+ & + & + & - & - & - & + & - & - & + & - & + \\
+ & + & + & + & - & - & - & + & - & - & + & - \\
+ & - & + & + & + & - & - & - & + & - & - & +
\end{bmatrix}
$$

we can obtain the following factorial designs after relabeling $+1$ with 1 and -1 with 0:

Treatment Combinations

$$
\Gamma_1=\begin{array}{cccccccccccc}
1 & 1 & 0 & 1 & 1 & 1 & 0 & 0 & 0 & 1 & 0 & 0 \\
1 & 0 & 1 & 0 & 1 & 1 & 1 & 0 & 0 & 0 & 1 & 0 \\
1 & 0 & 0 & 1 & 0 & 1 & 1 & 1 & 0 & 0 & 0 & 1 \\
1 & 1 & 0 & 0 & 1 & 0 & 1 & 1 & 1 & 0 & 0 & 0 \\
1 & 0 & 1 & 0 & 0 & 1 & 0 & 1 & 1 & 1 & 0 & 0 \\
1 & 0 & 0 & 1 & 0 & 0 & 1 & 0 & 1 & 1 & 1 & 0 \\
1 & 0 & 0 & 0 & 1 & 0 & 0 & 1 & 0 & 1 & 1 & 1 \\
1 & 1 & 0 & 0 & 0 & 1 & 0 & 0 & 1 & 0 & 1 & 1 \\
1 & 1 & 1 & 0 & 0 & 0 & 1 & 0 & 0 & 1 & 0 & 1 \\
1 & 1 & 1 & 1 & 0 & 0 & 0 & 1 & 0 & 0 & 1 & 0 \\
1 & 0 & 1 & 1 & 1 & 0 & 0 & 0 & 1 & 0 & 0 & 1
\end{array}
$$

Treatment Combinations

$$
\Gamma_2=\begin{array}{cccccccccccccccccccccccc}
0&0&0&0&0&0&0&0&0&0&0&0&1&1&1&1&1&1&1&1&1&1&1&1 \\
0&0&1&0&0&0&1&1&1&0&1&1&1&1&0&1&1&1&0&0&0&1&0&0 \\
0&1&0&1&0&0&0&1&1&1&0&1&1&0&1&0&1&1&1&0&0&0&1&0 \\
0&1&1&0&1&0&0&0&1&1&1&0&1&0&0&1&0&1&1&1&0&0&0&1 \\
0&0&1&1&0&1&0&0&0&1&1&1&1&1&0&0&1&0&1&1&1&0&0&0 \\
0&1&0&1&1&0&1&0&0&0&1&1&1&0&1&0&0&1&0&1&1&1&0&0 \\
0&1&1&0&1&1&0&1&0&0&0&1&1&0&0&1&0&0&1&0&1&1&1&0 \\
0&1&1&1&0&1&1&0&1&0&0&0&1&0&0&0&1&0&0&1&0&1&1&1 \\
0&0&1&1&1&0&1&1&0&1&0&0&1&1&0&0&0&1&0&0&1&0&1&1 \\
0&0&0&1&1&1&0&1&1&0&1&0&1&1&1&0&0&0&1&0&0&1&0&1 \\
0&0&0&0&1&1&1&0&1&1&0&1&1&1&1&1&0&0&0&1&0&0&1&0 \\
0&1&0&0&0&1&1&1&0&1&1&0&1&0&1&1&1&0&0&0&1&0&0&1
\end{array}
$$

It can be verified that Γ_1 is a factorial design for the 2^{11} factorial, that is an OA(12,11,2,2;3) array and Γ_2 is a factorial design for 2^{12} factorial that is an OA(24,12,2,3;3) array. Thus by Theorem 13.1 Γ_1 and Γ_2 have many useful statistical properties.

13.7 FACTORIAL DESIGNS BASED ON OTHER ARRAYS

Though the technique of constructing factorial designs based on orthogonal arrays produces designs with many desirable statistical properties, it has, unfortunately, its limitations and drawbacks. (i) It usually produces designs with many treatment combinations so that it may become uneconomical to utilize the produced designs for experimentations. (ii) It is applicable to situations in which all factors are at the same number of levels. Specialists in design theory have been able to devise techniques to deal with such problems. If we are willing to sacrifice some efficiency for the purpose of reducing the number of observations, then in many situations we can use factorial designs that are *balanced arrays* rather than orthogonal arrays.

Definition 13.2. A factorial design consisting of n treatment combinations from the s^t factorial is said to be based on a *balanced array* of strength d if, with respect to any d factors out of t factors, the treatment combination $(i_1, i_2, \ldots, i_d)$ and any other treatment combination that is a permutation of $i_1, i_2, \ldots, i_d$ appear the same number of times in the design. This number is called the *index* of the array with respect to $i_1, i_2, \ldots, i_d$ and is designated by $\lambda(i_1, i_2, \ldots, i_d)$. The set of all such numbers, as $i_1, i_2, \ldots, i_d$ runs over all s^t treatment combinations, is called the set of indices of the array, and is denoted by Λ.

It is customary to present a factorial design that is a balanced array in a $t \times n$ matrix and designate it by BA$(n, t, s, d; \Lambda)$. We observe that a BA$(n, t, s, d; \Lambda)$ array becomes an OA$(n, t, s, d; \lambda)$ array when the set of indices Λ contains a single number λ. For many choices of s, t, and d it is possible to construct balanced arrays that require fewer observations than their counterpart orthogonal arrays. For this and other aspects of balanced arrays (some also refer to them as *partially balanced arrays*), we refer the reader to the bibliography at the end of this chapter.

Unfortunately it is not necessarily true that a factorial design that is a balanced array of strength d is a factorial design of resolution $d+1$ as is the case for orthogonal arrays. What sets of parameters or functions of parameters are estimable depend largely on the set of indices. Thus we advise the reader to be aware of this fact before utilizing such factorial designs for practical purposes. Despite this drawback, the literature on balanced arrays contains many interesting and practically useful results.

Now we give several examples to illustrate the idea of a balanced array.

Example 13.15. The following array is a BA(24,9,3,2;{2,3}) array. For this array $\lambda(i_1, i_2) = 3$ if $i_1 \neq i_2$ and $= 2$ if $i_1 = i_2$.

$$\Gamma = \begin{matrix} 0&1&2&0&1&2&0&1&2&0&1&2&0&1&2&0&1&2&0&1&2&0&1&2 \\ 1&2&0&0&1&2&1&2&0&2&0&1&2&0&1&0&1&2&2&0&1&1&2&0 \\ 0&1&2&1&2&0&2&0&1&2&0&1&0&1&2&2&0&1&1&2&0&1&2&0 \\ 1&2&0&2&0&1&2&0&1&0&1&2&2&0&1&1&2&0&1&2&0&0&1&2 \\ 2&0&1&2&0&1&0&1&2&2&0&1&1&2&0&1&2&0&0&1&2&1&2&0 \\ 2&0&1&0&1&2&2&0&1&1&2&0&1&2&0&0&1&2&1&2&0&2&0&1 \\ 0&1&2&2&0&1&1&2&0&1&2&0&0&1&2&1&2&0&2&0&1&2&0&1 \\ 2&0&1&1&2&0&1&2&0&0&1&2&1&2&0&2&0&1&2&0&1&0&1&2 \\ 1&2&0&1&2&0&0&1&2&1&2&0&2&0&1&2&0&1&0&1&2&2&0&1 \end{matrix} .$$

If we consider the columns of this array as treatment combinations from the 3^9 factorial, then we have a factorial design with $n = 24$ treatment combinations from the 3^9 factorial.

Example 13.16. The following is an example of a BA(20,6,3,2;{1,3,4}) array.

$$\Gamma = \begin{matrix} 1&1&1&1&1&2&2&2&2&2&0&0&0&0&0&0&0&0&0&0 \\ 1&2&0&0&0&1&2&0&0&0&1&1&1&2&2&2&0&0&0&0 \\ 0&1&2&0&0&0&0&1&2&0&1&2&0&2&0&0&1&1&2&0 \\ 0&0&0&1&2&2&1&0&0&0&0&0&1&2&0&0&1&2&1&2 \\ 1&0&2&0&0&0&0&2&0&1&0&0&2&0&1&2&0&1&1&2 \\ 0&0&0&1&2&0&0&0&1&2&1&2&0&0&1&2&2&0&0&1 \end{matrix} .$$

As we can easily check, $\lambda(1,1) = \lambda(1,2) = \lambda(2,2) = 1$, $\lambda(0,1) = \lambda(0,2) = 3$, and $\lambda(0,0) = 4$.

It is customary to denote $\lambda(i_1, i_2, \ldots, i_d)$ in a balanced array of strength d based on two symbols coded by 0 and 1 by λ_w where w is equal to the number of ones in $(i_1, i_2, \ldots, i_d)$. w is called the *weight* of $(i_1, i_2, \ldots, i_d)$ (a term borrowed from the subject of binary codes). In this case the set of indices is completely specified by $\lambda_0, \lambda_1, \ldots, \lambda_d$. Now we give some examples of balanced arrays based on 0 and 1.

Example 13.17. The following is a BA$(20,4,2,3;\{1,3\})$ array.

$$\Gamma = \begin{matrix} 0 & 0 & 0 & 0 & 0 & 0 & 1 & 1 & 1 & 1 & 1 & 1 & 1 & 1 & 1 & 1 & 0 & 0 & 0 & 0 \\ 0 & 0 & 0 & 1 & 1 & 1 & 0 & 0 & 0 & 1 & 1 & 1 & 1 & 0 & 0 & 0 & 1 & 1 & 1 & 0 \\ 0 & 1 & 1 & 0 & 0 & 1 & 0 & 0 & 1 & 0 & 1 & 0 & 0 & 1 & 1 & 0 & 1 & 1 & 0 & 1 \\ 1 & 1 & 0 & 1 & 0 & 0 & 1 & 0 & 0 & 0 & 0 & 0 & 1 & 0 & 1 & 1 & 0 & 1 & 1 & 1 \end{matrix}.$$

For this array of strength 3, we have $\lambda_0 = \lambda_3 = 1$ and $\lambda_1 = \lambda_2 = 3$.

Factorial designs based on BA$(n,t,2,2r;\Lambda)$ arrays with $\lambda_0 = \lambda_1 = \cdots = \lambda_{2r-1} = \lambda$ and $\lambda_{2r} = \lambda + 1$ are balanced factorial designs of resolution $2r+1$, and they can be shown to be E-optimal over all factorial designs with n treatment combinations from 2^t factorial. Indeed, such factorial designs have further optimality properties. Such a balanced array can be easily obtained by adding one column consisting entirely of ones to an OA$(n-1,t,2,2r;\lambda)$ arrays. Thus all orthogonal arrays constructed by Hadamard matrices can be utilized for this purpose. The literature of balanced arrays contains many families of BA$(n,t,2,2r;\Lambda)$ arrays with $\lambda_1 = \lambda_2 = \cdots = \lambda_{2r} = \lambda$ and $\lambda_0 = \lambda + 1$. It is clear that if we change 0 to 1 and 1 to 0 in any such array we obtain a BA$(n,t,2,2r;\Lambda)$ array with $\lambda_0 = \lambda_1 = \cdots = \lambda_{2r-1} = \lambda$ and $\lambda_{2r} = \lambda + 1$.

For asymmetrical factorial designs some attempts have been made to produce arrays with variable numbers of symbols useful in such a context. Many more contributions are needed in this area.

With the help of the published results, the advanced reader is encouraged to solve the following interesting problems and to discover new results in the area of factorial designs based on orthogonal arrays and balanced arrays.

(i) Consider a factorial design based on an OA$(n,t,2,2r;\lambda)$ array from a 2^t factorial. Prove that by adding (or deleting) *any* treatment combination to (or from) such a factorial design the resulting design with $n+1$ (or $n-1$) treatment combinations is E-optimal over all resolution $2r+1$ factorial designs with $n+1$ (or $n-1$) treatment combinations from the 2^t factorial.

(ii) Similarly prove that a factorial design obtained by adding *any* two (or three) treatment combinations to a factorial design based on OA$(n,t,2,2t;\lambda)$ array is E-optimal over all resolution $2r+1$ factorial designs with $n+2$ (or $n+3$) treatment combinations.

(iii) Prove that factorial designs based on BA$(n,t,2,2r;\Lambda)$ arrays with indices

$$\lambda_0 = \lambda_1 = \cdots = \lambda_{2r-1} = \lambda, \qquad \lambda_{2r} = \lambda + 2$$

or

$$\lambda_1 = \lambda_2 = \cdots = \lambda_{2r-1} = \lambda, \qquad \lambda_0 = \lambda_{2r} = \lambda + 1$$

are E-optimal over factorial designs with n treatment combinations from the 2^t factorial.

13.8 FACTORIAL DESIGNS BASED ON COMPOSITION METHODS

In this category, for example, falls the *direct product method* of constructing factorial designs. Explicitly, the direct product method is as follows: Let Γ_1 and Γ_2 be two factorial designs based on $k_1^{t_1}$ and $k_2^{t_2}$ factorials, respectively. The design $\Gamma = \Gamma_1 \otimes \Gamma_2$ and consists of treatments $g = (g_1, g_2)$, where $g_1 \in \Gamma_1$ and $g_2 \in \Gamma_2$, and is called the direct product design. In many cases the useful statistical properties of Γ_1 and Γ_2 will be carried into Γ.

Example 13.18. To illustrate this method, the orthogonal main effect plan for the 2^7 factorial is as given previously, so that together with the orthogonal main effect plan for the 3^4 factorial:

	Factors			
	F_8	F_9	F_{10}	F_{11}
	0	0	0	0
	0	1	1	2
	0	2	2	1
Treatment	1	0	1	1
Combinations	1	1	2	0
	1	2	0	2
	2	0	2	2
	2	1	0	1
	2	2	1	0

we obtain the plan $\Gamma_1 \otimes \Gamma_2$, which is equal to

	F_1	F_2	F_3	F_4	F_5	F_6	F_7	F_8	F_9	F_{10}	F_{11}
	0	0	0	1	1	1	0	0	0	0	0
	0	0	0	1	1	1	0	0	1	1	2
$\Gamma_1 \otimes \Gamma_2 =$	⋯	⋯	⋯	⋯	⋯	⋯	⋯	⋯	⋯	⋯	⋯
	0	0	0	1	1	1	0	2	2	1	0
	⋯	⋯	⋯	⋯	⋯	⋯	⋯	⋯	⋯	⋯	⋯
	1	1	1	1	1	1	1	2	2	1	0

This design with $8 \times 9 = 72$ runs allows orthogonal estimation of main effects and the two-factor interactions of one of the factors $F_1, F_2, \ldots, F_7$ with one of the factors F_8, F_9, F_{10}, F_{11}.

This method can be generalized in several directions; that is, we may drop the orthogonality condition and/or increase the factorial to more than two different levels for the factors.

We have reached the end of this book and since the objectives were limited we could not cover *all* concepts arising in factorial design theory. Indeed, more concepts and a whole body of theorems and construction methods can be found in the published literature.

Our purpose has been to shed enough light on the subject of factorial design to enable the reader to pursue the finer and more specific details of the subject. In addition, a cohesive account of various facets of factorial design has been presented. This will provide the reader with the orientation that will be beneficial in further pursuit of knowledge about factorial design and in reading the starred references in the reading lists at the end of each chapter.

13.9 SELECTED ADDITIONAL READING

Starred references are recommended for a first reading.

1. Addelman, S. (1963). Techniques for constructing fractional replicate plans. *J. Am. Stat. Assoc.*, **58**, 45–71.

*2. Addelman, S. (1972). Recent developments in the design of factorial experiments. *J. Am. Stat. Assoc.*, **67**, 103–111.

3. Addelman, S., and Kempthorne, O. (1961). Some main-effect plans and orthogonal arrays of strength two. *Ann. Math. Stat.*, **32**, 1167–1176.

4. Anderson, D. A., and Federer, W. T. (1975). Possible absolute determinant values for square (0,1)-matrices useful in fractional replication. *Util. Math.*, **7**, 135–150.

5. Anderson, D. A., and Federer, W. T. (1976). Multidimensional balanced designs. *Commun. Stat.*, **A5**, 1193–1204.

6. Anderson, D. A., and Thomas, A. M. (1979). Resolution IV fractional factorial designs for the general asymmetric factorial. *Commun. Stat.*, **A8**, 931–943.

7. Bailey, N. T. J. (1959). The use of linear algebra in deriving prime power factorial designs with confounding and fractional replication. *Sankhyā*, **21**, 345–354.

8. Banerjee, K. S. (1975). *Weighing Designs*. Marcel Dekker, New York.

9. Berlekamp, E. R. (1968). *Algebraic Coding Theory*. McGraw-Hill, New York.

10. Blum, J. R., Schatz, J. A., and Seiden, E. (1970). On 2-level orthogonal arrays of odd index. *J. Comb. Theory*, **9**, 239–243.

11. Bose, R. C. (1963). Some ternary error correcting codes and fractionally replicated designs (with discussion). *Colloq. Int. CNRS.*, **110**, *Le Plan d'Expériences*, Paris, pp. 21–32.

12. Bose, R. C., and Bush, K. A. (1952). Orthogonal arrays of strength two and three. *Ann. Math. Stat.*, **23**, 508–524.

13. Bose, R. C., and Srivastava, J. N. (1964a). On a bound useful in the theory of factorial designs and error correcting codes. *Ann. Math. Stat.*, **35**, 408–414.

14. Bose, R. C., and Srivastava, J. N. (1964b). Mathematical theory of factorial designs. I. Analysis. II. Construction. *Bull. Int. Inst. Stat.*, **40**, 780–794.

15. Bose, R. C., and Srivastava, J. N. (1964c). Multidimensional partially balanced designs and their analysis, with applications to partially balanced factorial fractions. *Sankhyā*, **A26**, 145–167.

16. Box, G. E. P., and Hunter, J. S. (1957). Multi-factor experimental designs for exploring response surfaces. *Ann. Math. Stat.*, **28**, 195–241.

*17. Bush, K. A. (1952). Orthogonal arrays of index unity. *Ann. Math. Stat.* **23**, 426–434.

18. Carmichael, R. D. (1956). *Introduction to the Theory of Groups of Finite Order.* (First published in 1937). Dover, New York.

*19. Chakravarti, I. M. (1956). Fractional replication in asymmetrical factorial designs and partially balanced arrays. *Sankhyā*, **17**, 143–164.

20. Chakravarti, I. M. (1961). On some methods of construction of partially balanced arrays. *Ann. Math. Stat.*, **32**, 1181–1185.

21. Chakravarti, I. M. (1963). Orthogonal and partially balanced arrays and their applications in design of experiments. *Metrika*, **7**, 231–243.

22. Cheng, C.−S. (1980a). Optimality of some weighing and 2^n fractional factorial designs. *Ann. Stat.* **8**, 436–446.

*23. Cheng, C.−S. (1980b). Orthogonal arrays with variable number of symbols. *Ann. Stat.* **8**, 447–453.

24. Chopra, D. V. and Srivastava, J. N. (1973). Optimal balanced 2^7 fractional factorial designs of resolution V, $49 \leq N \leq 55$. *Commun. Stat.*, **2**, 59–84.

25. Chopra, D. V., and Srivastava, J. N. (1974). Optimal balanced 2^8 fractional factorial designs of resolution V, $37 \leq N \leq 51$. *Sankhyā*, **A36**, 41–52.

26. Crapo, H., and Rota, G. C. (1970). *Combinatorial Geometries.* MIT Press, Cambridge, Mass.

27. Daniel, C. (1973). One-at-a-time plans. *J. Am. Stat. Assoc.*, **68**, 353–360.

28. Das, M. N. (1961). Construction of rotatable designs from factorial designs. *J. Indian Soc. Agric. Stat.*, **13**, 169–194.

29. Das, M. N. (1964). A somewhat alternative approach for construction of symmetrical factorial designs and obtaining maximum number of factors. *Calcutta Stat. Assoc. Bull.*, **13**, 1–17.

30. Das, M. N., Sukla, G. K., and Kartha, C. P. (1967). On a method of construction of symmetrical fractional factorials and related error correcting codes. *Calcutta Stat. Assoc. Bull.*, **16**, 164–179.

31. Dembowski, P. (1968). *Finite Geometries.* Springer-Verlag, Berlin.

32. Dēnes, J., and Keedwell, A. D. (1974). *Latin Squares and Their Applications.* English Universities Press, London.

33. Draper, N. R., and Mitchell, T. J. (1967). The construction of saturated 2_R^{k-p} designs. *Ann. Math. Stat.*, **38**, 1110–1126.

34. Dykstra, O. (1971). The augmentation of experimental data to maximize $|X'X|$. *Technometrics*, **13**, 682–688.

35. Ehlich, H. (1964). Determinantenbeschatzungen für binäre matrizen. *Math. Z.*, **83**, 123–132.

36. Federer, W. T. (1955). *Experimental Design: Theory and Application*. Macmillan, New York.

37. Federer, W. T., Hedayat, A., Parker, E. T., Raktoe, B. L., Seiden, E., and Turyn, R. J. (1970). Some techniques for constructing mutually orthogonal Latin squares. *Proceedings of the 15th Conference on Design and Analysis of Experiments in Army Research Development and Testing No. 70-2*, U. S. Army Reseach Office, Research Triangle Park, N. C., pp. 673–796.

38. Finney, D. J. (1945). The fractional replication of factorial experiments. *Ann. Eugen.*, **12**, 291–301.

39. Fisher, R. A. (1942). The theory of confounding in factorial experiments in relation to the theory of groups. *Ann. Eugen.*, **11**, 341–353.

40. Fisher, R. A. (1945). A system of confounding for factors with more than two alternatives, giving completely orthogonal cubes and higher powers. *Ann. Eugen.*, **12**, 283–290.

41. Foata, D. C. (1962). Sur la construction des plans factoriels fractionnés et certains codes correcteurs à l'aide des caractères des groups abéliens. *Publ. Inst. Stat. Univ. Paris*, **11**, 57–66. Also in *Colloq. Int. CNRS,* **110** (1963), *Le Plan d'Expériences*, Paris, pp. 137–146.

42. Fry, R. E. (1961). Finding new fractions of factorial experimental designs. *Technometrics*, **3**, 359–370.

43. Fujii, Y. (1967). Geometrical association schemes and fractional factorial designs. *J. Sci. Hiroshima Univ. Ser. A-1*, **31**, 195–209.

44. Galil, Z., and Kiefer, J. (1980). D-optimum weighing designs. *Ann. Stat.*, **8**, 1293–1306.

45. Geramita, A. V., and Seberry, J. (1979). *Orthogonal Designs*. Marcel Dekker, New York.

46. Gulati, B. R. (1972). Construction of two-level saturated symmetrical factorial designs of resolution VI. *Ann. Math. Stat.*, **43**, 1652–1663.

47. Hall, M., Jr. (1967). *Combinatorial Theory*. Blaisdell, Waltham, Mass.

48. Hedayat, A., and Federer, W. T. (1970). On the equivalence of Mann's group automorphism method of constructing an $0(n, n-1)$ set and Raktoe's collineation method of constructing a balanced set of l-restrictional prime-powered lattice designs. *Ann. Math. Stat.*, **41**, 1530–1540.

49. Hedayat, A., Raghavarao, D., and Seiden, E. (1975). Further contributions to the theory of F-squares design. *Ann. Stat.*, **3**, 712–716.

50. Hedayat, A., and Shrikhande, S. S. (1977). Experimental designs and combinatorial systems associated with Latin squares and sets of mutually orthogonal Latin squares. *Sankhyā*, **A33**, 423–432.

51. Hedayat, A., and Wallis, W. D. (1978). Hadamard matrices and their applications. *Ann. Stat.*, **6**, 1184–1238.

52. John, P. W. M. (1971). *Statistical Design and Analysis of Experiments*. Macmillan, New York.

* 53. Joiner, J., Raktoe, B. L., and Federer, W. T. (1980). On sum composition of fractional factorial designs. *Commun. Stat.*, **A9**, 1875–1899.

54. Kempthorne, O. (1952). *The Design and Analysis of Experiments*. Wiley, New York.

55. Kiefer, J. and Galil, Z. (1980). Optimum weighing designs. In *Recent Developments in Statistical Inference and Data Analysis*, K. Matusita, Ed. North Holland, Amsterdam, pp. 183–189.

56. Kishen, K., and Srivastava, J. N. (1959). Mathematical theory of confounding in asymmetrical and symmetrical factorial designs. *J. Indian Soc. Agric. Stat.*, **11**, 73–110.

57. Kishen, K., and Tyagi, B. N. (1964). On the construction and analysis of some balanced asymmetrical factorial designs. *Calcutta Stat. Assoc. Bull.*, **13**, 123–149.

58. Kishen, K., and Tyagi, B. N. (1973). Recent developments in India in the construction of confounded asymmetrical factorial designs. In *A Survey of Combinatorial Theory*, J. N. Srivastava, F. Harary, C. R. Rao, G.C. Rota, and S. S. Shrikhande, Eds. North-Holland, Amsterdam, pp. 313–321.

59. Kitagawa, T., and Mitome, M. (1958). *Tables for the Design of Factorial Experiments.* Dover, New York.

60. Kurkjian, B., and Zelen, M. (1962). A calculus for factorial arrangements. *Ann. Math. Stat.*, **33**, 600–619.

61. Li, J. C. R. (1944). Design and statistical analysis of some confounded factorial experiments. *Iowa Agric. Exp. Sta. Res. Bull.*, No. 333, 452–492.

62. Margolin, B. H. (1969). Results on factorial designs of resolution IV for the 2^n and $2^n 3^m$ series. *Technometrics*, **11**, 431–444.

63. Mazumdar, S. (1967). On the construction of cyclic collineations for obtaining a balanced set of l-restrictional prime-powered lattice designs. *Ann. Math. Stat.*, **38**, 1293–1295.

64. Mitchell, T. J. (1972). An algorithm for the construction of "*D*-optimal" experimental designs applied to first-order models. Oak Ridge Natl. Laboratory, ORNL-4777.

65. Mitton, R. G., and Morgan, F. R. (1959). The design of factorial experiments: a survey of some schemes requiring not more than 256 treatment combinations. *Biometrika*, **46**, 251–259.

66. Narula, S. S. (1960). Augmented symmetrical and asymmetrical factorial designs. *J. Indian Soc. Agric. Stat.*, **12**, 57–87.

67. Okuno, T., and Shiyomi, M. (1965). On the construction of fractional factorial designs by using tables of orthogonal arrays. *Bull. Natl. Inst. Agric. Sci.*, **A12**, 23–75.

68. Patel, M. S. (1963). Partially duplicated fractional factorial designs. *Technometrics*, **5**, 71–83.

69. Pesotan, H., and Raghavarao, D. (1975). Embedded Hadamard matrices. *Util. Math.*, **8**, 99–110.

70. Pesotan, H., and Raktoe, B. L. (1975). Remarks on extensions of the Anderson-Federer (0,1)-matrix procedure for main effects to effects of a higher degree. *Commun. Stat.*, **4**, 797–811.

*71. Pesotan, H., Raktoe, B. L., and Federer, W. T. (1975). On complexes of Abelian groups with applications to fractional factorial designs. *Ann. Inst. Stat. Math.*, **27**, 117–142.

72. Pesotan, H., Raktoe, B. L., and Worthley, R. G. (1978). On main effect fold-over designs. *J. Stat. Plan. Inference*, 2, 277–291.

73. Peterson, W. W. (1961). *Error-Correcting Codes*. MIT Press, Cambridge, Mass.

74. Plackett, R. L., and Burman, J. P. (1946). The design of optimum multifactorial experiments. *Biometrika*, **33**, 305–325.

75. Preece, D. A. (1969). Near-cyclic representations for some resolution VI fractional factorial plans. *Ann. Math. Stat.*, **40**, 1840–1843.

*76. Rafter, J. A. and Seiden, E. (1974). Contributions to the theory and construction of balanced arrays. *Ann. Stat.*, **2**, 1256–1273.

77. Raghavarao, D. (1971). *Constructions and Combinatorial Problems in Design of Experiments*. Wiley, New York.

78. Raghavarao, D., and Pesotan, H. (1977). Embedded $(S_n + I_n)$-matrices. *Util. Math.*, **11**, 227–236.

79. Raktoe, B. L. (1967). Application of cyclic collineations to the construction of balanced l-restrictional prime-powered lattice designs. *Ann. Math. Stat.*, **38**, 1127–1141.

80. Raktoe, B. L. (1974). A geometrical formulation of an unsolved problem in fractional factorial designs. *Commun. Stat.*, **3**, 959–968.

81. Raktoe, B. L., and Federer, W. T. (1968). A unified approach for constructing a useful class of non-orthogonal main effect plans in k^n factorials. *J. R. Stat. Soc.*, **B30**, 371–380.

82. Raktoe, B. L., and Federer, W. T. (1970). A theorem on saturated plans and their complements. *Ann. Math. Stat.*, **41**, 2184–2185.

83. Raktoe, B. L., and Federer, W. T. (1973). Balanced optimal saturated main effect plans of the 2^n factorial and their relation to (v, k, λ) configurations. *Ann. Stat.*, **1**, 924–932.

84. Raktoe, B. L., Rayner, A. A., and Chalton, D. O. (1978). On construction of confounded mixed factorial and lattice designs. *Aust. J. Stat.*, **20**, 209–218.

85. Rao, C. R. (1946). Hypercubes of strength d leading to confounded designs in factorial experiments. *Calcutta Math. Soc. Bull.*, **38**, 67–78.

*86. Rao, C. R. (1947). Factorial experiments derivable from combinatorial arrangements of arrays. *J. R. Stat. Soc.*, **B9**, 128–139.

87. Rao, C. R. (1950). The theory of fractional replication in factorial experiments. *Sankhyā*, **10**, 81–86.

88. Rao, C. R. (1973). Some combinatorial problems of arrays and applications to design of experiments. In *A Survey of Combinatorial Theory*, J. N. Srivastava, F. Harary, C. R. Rao, G.-C. Rota, and S. S. Shrikhande., Eds. North-Holland, Amsterdam, pp. 349–359.

89. Ryser, H. J. (1963). *Combinatorial Mathematics (Carus Monograph No. 14)*. Mathematical Association of America, Oberlin, Ohio and Wiley, New York.

90. Saha, G. M., and Das, M. N. (1971). Construction of partially balanced incomplete block designs through 2^n factorials and some new designs of two associate classes. *J. Comb. Theory*, **11**, 282–295.

91. Saha, G. M., and Mohanty, S. (1970). On non-orthogonal main effect plans for asymmetrical factorials. *Ann. Inst. Stat. Math.*, **22**, 159–169.

92. Seiden, E. (1954). On the problem of construction of orthogonal arrays. *Ann. Math. Stat.* **25**, 151–156.

93. Seiden, E. (1955a). On the maximum number of constraints of an orthogonal array. *Ann. Math. Stat.* **26**, 132–135.

94. Seiden, E. (1955b). Further remarks on the maximum number of constraints of an orthogonal array. *Ann. Math. Stat.*, **26**, 759–763.

95. Seiden, E. (1973). On the problem of construction and uniqueness of saturated 2_R^{k-p} designs. In *A Survey of Combinatorial Theory*, J. N. Srivastava, F. Harary, C. R. Rao, G. C. Rota, and S. S. Shrikhande, Eds. North-Holland, Amsterdam, pp. 397–401.

*96. Seiden, E. and Zemach, R. (1966). On orthogonal arrays. *Ann. Math. Stat.* **37**, 1355–1370.

97. Shirakura, T., and Kuwada, M. (1975). Note on balanced 2^m factorial designs of resolution $2l+1$. *Ann. Inst. Stat. Math.*, **27**, 377–386.

98. Shrikhande, S. S. (1964). Generalized Hadamard matrices and orthogonal arrays of strength two. *Can. J. Math.*, **16**, 736–740.

99. Srivastava, J. N. (1968). Analysis and construction of fractional and confounded factorial designs with emphasis on the asymmetrical case. In *Contributions in Statistics and Agricultural Science and Indian Society of Agricultural Statistics.* New Delhi. Felicitation volume presented to Dr. V. G. Panse.

100. Srivastava, J. N. (1972). Some general existence conditions for balanced arrays of strength t and 2 symbols. *J. Comb. Theory*, **12**, 198–206.

101. Srivastava, J. N., and Anderson, D. A. (1970). Optimal fractional factorial plans for main effects orthogonal to two-factor interactions: 2^m series. *J. Am. Stat. Assoc.*, **65**, 828–843.

102. Srivastava, J. N., and Anderson, D. A. (1971). Factorial association schemes with application to the construction of multidimensional partially balanced designs. *Ann. Math. Stat.*, **42**, 1167–1181.

103. Srivastava, J. N., and Bose, R. C. (1966). Some economic partially balanced 2^m factorial fractions. *Ann. Inst. Stat. Math.*, **18**, 57–73.

104. Srivastava, J. N., and Chopra, D. V. (1971a). On the comparison of certain classes of balanced 2^8 fractional factorial designs of resolution V, with respect to the trace criterion. *J. Indian Soc. Agric. Stat.*, **23** (2), 124–131.

105. Srivastava, J. N., and Chopra, D. V. (1971b). On the characteristic roots of the information matrix of 2^m balanced factorial designs of resolution V, with applications. *Ann. Math. Stat.*, **42**, 722–734.

106. Srivastava, J. N., and Chopra, D. V. (1971c). Balanced optimal 2^m fractional factorial designs of resolution V, $m \leq 6$. *Technometrics*, **13**, 257–269.

*107. Srivastava, J. N., and Chopra, D. V. (1973). Balanced arrays and orthogonal arrays. In *A Survey of Combinatorial Theory*, J. N. Srivastava, F. Harary, C. R. Rao, G.-C. Rota, and S. S. Shrikhande., Eds. North-Holland Publishing Co., Amsterdam, pp. 411–428.

108. Srivastava, J. N., and Ghosh, S. (1977). Balanced 2^m factorial designs of resolution V which allow search and estimation of one extra unknown effect, $4 \leq m \leq 8$. *Commun. Stat.*, **A6**, 141–166.

109. Srivastava, J. N., Raktoe, B. L., and Pesotan, H. (1976). On invariance and randomization in fractional replication. *Ann. Stat.*, **4**, 423–430.

110. Webb, S. R. (1968a). Non-orthogonal designs of even resolution. *Technometrics*, **10**, 291–299.

111. Webb, S. R. (1968b). Saturated sequential factorial designs. *Technometrics*, **10**, 535–550.

112. Williams, E. J. (1949). Confounding and fractional replication in factorial experiments. *Aust. J. Agric. Sci.*, **15**, 145–153.

113. Yamamoto, S., Shirakura, T., and Kuwada, M. (1975). Balanced arrays of strength $2l$ and balanced fractional 2^m factorial designs. *Ann. Inst. Stat. Math.*, **27**, 143–157.

114. Yang, C. H. (1968). On design of maximal $(+1, -1)$-matrices of order $n \equiv 2 \pmod 4$. *Math. Comp.*, **22**, 174–180.

Appendix

The purpose of this appendix is to present some basic definitions and examples of finite mathematical structures, which have been mentioned in this book. For further reading, a list of references is provided as well.

A.1 BINARY RELATIONS

A binary relation on a set A to a set B is a subset $R \subset A \times B$. If $(a,b) \in R$, then we write aRb and if (a,b) is not in R we write $a\not R b$. By taking the Cartesian product of t sets we may extend the definition of binary relation to *t-ary relation*.

Example A1.1. Let $A = \{0,1\}$ and $B = \{0,1,2\}$ so that $A \times B = \{(0,0),(0,1),(0,2),(1,0),(1,1),(1,2)\}$. Let $R = \{(0,0),(0,1),(0,2)\}$ then R is a binary relation on A into B and $0R0$, $0R1$, $0R2$, $1\not R0$, $1\not R1$, $1\not R2$.

Let R be a binary relation on a set A to itself (i.e., R is a subset of $A \times A$). Then R is said to be:

(i) *Reflexive* if aRa for all $a \in A$.
(ii) *Symmetric* if a_1Ra_2 implies a_2Ra_1 for all a_1, $a_2 \in A$.
(iii) *Transitive* if a_1Ra_2 and a_2Ra_3 together imply a_1Ra_3 for all a_1, a_2, a_3 in A.

A binary relation on a set A (to itself) which is reflexive, symmetric and transitive is said to be an *equivalence relation on A*. A binary relation on a set A (to itself) is called *antisymmetric* if for all a_1, $a_2 \in A$, a_1Ra_2 and a_2Ra_1 together imply $a_1 = a_2$.

Example A1.2. Let R be the binary relation defined on the set of integers Z by aRb if and only if $2/(a-b)$ (i.e., 2 is a divisor of $a-b$). Then R is an equivalence relation, since it is easily verified that all three conditions are satisfied.

Given an equivalence relation E on a set A, we define the *equivalence class* under R of any element $a \in A$ to be the set

$$P_E(a) = \{x \in A : xEa\}.$$

Note that in the literature one uses the alternative notation $x = a(\text{mod } E)$ for xEa. One then calls the equivalence class $P_E(a)$ under E the *class of "a modulo E"*. The set of all equivalence classes under E is known as the *quotient set of A by E* and is usually denoted by A/E. Also note that the map $a \to P_E(a)$ which assigns to each $a \in A$ the class of a modulo E, determines a function $P_E\colon A \to A/E$. It is known as the *projection* of A onto its quotient A/E.

Let $\mathcal{P}(A)$ denote the *power set* of A, that is, the set of all possible subsets of A. Then we define a *partition* π of a set A to be a subset of $\mathcal{P}(A)$ such that: (i) for each $a \in A$, a is an element of exactly one $S \in \pi$, and (ii) $\phi \notin \pi$. The quotient set A/E is a partition of A.

Example A1.3. If a and b are in the set of integers Z, then a is said to be *congruent to b modulus k* (or modulo k) if $a-b$ is divisible by k. One writes this in abbreviated form as $a = b(\text{mod } k)$. This is an equivalence relation on Z defined by aEb if and only if $k/(a-b)$ (i.e., k is a divisor of $a-b$). Note that this example is a generalization of Example A1.2. The equivalence classes under E are

$$\begin{aligned}
P_E(0) &= P_{(k)}(0) = \{\ldots, -2k, -k, 0, k, 2k, \ldots\} \\
P_E(1) &= P_{(k)}(1) = \{\ldots, -2k+1, -k+1, 1, k+1, 2k+1, \ldots\} \\
P_E(2) &= P_{(k)}(2) = \{\ldots, -2k+2, -k+2, 2, k+2, 2k+2, \ldots\} \\
&\vdots \\
P_E(k-1) &= P_{(k)}(k-1) = \{\ldots, -2k+(k-1), -k+(k-1), k-1, k \\
&\qquad + (k-1), 2k+(k-1), \ldots\}.
\end{aligned}$$

The quotient set Z/E, also written as $Z/(k)$, is the set $\{P_E(i) = P_{(k)}(i) : i = 0, 1, 2, \ldots, (k-1)\}$ and is a partition of Z into k equivalence classes. The projection of Z onto $Z/(k)$ is the function $P_{(k)}\colon Z \to Z/(k)$, which maps every element of Z onto the equivalence class to which it belongs, for example, $P_{(k)}(k+2) = P_{(k)}(2)$. The numbers $0, 1, 2, \ldots, k-1$ in the equivalence classes are known as *least residues mod k* and are used as representatives of the classes. Thus an equivalence class with least residue i is called a *residue class i (mod k)*. For example, if $k=3$, then the three residue classes are

$$P_3(0)=\{\ldots\ldots,-6,-3,0,3,6,\ldots\ldots\}$$

$$P_3(1)=\{\ldots\ldots,-5,-2,1,4,7,\ldots\ldots\}$$

$$P_3(2)=\{\ldots\ldots,-4,-1,2,5,8,\ldots\ldots\}.$$

These three classes are often denoted as $\{0\}_3$, $\{1\}_3$, and $\{2\}_3$, and in general one uses $\{i\}_k$ for $P_k(i)$, $i=0,1,2,\ldots,k-1$.

A.2 GROUPS

A group is an ordered pair $(G,*)$ consisting of a nonempty set G and a binary operation $*\colon\ G\times G\to G[(x,y)\to x*y]$ satisfying the following axioms:

(i) $(x*y)*z=x*(y*z)$ for all $x, y, z\in G$ (*associativity*).

(ii) There exists an element $e\in G$ such that $x*e=x=e*x$ for all $x\in G$ (*identity element*).

(iii) For each $x\in G$, there exists an element $x'\in G$ such that $x*x'=e=x'*x$ (*inverse element*).

The operation $*$ is frequently written as a product; that is, $(x,y)\to xy$ or, as a sum, $(x,y)\to x+y$. We then say that $(G,*)$ is a *multiplicative* or an *additive group*, respectively.

Example A2.1. The set of residue classes $Z/(k)$ of Example A1.2 is an additive group and the reader may verify that the three axioms are satisfied if the addition of residue classes is defined by the addition table:

$+$	$\{0\}_k$	$\{1\}_k$	$\{2\}_k\cdots\cdots\cdots\{k-1\}_k$
$\{0\}_k$	$\{0\}_k$	$\{1\}_k$	$\{2\}_k\cdots\cdots\cdots\{k-1\}_k$
$\{1\}_k$	$\{1\}_k$	$\{2\}_k$	$\{3\}_k\cdots\cdots\cdots\{0\}_k$
$\{2\}_k$	$\{2\}_k$	$\{3\}_k$	$\{4\}_k\cdots\cdots\cdots\{1\}_k$
$\vdots$	$\vdots$	$\vdots$	$\vdots\qquad\vdots$
$\{k-1\}_k$	$\{k-1\}_k$	$\{0\}_k$	$\{1\}_k\cdots\cdots\cdots\{k-2\}_k$

Note that it is sufficient to define an addition table for the least residues.

A group G is said to be *commutative* or *Abelian* if the operation $*$ is commutative; that is, $x*y=y*x$ for all $x, y \in G$.

The example given in Example A2.1 is an additive Abelian group.

Algebraic structures that do not satisfy all the group axioms occur frequently in mathematics and the hierarchy with corresponding nomenclature is as follows:

Set with binary operation } *groupoid* } *semigroup* } *monoid* } *group*.
Associativity
Identity element
Inverse element

(Set with binary operation: *groupoid*; with Associativity: *semigroup*; with Identity element: *monoid*; with Inverse element: *group*.)

A group is called *finite* if the set G is finite, in which case the number of elements of G is known as the *order of the group*. If g is an element of the group $(G, *)$, then the *order* or *period of* g is the smallest positive integer n such that

$$\underbrace{g*g*\dots*g}_{n \text{ times}}=e.$$

For an additive group the left-hand side is written as ng and for a multiplicative group it is equal to g^n. If no such n exists, then g is said to be of *infinite order*. If p is a prime and every element of a group has order a power of p, then the group is called a *p-group*.

A group is said to be *cyclic* if it contains an element x such that every element of the group can be written as a power or multiple of x. One then says that the group is *generated* by x.

Example A2.2. The group of residue classes $Z/(k)$ of Example A2.1 can be depicted as $(\{0,1,2,\dots,k-1\},+)$ where $a+b=c$ and $c=a+b$ (mod k). For example if $k=6$, then $Z/(6)=(\{0,1,2,3,4,5\},+)$ a group of order 6. The identity element is 0 and the orders of $1,2,3,4$ and 5 are $6,3,2,3$ and 6, respectively. In fact $Z/(6)$ is cyclic and is generated by 1.

Given a set A of order n, then the set $S_n(A)$ of all permutations of A with the binary operation $(f,g)\to fog$ is the *group of permutations of the set* A. For the set $\Sigma_n=\{1,2,\dots,n\}$ the group of permutations of Σ_n is known as the *symmetric group* $S_n(\Sigma_n)$ of degree n. Note that the order of $S_n(\Sigma_n)$ is equal to $n!$.

A nonempty subset H of a group G is said to be a *subgroup* of G if for all $x, y \in H$: (i) $x*y \in H$, and (ii) x^{-1} (or $-x$) $\in H$. Subgroups of G other than G itself or $\{e\}$ are called *proper subgroups*. The order of a finite group is always a multiple of the order of any of its subgroups.

If H is a subgroup of G, then G can be partitioned into a set of equivalence classes by considering the equivalence relation E on G defined by

$$xEy \leftrightarrow x^{-1}y \in H \ (\text{or } -x+y \in H).$$

An equivalence class is then given by $E_x = \{y \in G: xEy\}$. Note that E_x is simply the set of all elements in G which is of the form xz (or $x+z$), where $z \in H$. In the literature, E_x is denoted by xH (or $x+H$) and it is referred to as the *left coset of* H in G. Similarly Hx (or $H+x$) is called the *right coset of* H in G.

Example A2.3. Take the group $Z/(6) = (\{0,1,2,3,4,5\},+)$ of Example A2.2. Then $H = \{0,3\}$ is a subgroup of $Z/(6)$ as can be easily verified. The right (or left) cosets of H are

$$H+0 = \{0,3\} = H,$$

$$H+1 = \{1,4\},$$

$$H+2 = \{2,5\}.$$

Hence $Z/(6)$ is partitioned into three equivalence classes.

If for all $g \in G$ the right coset Hg (or $H+g$) is equal to the left coset gH (or $g+H$) then H is called a *normal* or *invariant subgroup* of G. Thus in the preceding example H is a normal subgroup of $Z/(6)$.

A.3 RINGS

A *ring* is an ordered triple $(R,+,\cdot)$ consisting of a nonempty set R with two binary operations, $+$ = addition, and $\cdot$ = multiplication such that the following axioms are satisfied:

(i) The pair $(R,+)$ is a commutative group.
(ii) Multiplication is associative.
(iii) Multiplication is distributive (on both sides) over addition; that is, $a\cdot(b+c) = a\cdot b + a\cdot c$ and $(a+b)\cdot c = a\cdot c + b\cdot c$ for all $a, b, c \in R$.

If a ring admits a multiplicative identity (or unit) element, then it is referred to as a *ring with identity (or unit) element*. If multiplication in R is commutative, then R is said to be a *commutative ring*.

A *subring* of a ring, R, is a nonempty subset S of R that satisfies the conditions: (i) S is a subgroup of the additive group R and (ii) $x, y \in S$ imply $xy \in S$. For a ring with identity (or unit element), sometimes a third condition for a subset to be a subring is that the identity or (unit element) is in S.

Denoting the identity element by 1 in a ring R with identity element, we say that $x \in R$ is an *invertible element* with respect to multiplication if there exists a unique x^{-1} in R such that $x\,x^{-1} = x^{-1}x = 1$.

Example A3.1. Consider the group of residue classes $Z/(6) = \{0,1,2,3,4,5\}$. The reader may verify that $Z/(6)$ is a commutative ring with identity 1 with the following addition and multiplication tables:

+	0	1	2	3	4	5	·	0	1	2	3	4	5
0	0	1	2	3	4	5	0	0	0	0	0	0	0
1	1	2	3	4	5	0	1	0	1	2	3	4	5
2	2	3	4	5	0	1	2	0	2	4	0	2	4
3	3	4	5	0	1	2	3	0	3	0	3	0	3
4	4	5	0	1	2	3	4	0	4	2	0	4	2
5	5	0	1	2	3	4	5	0	5	4	3	2	1

The invertible elements in $Z/(6)$ are 1 and 5. The reader may verify that $\{0,2,4\}$ is a subring of $Z/(6)$ with unit element 4.

A.4 INTEGRAL DOMAINS

A commutative ring with identity element, K, is called an *integral domain* if $x, y \in \mathrm{K}$ and $xy = 0$ then $x = 0$ or $y = 0$. If in a ring R, $xy = 0$, $x \neq 0$ and $y \neq 0$, then x and y are called *zero divisors*. Thus an integral domain is a commutative ring that has an identity element but no zero divisors. In the literature some authors do not require commutativity for an integral domain.

Example A4.1. Consider the ring of residue classes $Z/(5) = \{0,1,2,3,4\}$ with addition and multiplication tables:

+	0	1	2	3	4	·	0	1	2	3	4
0	0	1	2	3	4	0	0	0	0	0	0
1	1	2	3	4	0	1	0	1	2	3	4
2	2	3	4	0	1	2	0	2	4	1	3
3	3	4	0	1	2	3	0	3	1	4	2
4	4	0	1	2	3	4	0	4	3	2	1

The reader may verify that $Z/(5)$ is an integral domain. As can be seen $Z/(5)$ does not contain zero divisors. Also, observe that $Z/(6)$ is not an integral domain since it contains zero divisors.

A.5 FIELDS

A *skew field* or *division ring* is a ring R with identity such that every nonzero element is invertible. For example, $Z/(5)$ is a skew field.

A *field* F is a commutative division ring. Well-known examples of a field are the rational numbers, real numbers, complex numbers, and the residue class ring $Z/(p)$, where p is a prime. For example, $Z/(5)$ is a field of order 5.

The *characteristic* of a field is the additive order of its multiplicative identity element; that is, F has finite characteristic m if m is the least positive integer for which $m1 = 0$. It has characteristic ∞, if no such multiple is 0. For example, the field $Z/(5)$ has characteristic 5 since $5 \cdot 1 = 0$.

It can be shown that a *finite field* has order equal to a prime or a power of a prime. Finite fields are known as *Galois fields* and the notation GF(s), where $s = p^u$, p a prime and u a positive integer has been used to denote them. For $u = 1$, one may show in a straightforward manner that the set of residue classes $Z/(p)$ is a field of order p, which is denoted by GF(p). If $u > 1$ a Galois field of order p^u is obtained using the following construction procedure. Let $R[x]$ be the ring of polynomials over GF(p) and let $P(x)$ be a fixed irreducible polynomial of degree u in $R[x]$. Define two polynomials $F(x)$ and $G(x)$ in $R[x]$ to be in the binary relation E if and only if $P(x)/[F(x) - G(x)]$ (i.e., $P(x)$ is a divisor of $F(x) - G(x)$ or $F(x) = G(x)$ mod $P(x)$). This is an equivalence relation on $R[x]$ which partitions $R[x]$ into p^u equivalence classes. Denote the set of equivalence classes by $R[x]/P(x)$ and use the set of p^u least residues as representatives of the equivalence classes. These residues are in the set $\{f(x) = a_0 + a_1 x + \cdots + a_{u-1}x^{u-1}\colon a_i \in \mathrm{GF}(p)\}$, which has order p^u. By defining addition and multiplication (that is, by presenting the addition and multiplication tables) of these residues in the usual way, we are led to addition and multiplication of the equivalence classes in $R[x]/P(x)$. It may then be established that $R[x]/P(x)$ is a field of order p^u, namely, GF(p^u).

Example A5.1. Let us construct a finite field of order 2^3, that is, GF(8). Take $P(x) = x^3 + x^2 + 1$. The least residues are $\{0, 1, x, x+1, x^2, x^2+1, x^2 + x, x^2 + x + 1\}$. The set of residue classes $R[x]/x^3 + x^2 + 1$ then forms a field with addition and multiplication defined by corresponding tables for the least residues. The arithmetic in these tables is performed over GF(2) modulo $x^3 + x^2 + 1$.

The crux of the matter in the construction of $GF(p^u)$ was the irreducible $P(x)$. There is a procedure for obtaining this and also tables. Examples are the following:

$$GF(2^2): \quad P(x) = x^2 + x + 1$$

$$GF(2^3): \quad P(x) = x^3 + x^2 + 1$$

$$GF(2^4): \quad P(x) = x^4 + x + 1$$

$$GF(3^2): \quad P(x) = x^2 + x + 2$$

$$GF(3^3): \quad P(x) = x^3 + 2x + 1$$

$$GF(5^2): \quad P(x) = x^2 + x + 2.$$

A.6 VECTOR SPACES

Let F be a field. Then a *vector space* V, over F is defined to be an additive Abelian group V together with a function $F \times V \to V$, given by $(\lambda, v) \to \lambda v$, called scalar multiplication, satisfying the following axioms:

(i) $\lambda(a+b) = \lambda a + \lambda b$,
(ii) $(\lambda + \mu)a = \lambda a + \mu a$,
(iii) $(\lambda\mu)a = \lambda(\mu a)$,
(iv) $1a = a$,

for all λ, μ in F and a, b in V.

In the literature the elements of F are called *scalars* and the elements of V are referred to as *vectors*.

A *vector subspace* of a vector space V is defined to be a subset V' of V such that: (i) V' is a subgroup of the additive group V, and (ii) $\lambda a \in V'$, for all $a \in V'$ and $\lambda \in F$.

Example A6.1. Let $F = GF(p^u)$, then the set of all t-tuples $(x_1, x_2, \ldots, x_t)$ where $x_i \in GF(p^u)$ forms a vector space over $GF(p^u)$, with addition and scalar multiplication defined as $(x_1, x_2, \ldots, x_t) + (y_1, y_2, \ldots, y_t) = (x_1 + y_1, x_2 + y_2, \ldots, x_t + y_t)$ and $\lambda(x_1, x_2, \ldots, x_t) = (\lambda x_1, \lambda x_2, \ldots, \lambda x_t)$, $\lambda \in GF(p^u)$.

A vector space V over F is said to be *finitely generated* if there exists a finite set of elements $v_1, v_2, \ldots, v_t$ in V which generate V; that is, every element of V is a linear combination of the form $\sum_{i=1}^{t} \lambda_i v_i$, $\lambda_i \in F$. The vectors $v_1, v_2, \ldots, v_t$ are said to be *linearly independent* if $\Sigma_{i=1}^{t} \lambda_i v_i = 0$ $(0 \in V)$ implies that $\lambda_1 = \lambda_2 = \cdots = \lambda_t = 0 (0 \in F)$. On the other hand, the vectors $v_1, v_2, \ldots, v_t$ are said to be *linearly dependent* if and only if there exists $\lambda_1, \lambda_2, \ldots, \lambda_t$ in F, not all zero, such that $\Sigma_{i=1}^{t} \lambda_i v_i = 0 (0 \in V)$. If $v_1, v_2, \ldots, v_t$ generate a vector space V and are linearly independent, then we say that $v_1, v_2, \ldots, v_t$ form a *basis* of V. If $v_1, v_2, \ldots, v_t$ form a basis of V and v is any vector of V then there exist a unique n-tuple $(\lambda_1, \ldots, \lambda_t)$ such that $v = \Sigma_{i=1}^{t} \lambda_i v_i$. One may show that every finitely generated vector space V has a basis. The number of vectors in a basis for V is called the dimension of V and V is called a *t-dimensional vector space*. Note that a t-dimensional vector space is isomorphic to the vector space of all t-tuples $(x_1, x_2, \ldots x_t)$, $x_i \in F$, which is denoted by F^t.

A.7 FINITE GEOMETRIES EG(t, s) AND PG$(t-1, s)$

The finite Euclidean and finite projective geometries can be developed in an axiomatic way just as was done for the other structures in the previous sections. For our purposes, it will be sufficient to present analytical representations of these.

Let F be the Galois field GF(s). The vector space F^t of all t-tuples $(x_1, x_2, \ldots, x_t)$, $x_i \in$ GF(s), is called a *finite dimensional Euclidean geometry*, EG(t, s) over GF(s). Since GF(s) contains s distinct elements it follows that EG(t, s) contains exactly s^t distinct vectors. A vector in EG(t, s) is known as a *point*. The subset of points in EG(t, s) which is a solution set to $t - m$ independent and consistent equations given by the matrix equation

$$\begin{bmatrix} a_{11} & a_{12} & \cdots & a_{1t} \\ a_{21} & a_{22} & \cdots & a_{2t} \\ & & \vdots & \\ a_{t-m1} & a_{t-m2} & \cdots & a_{t-mt} \end{bmatrix} \begin{bmatrix} x_1 \\ x_2 \\ \vdots \\ x_t \end{bmatrix} = \begin{bmatrix} y_1 \\ y_2 \\ \vdots \\ y_{t-m} \end{bmatrix}, \qquad y_i \in \mathrm{GF}(s),$$

is called an *m-flat* of EG(t, s). If all the y_i are zero, then the m-flat is a vector subspace and otherwise it is a coset of a vector subspace, when the subspace is viewed as a subgroup. An m-flat in EG(t, s) contains s^m points.

Example A7.1. Let $F = \mathrm{GF}(2) = Z/(2) = \{0, 1\}$. Then the vector space F^3 consisting of 3-tuples (x_1, x_2, x_3) is the finite Euclidean geometry EG(3,2) with the following 2^3 points: $(0,0,0)$, $(1,0,0)$, $(0,1,0)$, $(0,0,1)$, $(1,1,0)$, $(1,0,1)$, $(0,1,1)$, $(1,1,1)$. A 1-flat that is a vector subspace is

$\{(0,0,0),(0,0,1)\}$, which is the solution set to the equation system

$$\begin{bmatrix} 1 & 0 & 0 \\ 0 & 1 & 0 \end{bmatrix} \begin{bmatrix} x_1 \\ x_2 \\ x_3 \end{bmatrix} = \begin{bmatrix} 0 \\ 0 \end{bmatrix}.$$

A 2-flat in EG(3,2) of the coset type is $\{(1,0,0),(1,1,0),(1,0,1),(1,1,1)\}$, which is the solution set to

$$[1\ 0\ 0] \begin{bmatrix} x_1 \\ x_2 \\ x_3 \end{bmatrix} = [1].$$

Consider now the finite Euclidean geometry $\mathrm{EG}(t,s)$ over $\mathrm{GF}(s)$ and omit the point $\mathbf{0}' = (0,0,\ldots,0)$. Denote the resulting set by $\mathrm{EG}(t,s)-\mathbf{0}'$ and define two points in $\mathrm{EG}(t,s)-\mathbf{0}'$ to be in the relation $\mathrm{GF}(s)$ if and only if they are linearly dependent. This is an equivalence relation on $\mathrm{EG}(t,s)-\mathbf{0}'$, which partitions it into equivalence classes. Note that the equivalence classes are simply one-dimensional vector subspaces of $\mathrm{EG}(t,s)$ without $\mathbf{0}'$. The set of equivalence classes, that is, the quotient $[\mathrm{EG}(t,s)-\mathbf{0}']/\mathrm{GF}(s)$, is called the *finite projective geometry* of dimension $t-1$ over $\mathrm{GF}(s)$ and is denoted by $\mathrm{PG}(t-1,s)$.

A set of r linearly independent points of $\mathrm{PG}(t-1,s)$ generates an *(r − 1)-flat*; that is, it consists of the set of points $(x_1,x_2,\ldots,x_t)$ in $\mathrm{PG}(t-1,s)$ which are of the form $\Sigma_{i=1}^{r}\alpha_i\mathbf{Y}_i$, where $\mathbf{Y}_1,\mathbf{Y}_2,\ldots,\mathbf{Y}_r$ are r linearly independent points of $\mathrm{PG}(t-1,s)$ and the α_i are in $\mathrm{GF}(s)$ and not all simultaneously 0. A 0-flat is called a *point* and a 1-flat is called a *line* of $\mathrm{PG}(t-1,s)$ and so on. Note that an $(r-1)$-flat of $\mathrm{PG}(t-1,s)$ is simply a projective $(r-1)$-dimensional space, which is the quotient $[\mathrm{EG}(r,s)-\mathbf{0}']/\mathrm{GF}(s)$. The number of points in $\mathrm{PG}(t-1,s)$ is equal $(s^t-1)/(s-1)$ and the number of points in an $(r-1)$-flat is equal to $(s^r-1)/(s-1)$.

Example A7.2. Consider EG(2,3) over $F = \mathrm{GF}(3) = Z/(3) = \{0,1,2\}$. The points of PG(2, 2) are then the 3-tuples: (1, 0, 0), (0, 1, 0), (0,0,1), (1,1,0), (1,2,0), (1,0,1), (1,0,2), (0,1,1), (0,1,2), (1,1,1), (1,1,2), (1,2,1), (1,2,2). A 1-flat in PG(2,2) is for example generated by the two independent points (1,0,0) and (0,1,0) and constitutes the set {(1,0,0), (0,1,0), (1,1,0), (1,2,0)}.

A.8 SELECTED ADDITIONAL READING

1. Bose, R. C. (1947). Mathematical theory of the symmetrical factorial design. *Sankhyā*, **8**, 107–166.

2. Carmichael, R. D. (1956). *Introduction to the Theory of Groups of Finite Order*. (First published in 1937). Dover, New York.
3. Dembowski, P. (1968). *Finite Geometries*, Springer-Verlag, Berlin.
4. Kuiper, N. H. (1962). *Linear Algebra and Geometry*. North-Holland, Amsterdam.
5. Lederman, W. (1961). *Introduction to the Theory of Finite Groups*. Oliver and Boyd, London.
6. McCoy, N. H. (1972). *Fundamentals of Abstract Algebra*. Allyn & Bacon, Boston.

Index

Applied Probability and Statistics (Continued)

FLEISS • Statistical Methods for Rates and Proportions, *Second Edition*
GALAMBOS • The Asymptotic Theory of Extreme Order Statistics
GIBBONS, OLKIN, and SOBEL • Selecting and Ordering Populations: A New Statistical Methodology
GNANADESIKAN • Methods for Statistical Data Analysis of Multivariate Observations
GOLDBERGER • Econometric Theory
GOLDSTEIN and DILLON • Discrete Discriminant Analysis
GROSS and CLARK • Survival Distributions: Reliability Applications in the Biomedical Sciences
GROSS and HARRIS • Fundamentals of Queueing Theory
GUPTA and PANCHAPAKESAN • Multiple Decision Procedures: Theory and Methodology of Selecting and Ranking Populations
GUTTMAN, WILKS, and HUNTER • Introductory Engineering Statistics, *Second Edition*
HAHN and SHAPIRO • Statistical Models in Engineering
HALD • Statistical Tables and Formulas
HALD • Statistical Theory with Engineering Applications
HARTIGAN • Clustering Algorithms
HILDEBRAND, LAING, and ROSENTHAL • Prediction Analysis of Cross Classifications
HOEL • Elementary Statistics, *Fourth Edition*
HOLLANDER and WOLFE • Nonparametric Statistical Methods
JAGERS • Branching Processes with Biological Applications
JESSEN • Statistical Survey Techniques
JOHNSON and KOTZ • Distributions in Statistics
Discrete Distributions
Continuous Univariate Distributions—1
Continuous Univariate Distributions—2
Continuous Multivariate Distributions
JOHNSON and KOTZ • Urn Models and Their Application: An Approach to Modern Discrete Probability Theory
JOHNSON and LEONE • Statistics and Experimental Design in Engineering and the Physical Sciences, Volumes I and II, *Second Edition*
JUDGE, GRIFFITHS, HILL and LEE • The Theory and Practice of Econometrics
KALBFLEISCH and PRENTICE • The Statistical Analysis of Failure Time Data
KEENEY and RAIFFA • Decisions with Multiple Objectives
LAWLESS • Statistical Models and Methods for Lifetime Data
LEAMER • Specification Searches: Ad Hoc Inference with Nonexperimental Data
McNEIL • Interactive Data Analysis
MANN, SCHAFER and SINGPURWALLA • Methods for Statistical Analysis of Reliability and Life Data
MEYER • Data Analysis for Scientists and Engineers
MILLER • Survival Analysis
MILLER, EFRON, BROWN, and MOSES • Biostatistics Casebook
MONTGOMERY and PECK • Introduction to Linear Regression
NELSON • Applied Life Data Analysis
OTNES and ENOCHSON • Applied Time Series Analysis: Volume I, Basic Techniques
OTNES and ENOCHSON • Digital Time Series Analysis
POLLOCK • The Algebra of Econometrics
PRENTER • Splines and Variational Methods
RAO and MITRA • Generalized Inverse of Matrices and Its Applications
RIPLEY • Spatial Statistics
SCHUSS • Theory and Applications of Stochastic Differential Equations
SEAL • Survival Probabilities: The Goal of Risk Theory
SEARLE • Linear Models
SPRINGER • The Algebra of Random Variables
UPTON • The Analysis of Cross-Tabulated Data
WEISBERG • Applied Linear Regression
WHITTLE • Optimization Under Constraints